Student Solutions Manual & S

for

Serway and Vuille's

College Physics

Ninth Edition, Volume 2

John R. Gordon
Emeritus, James Madison University

Charles Teague
Emeritus, Eastern Kentucky University

Raymond A. Serway
Emeritus, James Madison University

Australia • Brazil • Japan • Korea • Mexico • Singapore • Spain • United Kingdom • United States

ISBN-13: 978-0-8400-6867-5

ISBN-10: 0-8400-6867-0

Brooks/Cole
20 Channel Center Street
Boston, MA 02210
USA

Cengage Learning is a leading provider of customized learning solutions with office locations around the globe, including Singapore, the United Kingdom, Australia, Mexico, Brazil, and Japan. Locate your local office at: **www.cengage.com/global**

Cengage Learning products are represented in Canada by Nelson Education, Ltd.

To learn more about Brooks/Cole, visit **www.cengage.com/brookscole**

Purchase any of our products at your local college store or at our preferred online store **www.cengagebrain.com**

Printed in the United States of America
2 3 4 5 6 7 14 13 12

PREFACE

This *Student Solutions Manual and Study Guide* has been written to accompany the textbook, *College Physics, Ninth Edition,* by Raymond A. Serway and Chris Vuille. The purpose of this ancillary is to provide students with a convenient review of the basic concepts and applications presented in the textbook, together with solutions to selected end-of-chapter problems. This is not an attempt to rewrite the textbook in a condensed fashion. Rather, emphasis is placed upon clarifying typical troublesome points and providing further practice in methods of problem solving.

Every textbook chapter has a matching chapter in this book, and each chapter is divided into several parts. Very often, reference is made to specific equations or figures in the textbook. Each feature of this Study Guide has been included to ensure that it serves as a useful supplement to the textbook. Most chapters contain the following components:

- **Notes from Selected Chapter Sections:** This is a summary of important concepts, newly defined physical quantities, and rules governing their behavior.

- **Equations and Concepts:** This is a review of the chapter, with emphasis on highlighting important concepts and describing important equations and formalisms.

- **Suggestions, Skills, and Strategies:** This section offers hints and strategies for solving typical problems that the student will often encounter in the course. In some sections, suggestions are made concerning mathematical skills that are necessary in the analysis of problems.

- **Review Checklist:** This is a list of topics and techniques the student should master after reading the chapter and working the assigned problems.

- **Solutions to Selected End-of-Chapter Problems:** Solutions are given for selected odd-numbered problems that were chosen to illustrate important concepts in each textbook chapter.

- **Tables:** A table of some Conversion Factors is provided on the inside front cover, and a list of selected Physical Constants is printed on the inside back cover.

An important note concerning significant figures: In preparing the problem solutions for this manual, we have attempted to carefully follow the rules regarding significant figures presented in Chapter 1 of the textbook. We sincerely hope that this *Student Solutions*

Manual and Study Guide will be useful to you in reviewing the material presented in the text and in improving your ability to solve problems and score well on exams. We welcome any comments or suggestions which could help improve the content of this study guide in future editions, and we wish you success in your study.

John R. Gordon
Harrisonburg, VA

Charles Teague
Richmond, KY

Raymond A. Serway
Leesburg, VA

Acknowledgments

We are indebted to everyone who contributed to this *Student Solutions Manual and Study Guide to Accompany College Physics, Ninth Edition*.

We wish to express our sincere appreciation to the staff of Cengage Learning who provided necessary resources and coordinated all phases of this project. Special thanks go to Brandi Kirksey (Associate Development Editor), Brendan Killion (Editorial Assistant), Holly Schaff (Associate Content Project Manager), Ed Dodd (Development Editor), and Charles Hartford (Publisher).

Our appreciation goes to our reviewer, Susan English, Durham Technical Community College. Her careful reading of the manuscript and checking the accuracy of the problem solutions contributed in an important way to the quality of the final product. Any errors remaining in the manual are the responsibility of the authors.

It is a pleasure to acknowledge the excellent work of the staff of MPS Limited, a Macmillan Company for assembling and typing this manual and preparing diagrams and page layouts. Their technical skills and attention to detail added much to the appearance and usefulness of this volume.

Finally, we express our appreciation to our families for their inspiration, patience, and encouragement.

Suggestions for Study

We have seen a lot of successful physics students. The question, "How should I study this subject?" has no single answer, but we offer some suggestions that may be useful to you.

1. Work to understand the basic concepts and principles before attempting to solve assigned problems. Carefully read the textbook before attending your lecture on that material. Jot down points that are not clear to you, take careful notes in class, and ask questions. Reduce memorization to a minimum. Memorizing sections of a text or derivations does not necessarily mean you understand the material.

2. After reading a chapter, you should be able to define any new quantities that were introduced and discuss the first principles that were used to derive fundamental equations. A review is provided in each chapter of the Study Guide for this purpose, and the marginal notes in the textbook (or the index) will help you locate these topics. You should be able to correctly associate with each physical quantity the symbol used to represent that quantity (including vector notation, if appropriate) and the SI unit in which the quantity is specified. Furthermore, you should be able to express each important principle or equation in a concise and accurate prose statement. Perhaps the best test of your understanding of the material will be your ability to answer questions and solve problems in the text or those given on exams.

3. Try to solve plenty of the problems at the end of the chapter. The worked examples in the text will serve as a basis for your study. This Study Guide contains detailed solutions to about twelve of the problems at the end of each chapter. You will be able to check the accuracy of your calculations for any odd-numbered problem, since the answers to these are given at the back of the text.

4. Besides what you might expect to learn about physics concepts, a very valuable skill you can take away from your physics course is the ability to solve complicated problems. The way physicists approach complex situations and break them down into manageable pieces is widely useful. Starting in Section 1.10, the textbook develops a general problem-solving strategy that guides you through the steps. To help you remember the steps of the strategy, they are called *Conceptualize, Categorize, Analyze,* and *Finalize*.

General Problem-Solving Strategy

Conceptualize

- The first thing to do when approaching a problem is to *think about* and *understand* the situation. Read the problem several times until you are confident you understand what is being asked. Study carefully any diagrams, graphs, tables, or photographs that accompany the problem. Imagine a movie, running in your mind, of what happens in the problem.

- If a diagram is not provided, you should almost always make a quick drawing of the situation. Indicate any known values, perhaps in a table or directly on your sketch.

- Now focus on what algebraic or numerical information is given in the problem. In the problem statement, look for key phrases such as "starts from at rest" ($v_i = 0$), "stops" ($v_f = 0$), or "freely falls" ($a_y = -g = -9.80$ m/s^2). Key words can help simplify the problem.

- Next focus on the expected result of solving the problem. Exactly what is the question asking? Will the final result be numerical or algebraic? If it is numerical, what units will it have? If it is algebraic, what symbols will appear in it?

- Incorporate information from your own experiences and common sense. What should a reasonable answer look like? What should its order of magnitude be? You wouldn't expect to calculate the speed of an automobile to be 5×10^6 m/s.

Categorize

- Once you have a really good idea of what the problem is about, you need to *simplify* the problem. Remove the details that are not important to the solution. For example, you can often model a moving object as a particle. Key words should tell you whether you can ignore air resistance or friction between a sliding object and a surface.

- Once the problem is simplified, it is important to *categorize* the problem. How does it fit into a framework of ideas that you construct to understand the world? Is it a simple *plug-in problem,* such that numbers can be simply substituted into a definition? If so, the problem is likely to be finished when this substitution is done. If not, you face what we can call an *analysis problem*—the situation must be analyzed more deeply to reach a solution.

- If it is an analysis problem, it needs to be categorized further. Have you seen this type of problem before? Does it fall into the growing list of types of problems that you have solved previously? Being able to classify a problem can make it much easier to lay out a plan to solve it. For example, if your simplification shows that the problem can be treated as a particle moving under constant acceleration and you have already solved such a problem (such as the examples in Section 2.6), the solution to the new problem follows a similar pattern.

Analyze

- Now, you need to analyze the problem and strive for a mathematical solution. Because you have already categorized the problem, it should not be too difficult to select relevant equations that apply to the type of situation in the problem. For example, if your categorization shows that the problem involves a particle moving under constant acceleration, Equations 2.9 to 2.13 are relevant.

- Use algebra (and calculus, if necessary) to solve symbolically for the unknown variable in terms of what is given. Substitute in the appropriate numbers, calculate the result, and round it to the proper number of significant figures.

Finalize

- This final step is the most important part. Examine your numerical answer. Does it have the correct units? Does it meet your expectations from your conceptualization of the problem? What about the algebraic form of the result—before you substituted numerical values? Does it make sense? Try looking at the variables in it to see whether the answer would change in a physically meaningful way if they were drastically increased or decreased or even became zero. Looking at limiting cases to see whether they yield expected values is a very useful way to make sure that you are obtaining reasonable results.

- Think about how this problem compares with others you have done. How was it similar? In what critical ways did it differ? Why was this problem assigned? You should have learned something by doing it. Can you figure out what? Can you use your solution to expand, strengthen, or otherwise improve your framework of ideas? If it is a new category of problem, be sure you understand it so that you can use it as a model for solving future problems in the same category.

When solving complex problems, you may need to identify a series of sub-problems and apply the problem-solving strategy to each. For very simple problems, you probably don't need this whole strategy. But when you are looking at a problem and you don't know what to do next, remember the steps in the strategy and use them as a guide.

Work on problems in this Study Guide yourself and compare your solutions with ours. Your solution does not have to look just like the one presented here. A problem can sometimes be solved in different ways, starting from different principles. If you wonder about the validity of an alternative approach, ask your instructor.

5. We suggest that you use this Study Guide to review the material covered in the text and as a guide in preparing for exams. You can use the sections Chapter Review, Notes from Selected Chapter Sections, and Equations and Concepts to focus in on any points which require further study. The main purpose of this Study Guide is to improve upon the efficiency and effectiveness of your study hours and your overall understanding of physical concepts. However, it should not be regarded as a substitute for your textbook or for individual study and practice in problem solving.

TABLE OF CONTENTS

15

Electric Forces and Electric Fields

NOTES FROM SELECTED CHAPTER SECTIONS

15.1 Properties of Electric Charges

Electric charge has the following important properties:

- There are two kinds of charges (positive and negative) in nature. Unlike charges attract one another, and like charges repel one another.
- The force between charges varies as the inverse square of the distance between them.
- Charge is always conserved.
- Charge comes in discrete packets (quantized) that are integral multiples of the electronic charge, $e = 1.602\,19 \times 10^{-19}$ C.

A neutral object becomes charged electrically due to the gain or loss of electrons. A negatively charged object has an excess of electrons (relative to protons), and a positively charged object has a deficiency of electrons.

15.2 Insulators and Conductors

Conductors are materials in which electrons move freely under the influence of an electric field; insulators are materials that do not readily transport charge.

15.3 Coulomb's Law

Experiments show that an electric force between two charges is:

- Inversely proportional to the square of the distance between the two charges and is directed along the line joining them.
- Proportional to the product of the magnitudes of the charges.
- Attractive if the charges are of opposite sign and repulsive if the charges have the same sign.

The electric interaction between two charges q_1 and q_2 obeys Newton's third law. The two charges experience forces that are equal in magnitude and opposite in direction. *This is true regardless of the relative magnitudes of the two charges.*

15.4 The Electric Field

An electric field exists at a point if a small test charge placed at that point experiences an electric force. The direction of the electric field is along the direction of the force on a positive test charge. The magnitude of the electric field is given by Equation 15.6. The SI units of the electric field are newtons per coulomb (N/C).

The net electric field at a point (e.g., point P in the figure at right), due to multiple charges, equals the *vector sum* of the electric fields at that point due to the individual charges. *This is called the superposition principle.*

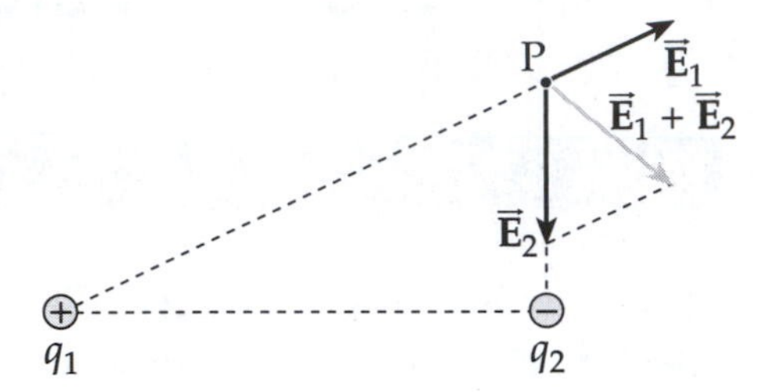

15.5 Electric Field Lines

A convenient aid for visualizing electric field patterns can be obtained by drawing a group of electric field lines. These lines indicate the direction and relative magnitude of the electric field at any point and are related to the electric field in any region of space in the following manner:

- The electric field vector $\vec{\mathbf{E}}$ is tangent to the electric field line at each point.
- The number of electric field lines per unit area through a surface perpendicular to the lines is proportional to the strength of the electric field in that region. Thus, $\vec{\mathbf{E}}$ is large when the field lines are close together and small when they are far apart.

The rules for drawing electric field lines for any charge distribution are as follows:

- The lines begin on positive charges and terminate on negative charges, or at infinity in the case of an excess of charge.
- The number of lines drawn leaving a positive charge or terminating on a negative charge is proportional to the magnitude of the charge. In the case of an isolated charge, the field lines are evenly distributed around the charge.
- No two field lines can cross.

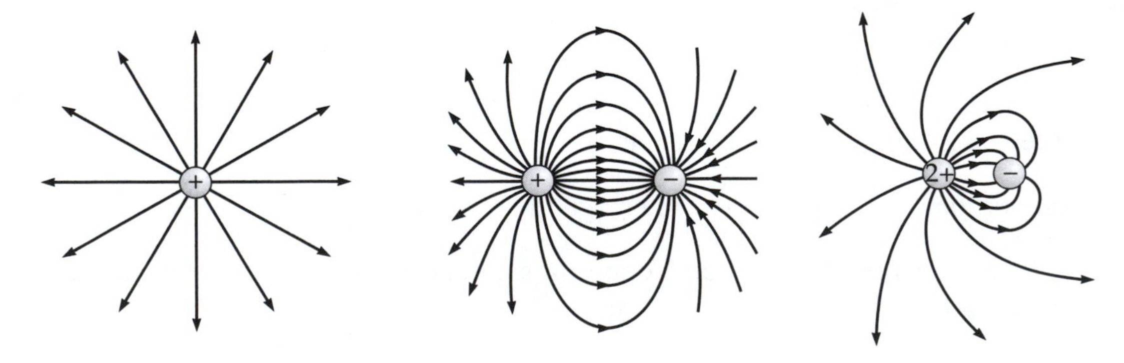

Representative Electric Field Line Patterns

15.6 Conductors in Electrostatic Equilibrium

A conductor in electrostatic equilibrium (no net motion of charge occurs within the conductor) has the following properties:

- The electric field is zero everywhere inside the conductor.
- Any excess charge on an isolated conductor resides entirely on its surface.
- The electric field just outside a charged conductor is perpendicular to the surface of the conductor.
- On an irregularly-shaped conductor, the charge per unit area is greatest at locations of sharpest curvature, that is, where the radius of curvature is smallest.

15.9 Electric Flux and Gauss's Law

There is an important general relation between the net electric flux through an imaginary closed surface (called a Gaussian surface) and the charge enclosed by the surface. This is known as Gauss's law and states that the net electric flux through any closed Gaussian surface is equal to the net charge inside the surface divided by ϵ_0. Although Gauss's law is valid for any closed surface, it is most useful for calculating the electric field in situations which have a high degree of symmetry in the charge distribution (for example, uniformly-charged spheres, cylinders, lines, or planes).

EQUATIONS AND CONCEPTS

Coulomb's law gives the magnitude of the electrostatic force between two point charges q_1 and q_2, separated by a distance r. This is also true for spherical charge distributions when r is the distance between their centers. In calculations the approximate value $(8.99 \times 10^9 \ \mathrm{N \cdot m^2/C^2})$ of the Coulomb constant, k_e, may be used. Like sign charges repel and unlike sign charges attract as illustrated in the figure at right.

$$F = k_e \frac{|q_1||q_2|}{r^2} \qquad (15.1)$$

$$k_e = 8.987\ 5 \times 10^9 \ \mathrm{N \cdot m^2/C^2} \qquad (15.2)$$

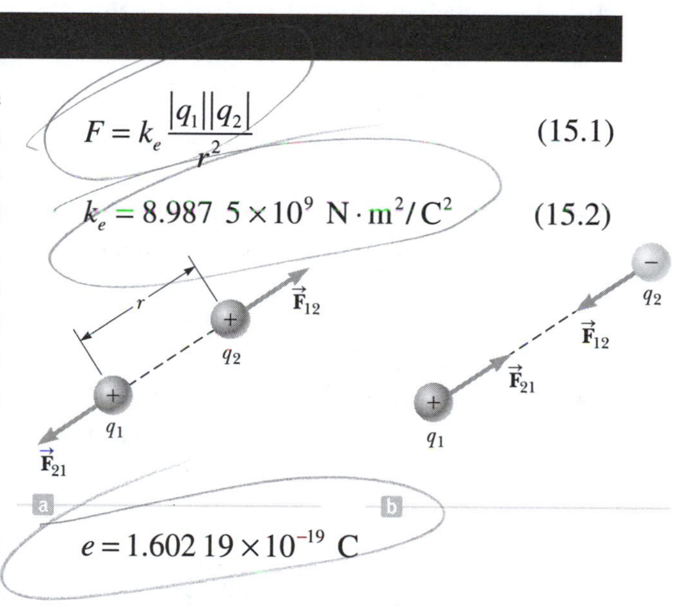

The **charge on the electron,** represented by the symbol e, is the fundamental quantity of charge in nature. The charge on the proton is positive and equal in magnitude to the charge on the electron.

$$e = 1.602\ 19 \times 10^{-19} \ \mathrm{C}$$

The definition of the electric field, $\vec{\mathbf{E}}$, at any point in space is the ratio of electric force per unit charge exerted on a small *positive test charge*, q_0, placed at the point where the field is to be determined. **The direction of $\vec{\mathbf{E}}$ is along the direction of $\vec{\mathbf{F}}$.** *An electric field exists in the vicinity of charge Q whether or not a test charge is present.*

$$\vec{\mathbf{E}} = \frac{\vec{\mathbf{F}}}{q_0} \qquad (15.3)$$

Q

q_0

$\vec{\mathbf{E}}$

Test charge

The **electric field due to a point charge** has a magnitude given by Equation 15.6. The direction of the electric field is radially outward from a positive point charge and radially inward toward a negative point charge. *The superposition principle holds when the electric field at a point is due to a number of point charges.*

$$E = k_e \frac{|q|}{r^2}\,\vec{\mathbf{F}} \qquad (15.6)$$

Electric flux through a surface is proportional to the number of electric lines that penetrate the surface. For a plane surface in a uniform field, the flux depends on the angle between the normal to the surface and the direction of the field. Electric flux has SI units of $\mathrm{N \cdot m^2/C}$. In the figure, the electric field is parallel to the normal to the plane surface, and therefore, in this case, $\theta = 0$ in Equation 15.9.

$$\Phi_E = EA\cos\theta \qquad (15.9)$$

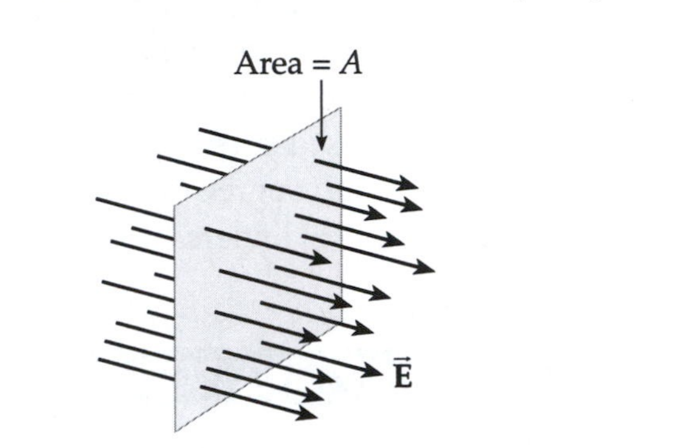

For the orientation shown above the electric flux through the area, A, is EA.

Gauss's law states that the net electric flux, Φ_E, out of a closed surface (Gaussian surface) is equal to the net charge enclosed by the surface divided by the constant ϵ_0. *By convention, for a closed surface, the flux lines passing into the interior of the volume are negative, and those passing out of the interior of the volume are positive.*

$$\Phi_E = \frac{Q_{\text{inside}}}{\epsilon_0} \qquad (15.11)$$

$$Q_{\text{inside}} = Q_{\text{positive}} - Q_{\text{negative}}$$

SUGGESTIONS, SKILLS, AND STRATEGIES

ELECTRIC FORCES AND FIELDS

1. **Units:** When performing calculations using Coulomb's law and the Coulomb constant $k_e = 8.99 \times 10^9\ \mathrm{N \cdot m^2/C^2}$, charge must be in coulombs and distances in meters. If they are given in other units, you must convert them to SI units.

2. **Applying Coulomb's law to point charges:** Remember to use the magnitude of the charges in Equation 15.1. The direction of the force is found by noting that the forces are repulsive between like charges and attractive between unlike charges. Use the superposition principle property when dealing with a collection of interacting point charges. When several charges are present, the resultant force on one of them is found by first calculating the force (magnitude and direction) that each of the other

charges exerts on it. Then, determine the vector sum of these forces. The magnitude of the force that any charged particle exerts on another is given by Coulomb's law.

3. **Calculating the electric field due to point charges:** The direction of the electric field is the direction of electric force on a positive "test" charge if placed at the point in question. The superposition principle can also be applied to electric fields, which, like electrostatic forces, are also vector quantities. To find the electric field at a given point, first calculate the electric field at the point due to each individual charge. The resultant field at the point is the vector sum of the fields due to the individual charges.

4. **Electric flux and Gauss's law:** When using Gauss's law to calculate the electric field in the vicinity of a charge distribution, you must anticipate the expected symmetry of the electric field by considering the shape of the charge distribution. Choose a Gaussian surface which has a shape matching the symmetry of the field (spherical surface for a point charge, cylindrical surface for a line of charge, etc.). The Gaussian surface must be closed and include the point at which the field is to be calculated.

   An important factor in your choice of a shape should be the ease with which you can calculate the flux through the closed surface. This will be the case when the surface can be divided into regions so that over a particular region either (1) the flux is zero, or (2) the electric field is constant. Consider, for example, a charge distributed uniformly along a line. In order to calculate the electric field a distance, r, from the line, choose as a Gaussian surface a closed cylinder of radius r whose axis is along the line of charge. In this case there will be zero flux through the ends of the cylinder (the direction of the electric field will be perpendicular to the surface), and the field will have a constant magnitude over the wall of the cylinder (all points on the curved surface are equidistant from the line of charge).

REVIEW CHECKLIST

You should be able to:

- Use Coulomb's law to determine the net electrostatic force (magnitude and direction) on a point electric charge due to a known distribution of a finite number of point charges. (Section 15.3)

- Calculate the electric field $\vec{\mathbf{E}}$ (magnitude and direction) at a specified location in the vicinity of a group of point charges. (Section 15.4)

- Describe the pattern of electric field lines associated with charge distributions such as (i) single point charges, (ii) pairs of charges with like or unlike sign and equal or unequal magnitudes, (iii) charged metallic spheres, etc. (Section 15.5)

- State and justify the conditions for the charge distribution on a conductor in electrostatic equilibrium. (Section 15.6)

- Calculate the electric flux through a surface and use Gauss's law to determine the electric field due to symmetric charge distributions (point, sphere, line, cylinder, plane). (Section 15.9)

SOLUTIONS TO SELECTED END-OF-CHAPTER PROBLEMS

2. A charged particle *A* exerts a force of 2.62 N to the right on charged particle *B* when the particles are 13.7 mm apart. Particle *B* moves straight away from *A* to make the distance between them 17.7 mm. What vector force does particle *B* then exert on *A*?

Solution

The force that charged particle *A* exerts on charged particle *B* is directed along the line connecting the positions of the two particles. If the two particles have charges of the same sign, this force is repulsive and directed away from *A*, while the force is attractive and directed toward *A* if the two particles have charges of opposite signs. The magnitude of the force is given by Coulomb's law as $F = k_e |q_A||q_B|/r^2$, where r is the distance separating the two particles, and the constant k_e is given by $k_e = 1/4\pi \in_0 = 8.99 \times 10^9 \text{ N} \cdot \text{m}^2/\text{C}^2$. Thus, with particle *B* initially located at $r_i = 13.7$ mm from particle *A*, we have

$$\vec{\mathbf{F}}_i = \frac{k_e |q_A||q_B|}{r_i^2} \text{ to the right}$$

with $|\vec{\mathbf{F}}_i| = F_i = 2.62$ N. If Particle *B* now moves straight away from *A*, so the line connecting their positions has the same orientation as before, the new force that particle *A* exerts on *B* will have the same direction as before but a smaller magnitude given by

$$|\vec{\mathbf{F}}_f| = F_f = \frac{k_e |q_A||q_B|}{r_f^2}$$

where $r_f = 17.7$ mm is the new distance separating the two particles. The ratio of the magnitudes of the initial and final forces of interaction is

$$\frac{F_f}{F_i} = \frac{\cancel{k_e |q_A||q_B|}/r_f^2}{\cancel{k_e |q_A||q_B|}/r_i^2} = \left(\frac{r_i}{r_f}\right)^2 \quad \text{or} \quad F_f = F_i\left(\frac{r_i}{r_f}\right)^2 = (2.62 \text{ N})\left(\frac{13.7 \text{ mm}}{17.7 \text{ mm}}\right)^2 = 1.57 \text{ N}$$

and $\vec{\mathbf{F}}_f = 1.57$ N to the right. The final reaction force that particle *B* exerts back on *A* is then given by Newton's third law as $\vec{\mathbf{F}}_f' = 1.57$ N to the left. ◊

7. A small sphere of charge 0.800 μC hangs from the end of a spring as in Figure P15.7a. When another small sphere of charge –0.600 μC is held beneath the first sphere as in Figure P15.7b, the spring stretches by $d = 3.50$ cm from its original length and reaches a new equilibrium position with a separation between the charges of $r = 5.00$ cm. What is the force constant of the spring?

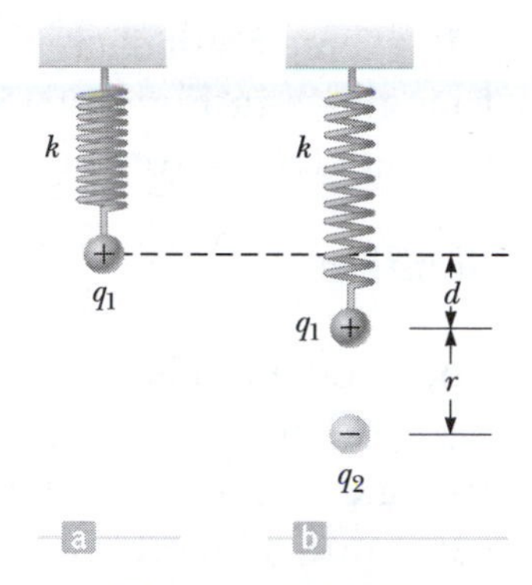

Figure P15.7

Solution

We assume the weight of the sphere attached to the spring is negligible in comparison to the magnitude of the other forces acting on this sphere. Then, we must consider two forces that act on this object. One is an upward force exerted on the sphere by the spring, with magnitude $F_s = kd$, where k is the force constant of the spring and d is the distance the spring has been stretched from its original length. The other force is a downward electrical force exerted on the upper positively charged sphere by the negatively charged sphere below it. The magnitude of this force is given by Coulomb's law as

$$F_e = \frac{|q_1||q_2|}{4\pi \in_0 r^2} = k_e \frac{|q_1||q_2}{r^2}$$

where $k_e = 8.99 \times 10^9 \text{ N} \cdot \text{m}^2/\text{C}^2$ and r is the distance separating the centers of the two small spheres. When the upper sphere comes to equilibrium, the net force acting on it is zero, so the upward spring force must equal the downward electrical force. Equating the magnitudes of these two forces gives the force constant of the spring as

$$k = \frac{F_e}{d} = \frac{k_e|q_1||q_2|/r^2}{d}$$

With $|q_1| = 0.800\,\mu\text{C}$, $|q_2| = 0.600\,\mu\text{C}$, $d = 3.50\,\text{cm}$, and $r = 5.00\,\text{cm}$, the force constant is found to be

$$k = \frac{(8.99\times10^9 \text{ N}\cdot\text{m}^2/\text{C}^2)(0.800\times10^{-6}\text{ C})(0.600\times10^{-6}\text{ C})}{(3.50\times10^{-2}\text{ m})(5.00\times10^{-2}\text{ m})^2} = 49.3 \text{ N/m} \qquad \Diamond$$

13. Three point charges are located at the corners of an equilateral triangle as in Figure P15.13. Find the magnitude and direction of the net electric force on the 2.00-μC charge.

y
7.00 μC
0.500 m
60.0°
x
2.00 μC
−4.00 μC

Figure P15.13

Solution

The 2.00-μC charge experiences an electrical force exerted by each of the charges on the other two corners of the equilateral triangle as shown in the force diagram at the lower right. Since each side of the triangle has length $r = 0.500$ m, Coulomb's law gives the magnitudes of these forces as

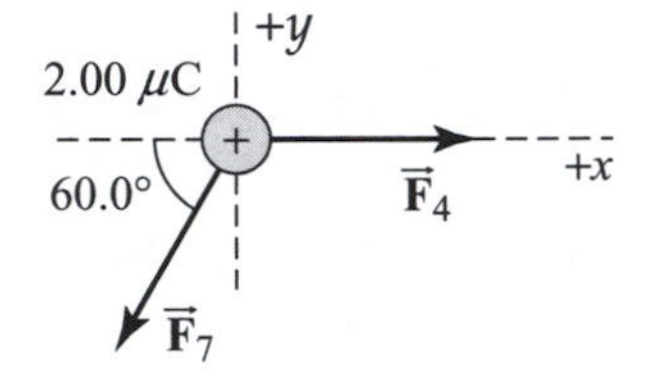

$$F_7 = \frac{(8.99\times 10^9\ \text{N}\cdot\text{m}^2/\text{C}^2)(7.00\times 10^{-6}\ \text{C})(2.00\times 10^{-6}\ \text{C})}{(0.500\ \text{m})^2}$$

$$= 0.503\ \text{N}$$

and $$F_4 = \frac{(8.99\times 10^9\ \text{N}\cdot\text{m}^2/\text{C}^2)(4.00\times 10^{-6}\ \text{C})(2.00\times 10^{-6}\ \text{C})}{(0.500\ \text{m})^2} = 0.288\ \text{N}$$

Therefore, the components of the net force on the 2.00-μC charge are

$$\Sigma F_x = F_4 - F_{7,\,x} = 0.288\ \text{N} - (0.503\ \text{N})\cos 60.0° = 3.65\times 10^{-2}\ \text{N}$$

and $$\Sigma F_y = 0 - F_{7,\,y} = 0 - (0.503\ \text{N})\sin 60.0° = -0.436\ \text{N}$$

The magnitude and direction of the resultant force are then

$$F_R = \sqrt{(\Sigma F_x)^2 + (\Sigma F_y)^2} = \sqrt{(3.65\times 10^{-2}\ \text{N})^2 + (-0.436\ \text{N})^2} = 0.438\ \text{N}$$

and $$\theta = \tan^{-1}\left(\frac{\Sigma F_y}{\Sigma F_x}\right) = \tan^{-1}\left(\frac{-0.436\ \text{N}}{3.65\times 10^{-2}\ \text{N}}\right) = -85.2°$$

The resultant electrical force acting on the 2.00-μC charge is

$\vec{\mathbf{F}}_R = 0.438$ N directed to the right and 85.2° below the $+x$-axis ◊

18. (a) Determine the electric field strength at a point 1.00 cm to the left of the middle charge shown in Figure P15.10. (b) If a charge of $-2.00\ \mu C$ is placed at this point, what are the magnitude and direction of the force on it?

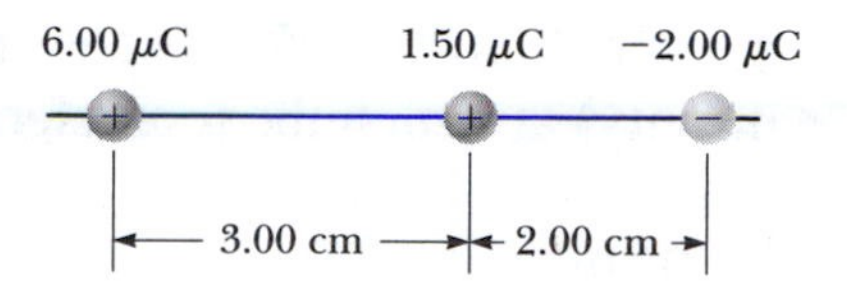

Figure P15.10

Solution

(a) The electric field due to a point charge points away from the charge if the charge is positive and toward the charge if the charge is negative. The electric field at distance r from a point charge Q has magnitude $E = k_e Q/r^2$. With this in mind, we sketch the contributions of each of the charges in Figure P15.10 to the resultant electric field at the point of interest (1.00 cm to the left of the charge $q_2 = +1.50\ \mu C$):

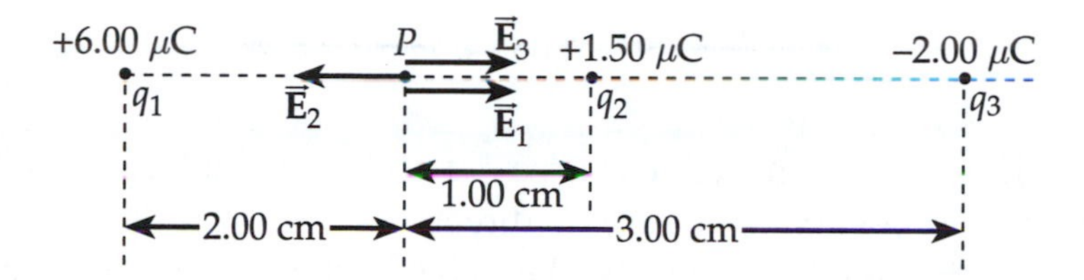

The contribution $\vec{\mathbf{E}}_1$ is due to the charge $q_1 = +6.00\ \mu C$, $\vec{\mathbf{E}}_2$ is due to $q_2 = +1.50\ \mu C$, while $\vec{\mathbf{E}}_3$ is due to $q_3 = -2.00\ \mu C$. In this case, all the contributions to the resultant field at point P lie along the same line. This means that the magnitude of the resultant field is the same as the algebraic sum of these contributions, or $E_R = +E_1 - E_2 + E_3$, where toward the right is taken as the positive direction. Thus, $E_R = k_e\left(|q_1|/r_1^2 - |q_2|/r_2^2 + |q_3|/r_3^2\right)$, which yields

$$E_R = \left(8.99\times 10^9\ \text{N}\cdot\text{m}^2/\text{C}^2\right)\left[\frac{6.00\ \mu\text{C}}{(0.020\,0\ \text{m})^2} - \frac{1.50\ \mu\text{C}}{(0.010\,0\ \text{m})^2} + \frac{2.00\ \mu\text{C}}{(0.030\,0\ \text{m})^2}\right]\left(\frac{10^{-6}\ \text{C}}{1\ \mu\text{C}}\right)$$

$$= +2.00\times 10^7\ \text{N/C}$$

and the resultant electric field is $\vec{\mathbf{E}}_R = +2.00\times 10^7\ \text{N/C}$ toward the right. ◊

(b) If a charge $q = -2.00\ \mu C$ is placed at point P, it will experience a force of

$$\vec{\mathbf{F}}_e = q\vec{\mathbf{E}}_R = -\left(2.00\times 10^{-6}\ \text{C}\right)\left(2.00\times 10^7\ \text{N/C}\right) = -40.0\ \text{N}$$

or $\quad \vec{\mathbf{F}}_e = 40.0\ \text{N}$ directed toward the left ◊

27. In Figure P15.27, determine the point (other than infinity) at which the total electric field is zero.

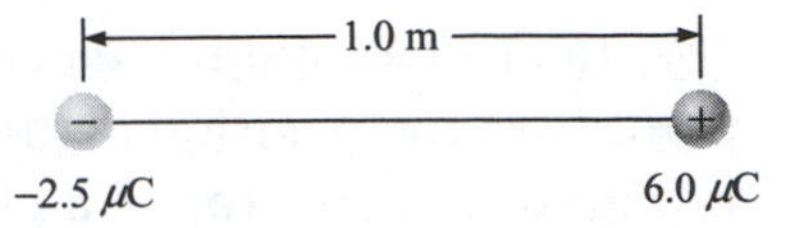

Figure P15.27

Solution

At each point in space, the electric field generated by the charge distribution shown in Figure P15.27 is the vector sum of two contributions, one due to each of the charges q_1 and q_2. In order for this field to be zero, the two contributions must have equal magnitudes and opposite directions.

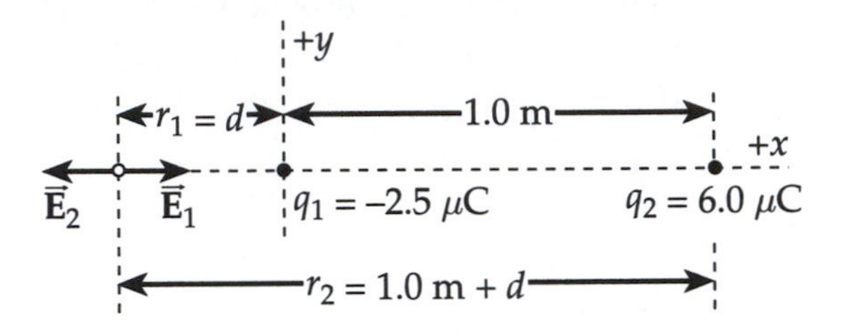

The contribution $\vec{\mathbf{E}}_1$ due to the negative charge q_1 will be directed toward q_1, while the contribution $\vec{\mathbf{E}}_2$ due to the positive charge q_2 is directed away from that charge. Only on the straight line passing through both charges (the x-axis in the above sketch) will these contributions lie along the same line. In the region of this line between the two charges, the two contributions are in the same direction and cannot add to zero. Thus, the point of interest must lie on the x-axis, either to the left of q_1 or to the right of q_2. If the two contributions to the resultant field are to have equal magnitudes, the point of interest must be located nearest to the smaller charge q_1. Thus, it must be located on the portion of the x-axis to the left of q_1 as shown in the sketch given above.

Using the notation given in the sketch, and requiring that $E_1 = E_2$, or $k_e \dfrac{|q_1|}{r_1^2} = k_e \dfrac{|q_2|}{r_2^2}$, we obtain

$$k_e \frac{2.5\ \mu\text{C}}{d^2} = k_e \frac{6.0\ \mu\text{C}}{(1.0\text{ m} + d)^2} \qquad \text{or} \qquad 1.0\text{ m} + d = \pm d\sqrt{\frac{6.0}{2.5}} = \pm 1.55d$$

Using the upper sign gives $\quad 1.0\text{ m} = 0.55d \quad$ and $\quad d = \dfrac{+1.0\text{ m}}{0.55} = +1.8\text{ m}$

so the point where the field is zero is on the x-axis, 1.8 m to the left of q_1. ◊

Note that using the lower sign would have given $d = \dfrac{-1.0\text{ m}}{2.55} = -0.39\text{ m}$, which is located between the two charges (since $d > 0$ is to the left of q_1 according to the notation adopted in the sketch). Here E_1 and E_2 have the same magnitudes but also the same direction and cannot add to zero. Thus, this is not an acceptable solution.

33. Two point charges are a small distance apart. (a) Sketch the electric field lines for the two if one has a charge four times that of the other and both charges are positive. (b) Repeat for the case in which both charges are negative.

Solution

In sketching the electric field patterns for given charge distributions, there are several points one should keep in mind.

Some of these are:

(1) When point charges are isolated in space, the field lines exhibit radial symmetry. That is, they are uniformly spaced around the charge and either radially outward away from the charge or radially inward toward the charge. When in the presence of other charges, the field lines will still be nearly radial at points very close to the charge but will become distorted farther out.

(2) Field lines originate on positive charges and terminate on negative charges.

(3) The number of field lines one should draw leaving a positive charge or approaching a negative charge is proportional to the charge. Therefore, the number of lines leaving (or approaching) a charge of magnitude $4q$ should be four times the number leaving (or approaching) a charge of magnitude q.

(4) Field lines never cross each other.

(5) If the charge distribution producing the electric field is symmetric about some line in space, the electric field must also be symmetric about that line.

(a) In the drawing below, for two point charges of magnitudes q_2 and $4q_2$ located near each other, not all of the lines leaving the charge $4q_2$ are shown for the sake of clarity in the drawing. ◊

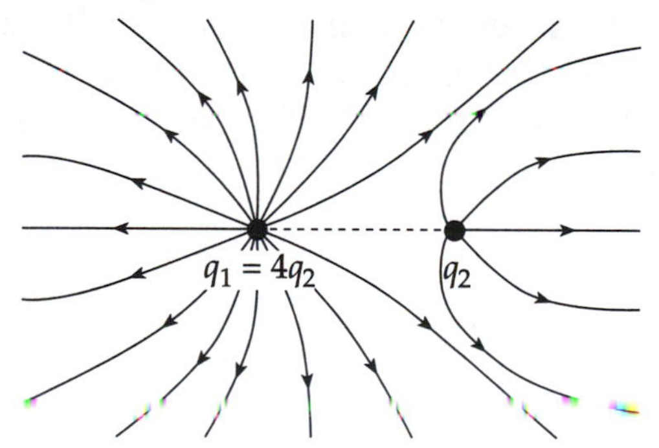

(b) If the charges q_1 and q_2 were negative instead of positive, the pattern of field lines would be as shown above with the exception that the arrows would point in the opposite direction along the lines (that is, in toward the negative charges). ◊

39. A Van de Graaff generator is charged so that a proton at its surface accelerates radially outward at 1.52×10^{12} m/s^2. Find (a) the magnitude of the electric force on the proton at that instant and (b) the magnitude and direction of the electric field at the surface of the generator.

Solution

(a) The proton is a positively charged particle with charge $q=+e=1.60\times10^{-19}$ C and mass $m_p=1.67\times10^{-27}$ kg. If the proton experiences a radially outward acceleration of magnitude $a=1.52\times10^{12}$ m/s^2, Newton's second law gives the magnitude of the electrical force producing this acceleration as

$$F_e=m_pa=\left(1.67\times10^{-27}\ \text{kg}\right)\left(1.52\times10^{12}\ \text{m/s}^2\right)=2.54\times10^{-15}\ \text{N}$$ ◊

(b) When a particle having charge q is in an electric field, the electrical force exerted on the particle is given by $\vec{\mathbf{F}}_e=q\vec{\mathbf{E}}$. Note that this means that the force and the electric field are in the same direction if the particle has positive charge and are in opposite directions if the particle is negatively charged. If the acceleration of the proton is radially outward, the resultant force acting on it must also be radially outward according to Newton's second law, $\vec{\mathbf{F}}_R=m\vec{\mathbf{a}}$. Since the proton is positively charged, the electric field is in the same direction as the force. The magnitude of this electric field is

$$E=\frac{F_e}{q}=\frac{F_e}{e}=\frac{2.54\times10^{-15}\ \text{N}}{1.60\times10^{-19}\ \text{C}}=1.59\times10^{4}\ \text{N/C}$$

Thus, the electric field at the surface of this Van de Graaff generator is $\vec{\mathbf{E}}=1.59\times10^{4}$ N/C in the outward radial direction. ◊

---

47. Suppose the conducting spherical shell of Figure 15.29 carries a charge of 3.00 nC and that a charge of -2.00 nC is at the center of the sphere. If $a=2.00$ m and $b=2.40$ m, find the electric field at (a) $r=1.50$ m, (b) $r=2.20$ m, and (c) $r=2.50$ m. (d) What is the charge distribution on the sphere?

Solution

With the positive point change at the center of the spherical shell, the charge on the surrounding shell will distribute itself uniformly around that shell, yielding complete spherical symmetry. Thus, anywhere that an electric field due to this charge distribution exists, that field will be in the radial direction, and its magnitude will depend only on the distance r from the center of the spherical shell.

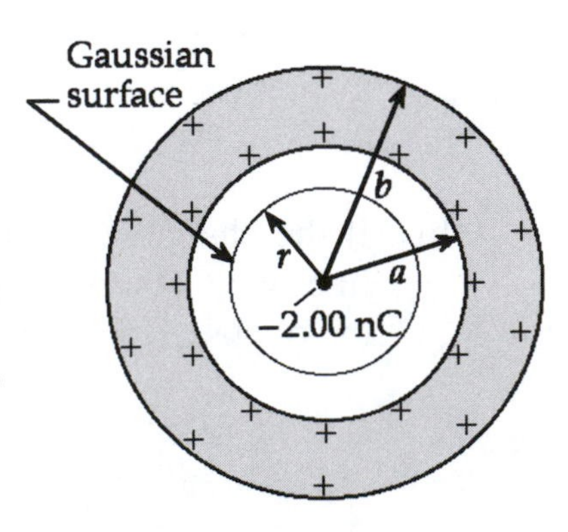

To take advantage of the symmetry known to exist in the field, we shall employ a spherical Gaussian surface that is concentric with the spherical shell. The flux through this surface (with outward, away from the interior, taken as the positive direction) is then

$$\Phi_E = EA\cos\theta = E\left(4\pi r^2\right)\cos 0° = 4\pi r^2 E$$

where r is the radius of the Gaussian surface, and E is the magnitude of the electric field at all points on this surface.

Gauss's law also tells us that the flux through our Gaussian surface is equal to the net charge enclosed within that surface divided by ϵ_0. Thus, we have

$$\Phi_E = 4\pi r^2 E = Q_{\text{inside}}/\epsilon_0 \qquad \text{or} \qquad E = Q_{\text{inside}}/4\pi \epsilon_0 r^2 = k_e Q_{\text{inside}}/r^2$$

(a) For $r < a$, the only charge inside the Gaussian surface is the negative point charge at the center of the shell, $Q_{\text{inside}} = -2.00\text{ nC}$. Hence, the electric field at $r = 1.50\text{ m}$ from the center of the spherical shell is

$$E = \frac{\left(8.99\times10^9\ \text{N}\cdot\text{m}^2/\text{C}^2\right)\left(-2.00\times10^{-9}\ \text{C}\right)}{(1.50\ \text{m})^2} = -7.99\ \text{N/C}$$

The negative sign means that the field is directed inward toward the interior of our Gaussian surface. The electric field at $r = 1.50\text{ m}$ is then

$$\vec{\mathbf{E}} = 7.99\ \text{N/C directed radially inward}$$ ◊

(b) All points located at distance $r = 2.20\text{ m}$ from the center of the shell are in the range $a < r < b$ or are located within the conducting material making up the shell. Since a static electric field cannot exist inside a conducting material, we must conclude that $\vec{\mathbf{E}} = 0$ at $r = 2.20\text{ m}$. ◊

(c) At $r = 2.50\text{ m}$, the Gaussian surface encloses both the point charge at the center of the shell and all of the charge on the shell itself. Thus,

$$Q_{\text{inside}} = +3.00\text{ nC} - 2.00\text{ nC} = +1.00\text{ nC},$$

and the field is

$$E = \frac{\left(8.99\times10^9\ \text{N}\cdot\text{m}^2/\text{C}^2\right)\left(+1.00\times10^{-9}\ \text{C}\right)}{(2.50\ \text{m})^2} = +1.44\ \text{N/C}$$

The positive sign indicates that the field at this distance from the center is directed outward, away from the interior of the closed Gaussian surface. The field at $r = 2.50\text{ m}$ is $\vec{\mathbf{E}} = 1.44\ \text{N/C}$ directed radially outward. ◊

(d) Let our Gaussian surface now have a radius that is infinitesimally greater than $a = 2.00\text{ m}$. This places all points on the Gaussian surface inside the conducting material of the shell where the electric field is zero. Thus, the flux through this surface, and the net charged enclosed by this surface, must both be zero. Since there is a point charge $q = -2.00$ nC at the center of the shell and no other charge in the space enclosed by the shell, the only way the net charge enclosed by the Gaussian surface can be zero is if 2.00 nC of positive charge resides on the inside surface of the shell at $r = a$. Because no point inside a conducting material can have a net charge under static conditions, the remainder of the 3.00 nC of net charge carried by the shell must reside on the outer surface at $r = b$. Our conclusion is that all the charge carried by the shell must reside on its two surfaces with the total amount on each surface being $q_{\text{inner}} = +2.00$ nC and $q_{\text{outer}} = +1.00$ nC. ◊

---

48. A very large nonconducting plate lying in the *xy*-plane carries a charge per unit area of σ. A second such plate located at $z = 2.00$ cm and oriented parallel to the *xy*-plane carries a charge per unit area of -2σ. Find the electric field (a) for $z < 0$, (b) $0 < z < 2.00$ cm, and (c) $z > 2.00$ cm.

Solution

Before solving this problem, the reader should carefully review Example 15.8 in the textbook. There, Gauss's law is used to determine the electric field produced by a nonconducting plane sheet of charge. The results of that analysis are that the field has a constant magnitude given by $E = |\sigma_{\text{sheet}}|/2\epsilon_0$ (where σ_{sheet} is the charge per unit area on the sheet), is everywhere perpendicular to the sheet, is directed away from the sheet if $\sigma_{\text{sheet}} > 0$, and is directed toward the sheet if $\sigma_{\text{sheet}} < 0$.

In this problem, we have two plane sheets of charge, both parallel to the *xy*-plane and separated by a distance of 2.00 cm. The upper sheet has charge density $\sigma_{\text{sheet}} = -2\sigma$, while the lower sheet has $\sigma_{\text{sheet}} = +\sigma$. At any location, the resultant electric field is the vector sum of the fields due to the two individual sheets of charge. Taking upward as the positive *z*-direction, the fields due to each of the sheets in the three regions of interest are:

	Lower sheet (at $z = 0$)	**Upper sheet (at $z = 2.00$ cm)**
Region	Electric Field	Electric Field
$z < 0$	$E_z = -\dfrac{\lvert+\sigma\rvert}{2\epsilon_0} = -\dfrac{\sigma}{2\epsilon_0}$	$E_z = +\dfrac{\lvert-2\sigma\rvert}{2\epsilon_0} = +\dfrac{\sigma}{\epsilon_0}$
$0 < z < 2.00$ cm	$E_z = +\dfrac{\lvert+\sigma\rvert}{2\epsilon_0} = +\dfrac{\sigma}{2\epsilon_0}$	$E_z = +\dfrac{\lvert-2\sigma\rvert}{2\epsilon_0} = +\dfrac{\sigma}{\epsilon_0}$
$z > 2.00$ cm	$E_z = +\dfrac{\lvert+\sigma\rvert}{2\epsilon_0} = +\dfrac{\sigma}{2\epsilon_0}$	$E_z = -\dfrac{\lvert-2\sigma\rvert}{2\epsilon_0} = -\dfrac{\sigma}{\epsilon_0}$

The resultant electric field in each of the regions of interest is then:

(a) For $z<0$: $E_z = E_{z,\text{lower}} + E_{z,\text{upper}} = -\dfrac{\sigma}{2\epsilon_0} + \dfrac{\sigma}{\epsilon_0} = +\dfrac{\sigma}{2\epsilon_0}$ ◊

(b) For $0<z<2.00$ cm : $E_z = E_{z,\text{lower}} + E_{z,\text{upper}} = +\dfrac{\sigma}{2\epsilon_0} + \dfrac{\sigma}{\epsilon_0} = +\dfrac{3\sigma}{2\epsilon_0}$ ◊

(c) For $z>2.00$ cm : $E_z = E_{z,\text{lower}} + E_{z,\text{upper}} = +\dfrac{\sigma}{2\epsilon_0} - \dfrac{\sigma}{\epsilon_0} = -\dfrac{\sigma}{2\epsilon_0}$ ◊

---

53. (a) Two identical point charges $+q$ are located on the y-axis at $y=+a$ and $y=-a$. What is the electric field along the x-axis at $x=b$? (b) A circular ring of charge of radius a has a total positive charge Q distributed uniformly around it. The ring is in the $x=0$ plane with its center at the origin. What is the electric field along the x-axis at $x=b$ due to the ring of charge? (*Hint*: Consider the charge Q to consist of many pairs of identical point charges positioned at the ends of diameters of the ring.)

Solution

(a) Any point on the x-axis is equidistant from the two identical point charges. Thus, the contributions by the two charges to the resultant field at this point have equal magnitudes given by

$$E_1 = E_2 = \frac{k_e q}{r^2}$$

The components of the resultant field are

$$E_y = E_{1y} - E_{2y}: \qquad E_y = \left(\frac{k_e q}{r^2}\right)\sin\theta - \left(\frac{k_e q}{r^2}\right)\sin\theta = 0$$

and

$$E_x = E_{1x} + E_{2x}: \qquad E_x = \left(\frac{k_e q}{r^2}\right)\cos\theta + \left(\frac{k_e q}{r^2}\right)\cos\theta = \left[\frac{k_e(2q)}{r^2}\right]\cos\theta$$

Since $r=\sqrt{a^2+b^2}$ and $\cos\theta = b/r$, the net x-component becomes

$$E_x = \frac{k_e(2q)b}{r^3} = \frac{k_e b(2q)}{\left(a^2+b^2\right)^{3/2}}$$

The resultant electric field on the axis at $x=b$ is

$$\vec{\mathbf{E}}_R = \frac{k_e b(2q)}{\left(a^2+b^2\right)^{3/2}} \text{ in the } +x\text{- direction}$$ ◊

(b) Note that the result of part (a) is $\vec{\mathbf{E}}_R = k_e b(Q)/\left(a^2+b^2\right)^{3/2}$ in the $+x$-direction, where $Q = 2q$ is the total charge in the charge distribution generating the field. In the case of a uniformly-charged circular ring, consider the ring to consist of many pairs of identical charges, located on opposite ends of a diameter of the ring and uniformly spaced around the ring. Each pair of charges has a total charge Q_i. At a point on the axis of the ring, and distance b from the center, this pair of charges generates an electric field contribution that is parallel to the axis and has magnitude

$$E_i = \frac{k_e b Q_i}{\left(a^2+b^2\right)^{3/2}}$$

The resultant electric field of the ring is the summation of the contributions by all such pairs of charges, or

$$E_R = \Sigma E_i = \left[\frac{k_e b}{\left(a^2+b^2\right)^{3/2}}\right]\Sigma Q_i = \frac{k_e b Q}{\left(a^2+b^2\right)^{3/2}}$$

where $Q = \Sigma Q_i$ is the total charge on the ring.

Thus, the electric field at a point on the axis of the ring (radius a), and distance b from the plane of the ring, is directed parallel to the axis, and if $Q > 0$, away from the ring. Its magnitude is

$$\left|\vec{\mathbf{E}}_R\right| = \frac{k_e b Q}{\left(a^2+b^2\right)^{3/2}}$$

◊

59. Two hard rubber spheres, each of mass $m = 15.0$ g, are rubbed with fur on a dry day and are then suspended with two insulating strings of length $L = 5.00$ cm whose support points are a distance $d = 3.00$ cm from each other as shown in Figure P15.59. During the rubbing process, one sphere receives exactly twice the charge of the other. They are observed to hang at equilibrium, each at an angle of $\theta = 10.0°$ with the vertical. Find the amount of charge on each sphere.

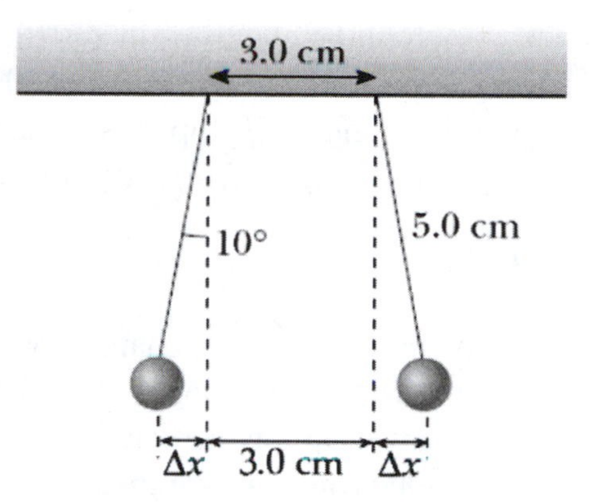

Figure P15.59 (modified)

Solution

Each sphere exerts a repulsive force of magnitude $F_e = k_e q_1 q_2 / r^2$, where $q_2 = 2q_1$, on the other. Consider the force diagram of one of the spheres given at the right. Here, T is the tension in the string supporting the sphere. With the spheres in equilibrium in the positions shown, Newton's second law gives

$$\Sigma F_y = T\cos 10.0° - mg = 0 \qquad \text{or} \qquad T = \frac{mg}{\cos 10.0°}$$

and $$\Sigma F_x = T\sin 10.0° - F_e = 0 \qquad \text{or} \qquad F_e = T\sin 10° = \left(\frac{mg}{\cos 10.0°}\right)\sin 10.0° = (mg)\tan 10.0°$$

Thus, $$k_e \frac{q_1(2q_1)}{r^2} = (mg)\tan 10.0° \quad \text{and} \quad q_1 = r\sqrt{\frac{(mg)\tan 10.0°}{2k_e}}$$

where the equilibrium distance between the centers of the two spheres is

$$r = d + 2(\Delta x) = d + 2(L\sin 10.0°) = 3.00 \text{ cm} + 2[(5.00 \text{ cm})\sin 10.0°] = 4.74 \text{ cm}$$

Our estimate of the charge q_1 on one sphere is

$$q_1 = (0.0474 \text{ m})\sqrt{\frac{(15.0\times 10^{-3} \text{ kg})(9.80 \text{ m/s}^2)\tan 10.0°}{2(8.99\times 10^9 \text{ N}\cdot\text{m}^2/\text{C}^2)}} = 5.69\times 10^{-8} \text{ C} \qquad \lozenge$$

The charge on the other sphere is then $q_2 = 2q_1 = 2(5.69\times 10^{-8} \text{ C}) = 1.14\times 10^{-7} \text{ C}$. ◊

63. Each of the electrons in a particle beam has a kinetic energy of 1.60×10^{-17} J. (a) What is the magnitude of the uniform electric field (pointing in the direction of the electrons' movement) that will stop these electrons in a distance of 10.0 cm? (b) How long will it take to stop the electrons? (c) After the electrons stop, what will they do? Explain.

Solution

(a) When a negatively charged particle, such as an electron, is in an electric field, it experiences a force of magnitude $F_e = |q|E$ in the direction opposite to that of the field. Thus, the electrons in this particle beam experience a retarding force of magnitude $F_e = eE$. When an electron moves a distance $\Delta x = 0.100$ m against this retarding force, the work done on it by the field is $W = F_e(\Delta x)\cos 180° = -eE(\Delta x)$. If the electrons are brought to rest $(KE_f = 0)$ in this distance, the work-energy theorem gives $W = \Delta KE$, or

$$-eE(\Delta x) = 0 - KE_i \qquad \text{and} \qquad E = \frac{KE_i}{e(\Delta x)}$$

The required magnitude of the field is then

$$E = \frac{1.60 \times 10^{-17}\ \text{J}}{(1.60 \times 10^{-19}\ \text{C})(0.100\ \text{m})} = 1.00 \times 10^{3}\ \text{N/C} \qquad \Diamond$$

(b) The acceleration of the electrons is $a_x = -F_e/m_e = -eE/m_e$. The time required for them to come to rest is

$$t = \frac{v_f - v_i}{a_x} = \frac{0 - v_i}{-eE/m_e} = \frac{m_e v_i}{eE} = \frac{\sqrt{2m_e(KE_i)}}{eE}$$

or $$t = \frac{\sqrt{2(9.11 \times 10^{-31}\ \text{kg})(1.60 \times 10^{-17}\ \text{J})}}{(1.60 \times 10^{-19}\ \text{C})(1.00 \times 10^{3}\ \text{N/C})} = 3.37 \times 10^{-8}\ \text{s} = 33.7\ \text{ns} \qquad \Diamond$$

(c) After bringing the electron to rest, the electric force continues to act on it, causing it to accelerate in the direction opposite to the field (hence, opposite to the direction of the electron's original motion) at a rate of

$$|a_x| = \frac{eE}{m_e} = \frac{(1.60 \times 10^{-19}\ \text{C})(1.00 \times 10^{3}\ \text{N/C})}{9.11 \times 10^{-31}\ \text{kg}} = 1.76 \times 10^{14}\ \text{m/s}^2 \qquad \Diamond$$

16

Electrical Energy and Capacitance

NOTES FROM SELECTED CHAPTER SECTIONS

16.1 Potential Difference and Electric Potential

The electrostatic force is conservative; it is possible to define an electric potential energy function associated with this force. *The potential difference between two points is proportional to the change in potential energy of a charge as it moves between the two points.*

Electrical potential difference (a scalar quantity) is the negative of the work done by the electric field (the conservative force) in moving a charge from point A to point B divided by the magnitude of the charge. The SI unit of potential is the joule per coulomb or volt (V).

A positive charge gains electrical potential energy when it is moved opposite to the direction of the electric field. When a positive charge is placed in an electric field, it moves in the direction of the field, from a point of high potential to a point of lower potential.

16.2 Electric Potential and Potential Energy Due to Point Charges

In electric circuits, a point of zero potential is often defined by grounding (connecting to Earth) some point in the circuit. In the case of a point charge, the point of zero potential is taken to be at an infinite distance from the charge. The electric potential at a given point in space due to a point charge q depends only on the value of the charge and the distance, r, from the charge to the specified point in space. An electric potential can exist at a point in space whether or not a test charge exists at that point.

When using the superposition principle to determine the value of the electric potential at a point due to several point charges, the **algebraic sum** of the individual electric potentials must be used. *Be careful not to confuse electric potential difference with electric potential energy.*

Electric potential is a scalar property of the region surrounding an electric charge. It does not depend on the presence of a test charge in the field.

Electric potential energy is a characteristic of a charge-field system. It is due to the interaction of the field and a charge located within the field.

In the figure to the right an electric field (due to the presence of $+Q$) is present in the region surrounding the charge. This field is due to a point charge and has the properties described in Chapter 15.

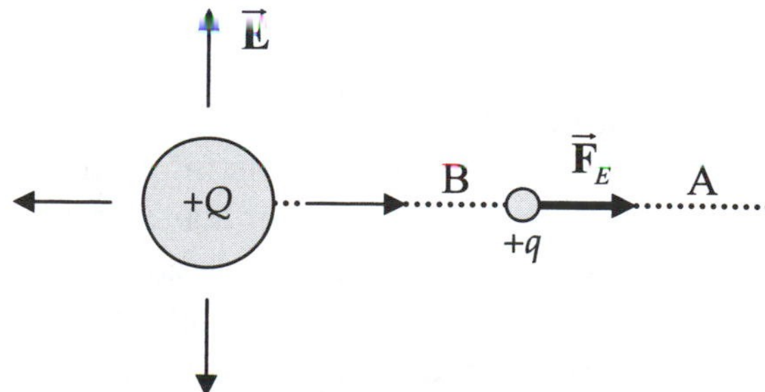

Refer to the figure as you consider the following statements about electric potential difference and electric potential energy.

- Electric field lines, four of which are shown, are directed along outward radials from the charge Q. Points "A" and "B" are shown along one of the field lines.
- If a positive test charge q is placed at some point between A and B, the electric field will exert a force $\vec{\mathbf{F}}_E = q\vec{\mathbf{E}}$ on q directed along the field line and away from Q.
- If the test charge q is moved from point A to point B (where the potential is greater), the electric field will do work, $W = -q\Delta V = -q(V_B - V_A)$, and the electric potential energy of the two-charge system (Q and q) increases by an amount equal to $q\Delta V$. ***The electric potential energy of a positive charge increases when the charge moves to a point closer to another isolated positive point charge (or along a direction opposite the direction of the electric field).***
- ***Since the electrostatic force is a conservative force, the work done by the electric field as the test charge q moves from point A to point B will be the same regardless of the actual path taken.***
- You should consider the result of changing either or both Q and q from a positive to a negative charge.

16.3 Potentials and Charged Conductors
16.4 Equipotential Surfaces

A surface on which every point has the same value of electric potential is an **equipotential surface**. The electric field at every point on an equipotential surface is perpendicular to the surface because a parallel component of $\vec{\mathbf{E}}$ would imply work would be required to move a charge between two points on the surface.

No work is required to move a charge between two points that are at the same potential. That is, $W = 0$ when $V_B = V_A$.

The electric potential is a constant everywhere on the surface of a charged conductor in electrostatic equilibrium.

The electric potential is constant everywhere inside a conductor and equal to its value at the surface. This is true because the electric field is zero inside a conductor, and no work is required to move a charge between any two points inside the conductor.

The electron volt (eV) is defined as the kinetic energy that an electron (or proton) gains (or loses) when accelerated through a potential difference of 1 V.

16.6 Capacitance

A capacitor is a device consisting of a pair of conductors (plates) separated by insulating material. A charged capacitor acts as a storehouse of charge and energy in an electric field. The capacitance of a capacitor depends on the following physical characteristics of the device: size, shape, plate separation, and the nature of the dielectric medium filling the region between the plates.

The capacitance, C, of a capacitor is defined as the ratio of the magnitude of the charge on either conductor to the magnitude of the potential difference between the conductors. Capacitance has SI units of coulombs per volt, called **farads** (F). The farad is a very large unit of capacitance. In practice, most capacitors have capacitances ranging from picofarads to microfarads.

An **air-filled parallel-plate capacitor** has a capacitance that is proportional to the area of overlap between the plates and inversely proportional to the separation of the plates.

$$C = \frac{\varepsilon_0 A}{d} \quad (16.9)$$

16.8 Combinations of Capacitors

Two or more capacitors can be connected in a circuit in several possible combinations. For example, three capacitors (each assumed to have the same value of capacitance, C) can be combined as illustrated in the figure at right to achieve four different values of equivalent capacitance.

(a) all in parallel — $C_{eq} = 3C$

(b) all in series — $C_{eq} = \frac{1}{3}C$

(c) 2 in series, in parallel with a third — $C_{eq} = \frac{3}{2}C$

(d) 2 in parallel, in series with a third — $C_{eq} = \frac{2}{3}C$

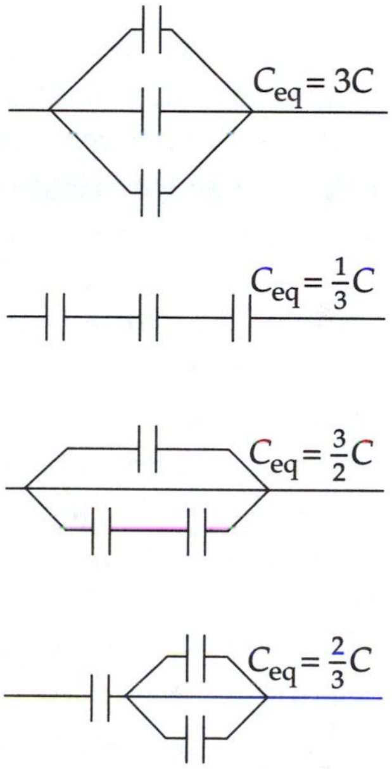

Can you determine how many different values of equivalent capacitance could be achieved by combining the three capacitors if they each had a different capacitance, C_1, C_2, and C_3?

Note in the figure above that when a group of capacitors is connected in series, they are arranged "end-to-end," and only adjacent capacitors have a circuit point in common. When connected in parallel, each capacitor in the group has two circuit points in common with each of the others.

For capacitors connected in series:

- The reciprocal of the equivalent capacitance equals the sum of the reciprocals of the individual capacitances. See Equation 16.15.
- The potential difference across the series group equals the sum of the potential differences across the individual capacitors.
- Each capacitor of a group in series has the same charge (even if their capacitances are not the same), and this is also the value of the total charge on the series group.

For capacitors connected in parallel:

- The equivalent capacitance equals the sum of the individual capacitances. See Equation 16.12.
- The potential difference across each capacitor has the same value.
- The total charge on the parallel group equals the sum of the charges on the individual capacitors.

16.9 Energy Stored in a Charged Capacitor

The energy stored in a given charged capacitor is proportional to the square of the potential difference between the two plates.

Each capacitor has a limiting voltage that depends on the capacitor's physical characteristics. When the potential difference between the plates of a capacitor exceeds that limiting voltage, a discharge will occur through the insulating material between the two plates. This is known as electrical breakdown. Therefore, the maximum energy which can be stored in a capacitor is limited by the breakdown voltage.

16.10 Capacitors with Dielectrics

A dielectric is an insulating material, such as rubber, glass, waxed paper, or air. When a dielectric is inserted between the plates of a capacitor, the capacitance increases. If the dielectric completely fills the space between the plates, the capacitance is multiplied by a dimensionless factor κ, called the **dielectric constant.** The dielectric constant is a property of the dielectric material and has a value of 1.000 for a vacuum.

The smallest plate separation for a capacitor is limited by the electric discharge that can occur through the dielectric material separating the plates. For any given plate separation, there is a maximum electric field that can be produced in the dielectric before it breaks down and begins to conduct. This maximum electric field is called the **dielectric strength** and has SI units of V/m.

EQUATIONS AND CONCEPTS

The **change in electric potential energy** of an electric charge in moving between two points in an electric field is equal to the negative of the work done by the electric force.

$$\Delta PE = -W_{AB} = -(qE_x)\Delta x \quad (16.1)$$

In a **uniform electric field** the change in electric potential energy can be expressed in terms of the charge, the magnitude of the field, and the distance moved parallel to the direction of the field.

$$\Delta PE = -qEd$$

(in a uniform electric field)

The **electric potential difference** between points A and B is defined as the change in electric potential energy per unit charge as a charge is moved from point A to point B. The SI unit of electric potential is the volt, V.

$$\Delta V = V_B - V_A = \frac{\Delta PE}{q} \quad (16.2)$$

$$1\ \text{V} \equiv 1\ \text{J/C}$$

The **potential difference in a uniform field** (constant in magnitude and direction) depends only on the displacement Δx in a direction parallel to the field. *Electric field lines always point along the direction of decreasing potential.*

$$\Delta V = -E_x \Delta x \quad (16.3)$$

The **electric potential due to a point charge** is inversely proportional to the distance from the charge. *This equation assumes that the potential at infinity is zero. Remember that electric potential is a scalar, and the correct algebraic sign for the charge, q, must be used in Equation 16.4.*

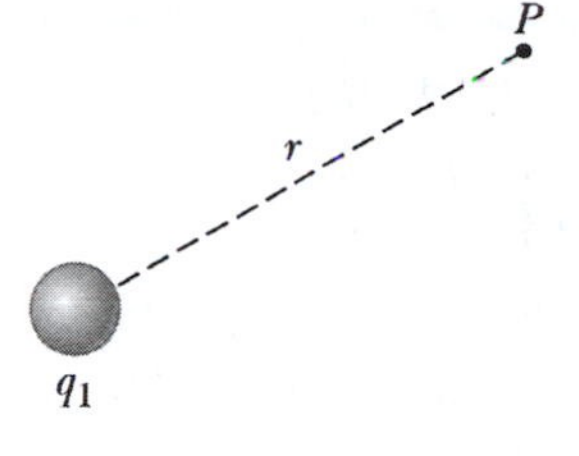

$$V = k_e \frac{q}{r} \quad (16.4)$$

The **total electric potential due to several point charges** is the algebraic sum of the potentials due to the individual charges.

$$V = k_e \sum_i \frac{q_i}{r_i}$$

The potential energy of a pair of charges separated by a distance, r, represents the minimum work required to assemble the charges from an infinite separation. *The potential energy of the two charges is positive if the two charges have the same sign; and it is negative if the two charges are of opposite sign.*

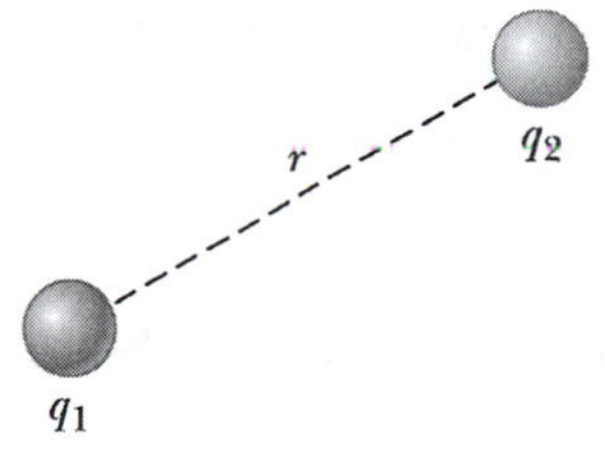

$$PE = k_e \frac{q_1 q_2}{r} \quad (16.5)$$

The **work required to move a charge** q from point A to point B in an electric field depends on the magnitude and sign of the charge. *Zero work is required to move a charge between two points that are at the same potential.*

$$W = -q(V_B - V_A) \quad (16.6)$$

$$1\ \text{eV} = 1.60 \times 10^{-19}\ \text{J} \quad (16.7)$$

Important characteristics of a charged conductor in electrostatic equilibrium:

- All excess charge resides on the surface.
- $\vec{\mathbf{E}} = 0$ at all interior points.
- $\vec{\mathbf{E}}$ is perpendicular to the surface on the outside.
- Electric potential is constant in the interior and equal to its value on the surface.

The **capacitance** (C) **of a capacitor** is defined as the ratio of the magnitude of the charge on either plate (conductor) to the potential difference between the plates (C is always positive). The SI unit of capacitance is the farad (F).

$$C \equiv \frac{Q}{\Delta V} \qquad (16.8)$$

$$1\text{ F} \equiv 1\text{ C/V}$$

1 microfarad = 10^{-6} F

1 picofarad = 10^{-12} F

The **capacitance of an air-filled parallel-plate capacitor** is proportional to the area of one of the plates and inversely proportional to the separation of the plates. *Note*: If the two plates do not exactly overlap, A in Equation 16.9 is equal to the actual area of overlap between the plates.

$$C = \epsilon_0 \frac{A}{d} \qquad (16.9)$$

When a **dielectric material** (or insulator) is inserted between the plates of a capacitor, the capacitance of the device increases by a factor κ. Kappa, called the dielectric constant, is dimensionless and is characteristic of a specific material.

$$C = \kappa \epsilon_0 \left(\frac{A}{d}\right) \qquad (16.19)$$

The **permittivity** of free space is expressed in SI units.

$$\epsilon_0 = \frac{1}{4\pi k_e} = 8.85 \times 10^{-12}\ \text{C}^2/\text{N}\cdot\text{m}^2$$

For a **parallel combination of capacitors**:

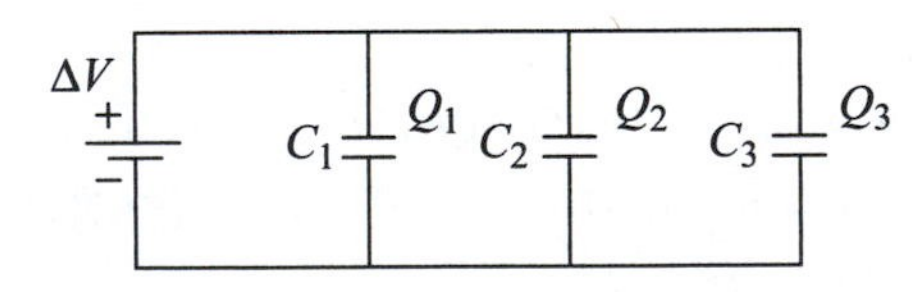

Capacitors in parallel

- *Each capacitor has two circuit points in common* with each of the other capacitors in the group.
- *The equivalent capacitance* equals the sum of the individual capacitances.

$$C_{eq} = C_1 + C_2 + C_3 + \ldots \qquad (16.12)$$

- The *total charge* is the sum of the charges on the individual capacitors.

$$Q_{total} = Q_1 + Q_2 + Q_3 + \ldots$$

- The *potential difference* is the same for each capacitor.

$$\Delta V_{total} = \Delta V_1 = \Delta V_2 = \Delta V_3 = \ldots$$

For **series combination of capacitors**:

- Adjacent capacitors have one circuit point in common.

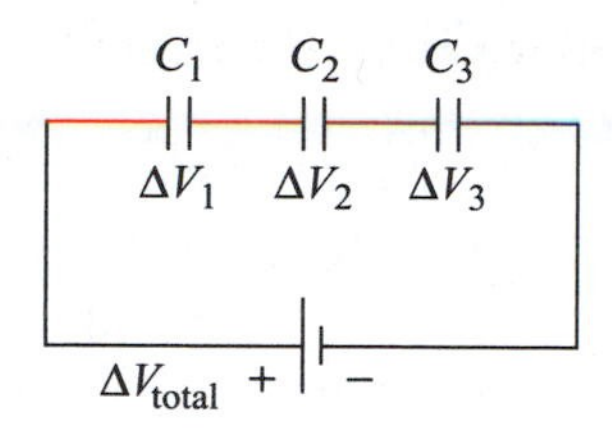

Capacitors in series

- The reciprocal of the equivalent capacitance is equal to the sum of the reciprocals of the individual capacitances.

$$\frac{1}{C_{eq}} = \frac{1}{C_1} + \frac{1}{C_2} + \frac{1}{C_3} + \ldots \qquad (16.15)$$

- The potential difference across the group equals the sum of the potential differences across each capacitor.

$$\Delta V = \Delta V_1 + \Delta V_2 + \Delta V_3 + \ldots$$

- The total charge on the series combination is equal to the charge on each capacitor, each capacitor having the same charge.

$$Q_{total} = Q_1 = Q_2 = Q_3 = \ldots$$

The **electrostatic potential energy** stored in the electric field of a charged capacitor is equal to the work done by a battery (or other source of emf) in charging the capacitor from $q = 0$ to $q = Q$.

$$W = \tfrac{1}{2} Q \Delta V \qquad (16.16)$$

$$\text{Energy stored} = \tfrac{1}{2} Q\Delta V = \tfrac{1}{2} C(\Delta V)^2 = \frac{Q^2}{2C} \qquad (16.17)$$

SUGGESTIONS, SKILLS, AND STRATEGIES

A STRATEGY FOR PROBLEMS INVOLVING ELECTRIC POTENTIAL

1. When working problems involving electric potential, remember that electric potential is a scalar quantity (rather than a vector quantity like the electric field), so there are no components to worry about. Therefore, when using the superposition principle to evaluate the electric potential at a point due to a system of point charges, you simply take the algebraic sum of the electric potentials due to each charge. However, you must keep track of signs. The electric potential due to each positive charge is positive; the electric potential due to each negative charge is negative. The basic equation to use is $V = k_e q / r$.

2. As in mechanics, only changes in potential energy are significant; hence, the point where you choose the potential energy to be zero is arbitrary. However, it is common practice to set $V = 0$ at infinity when this choice suits the particular situation.

PROBLEM-SOLVING HINTS FOR CAPACITANCE

- When analyzing a series-parallel combination of capacitors to determine the equivalent capacitance, you should make a sequence of circuit diagrams which show the successive steps in the simplification of the circuit. At each step, combine those capacitors which are in simple-parallel or simple-series relationship to each other, and use appropriate equations for series or parallel capacitors at each step of the simplification. At each step, you know two of the three quantities: Q, ΔV, and C. You will be able to determine the remaining quantity using the relation $Q = C\Delta V$.

- When calculating capacitance, be careful with your choice of units. To calculate capacitance in farads, make sure that distances are in meters and use the SI value of ϵ_0. When checking consistency of units, remember that the electric field has units of newtons per coulomb (N/C) or volts per meter (V/m).

- When two or more unequal capacitors are connected in series, they carry the same charge, but their potential differences are not the same. The capacitances add as reciprocals, and the equivalent capacitance of the combination is always less than that of the smallest individual capacitor.

- When two or more capacitors are connected in parallel, the potential difference is the same for each capacitor. The charge on each capacitor is proportional to its capacitance; the capacitances add directly to give the equivalent capacitance of the parallel combination.

- A dielectric increases capacitance by the factor κ (the dielectric constant). This occurs because induced surface charges on the dielectric reduce the electric field inside the material from E to (E/κ).

- Be careful about problems in which you may be connecting or disconnecting a battery to a capacitor. It is important to note whether modifications to the capacitor are being made while the capacitor is connected to the battery or after it is disconnected. If the capacitor remains connected to the battery, the potential difference across the capacitor necessarily remains the same (equal to that of the battery), and the charge is proportional to the capacitance. However, it may be modified, for example, by insertion of a dielectric. On the other hand, if you disconnect the capacitor from the battery before making any modifications to the capacitor, then its charge remains the same. In this case, as you vary the capacitance, the potential difference across the plates changes in inverse proportion to capacitance, according to $\Delta V = Q/C$.

REVIEW CHECKLIST

- Understand that each point in the vicinity of a charge distribution can be characterized by a scalar quantity called the electric potential, V; and define the quantity, electrical potential difference. (Section 16.1)
- Calculate the electric potential difference between any two points in a uniform electric field and calculate the electric potential difference between any two points in the vicinity of a group of point charges. (Sections 16.1–16.2)
- Calculate the electric potential energy associated with a group of point charges and define the unit of energy, the electron volt. (Section 16.3)
- Justify the claims that (i) all points on the surface and within a charged conductor are at the same potential and (ii) the electric field within a charged conductor is zero. (Sections 16.3–16.4)
- Define the quantity capacitance; evaluate the capacitance of a parallel-plate capacitor of given area and plate separation. (Sections 16.6–16.7)
- Determine the equivalent capacitance of a network of capacitors in series-parallel combination and calculate the final charge on each capacitor and the potential difference across each when a known potential is applied across the combination. (Section 16.8)
- Calculate the energy stored in a charged capacitor. (Section 16.9)

SOLUTIONS TO SELECTED END-OF-CHAPTER PROBLEMS

1. A uniform electric field of magnitude 375 N/C pointing in the positive x-direction acts on an electron, which is initially at rest. After the electron has moved 3.20 cm, what is (a) the work done by the field on the electron, (b) the change in potential energy associated with the electron, and (c) the velocity of the electron?

Solution

(a) A particle of charge q in an electric field $\vec{\mathbf{E}}$ experiences an electrical force $\vec{\mathbf{F}}_e = q\vec{\mathbf{E}}$. Since the electron has a negative charge, $q = -e$, the electrical force (and the displacement of the electron after it is released from rest) is in the negative x-direction direction, opposite to the direction of the field. Taking to the right as the positive x-direction, the work done on the electron after it has moved 3.20 cm to the left is

$$W = F_e(\Delta x)\cos\theta = qE(\Delta x)\cos\theta$$

giving $W = (-1.60\times10^{-19}\ \text{C})(+375\ \ \text{N/C})(-3.20\times10^{-2}\ \text{m})\cos 0° = +1.92\times10^{-18}\ \text{J}$ ◊

(b) As a charged particle moves through an electric field, the change in the electrical potential energy of the particle is the negative of the work done on the particle by that field. Hence, the change in the electrical potential energy of the electron when it moves 3.20 cm to the left is

$$\Delta PE = PE_f - PE_i = -W = -\left(+1.92\times10^{-18}\text{ J}\right) = -1.92\times10^{-18}\text{ J}$$ ◊

(c) Electrical forces are conservative forces, so the mechanical energy of the electron is conserved, or $KE_f + PE_f = KE_i + PE_i$. Since the electron starts from rest ($KE_i = 0$), the final kinetic energy of the electron is $KE_f = PE_i - PE_f = -\Delta PE$, and the velocity of the electron after moving 3.20 cm in the negative x-direction is

$$v = -\sqrt{\frac{2KE_f}{m}} = -\sqrt{\frac{2(-\Delta PE)}{m}} = -\sqrt{\frac{2\left[-\left(-1.92\times10^{-18}\text{ J}\right)\right]}{9.11\times10^{-31}\text{ kg}}} = -2.05\times10^{6}\text{ m/s}$$

or $\vec{\mathbf{v}} = 2.05\times10^{6}$ m/s in the negative x-direction. ◊

---

10. On planet Tehar, the free-fall acceleration is the same as that on Earth, but there is also a strong downward electric field that is uniform close to the planet's surface. A 2.00-kg ball having a charge of 5.00 μC is thrown upward at a speed of 20.1 m/s. It hits the ground after an interval of 4.10 s. What is the potential difference between the starting point and the top point of the trajectory?

Solution

Since both the gravitational and electric fields this charged ball moves in may be considered uniform over the range of motion, the ball will have a constant vertical acceleration. We use the kinematics equation $y = v_{0y}t + a_y t^2/2$, with $y = 0$ at the time $t = t_f$, to obtain

$$a_y = -\frac{2v_{0y}}{t_f} \qquad \textbf{[1]}$$

We also make use of Newton's second law to find

$$a_y = \frac{\Sigma F_y}{m} = \frac{-mg + qE_y}{m} \quad \text{or} \quad a_y = -\left(g - \frac{qE_y}{m}\right) \qquad \textbf{[2]}$$

Equating the results in Equations [1] and [2] gives $g - \dfrac{qE_y}{m} = \dfrac{2v_{0y}}{t_f}$,

or the electric field strength is

$$E_y = \frac{m}{q}\left(g - \frac{2v_{0y}}{t_f}\right) = \frac{2.00\text{ kg}}{5.00\times10^{-6}\text{ C}}\left[9.80\text{ m/s}^2 - \frac{2(20.1\text{ m/s})}{4.10\text{ s}}\right] = -1.95\times10^{3}\text{ N/C}$$

The negative sign in this result simply means the electric field is directed downward. Next, we use the kinematics equation $v_y^2 = v_{0y}^2 + 2a_y(\Delta y)$, with $v_y = 0$ at $(\Delta y) = (\Delta y)_{\text{max}}$, and make use of Equation [1] again to find

$$(\Delta y)_{max} = \frac{0 - v_{0y}^2}{2a_y} = \frac{-v_{0y}^2}{2}\left(-\frac{t_f}{2v_{0y}}\right) = \frac{v_{0y}t_f}{4} = \frac{(20.1 \text{ m/s})(4.10 \text{ s})}{4} = 20.6 \text{ m}$$

The electrical potential difference between the starting point and highest point in the trajectory is then

$$\Delta V = V_{top} - V_{ground} = \frac{\Delta PE_e}{q} = \frac{-(\text{work done by field})}{q} = \frac{-\left[qE_y(\Delta y)_{max}\right]}{q} = -E_y(\Delta y)_{max}$$

or $\Delta V = -\left(-1.95\times10^3 \text{ N/C}\right)(+20.6 \text{ m}) = 4.02\times10^4 \text{ J/C} = 40.2\times10^3 \text{ V} = 40.2 \text{ kV}$ ◊

---

17. The three charges in Figure P16.17 are at the vertices of an isosceles triangle. Let $q = 7.00$ nC and calculate the electric potential at the midpoint of the base.

Solution

The electrical potential V_i at distance r_i from an isolated point charge Q_i is given by

$$V_i = \frac{Q_i}{4\pi \in_0 r_i} = \frac{k_e Q_i}{r_i}$$

where $k_e = 8.99\times10^9 \text{ N}\cdot\text{m}^2/\text{C}^2$. The charge distribution shown in Figure P16.17 consists of a set of three individual point charges, each having a magnitude or absolute value of q, with two negative charges at the corners of the base of the triangle and a positive charge at the apex of the triangle. Each of these charges makes its own contribution V_i to the electrical potential at point P, with the total electrical potential at this point being the algebraic sum of these contributions, or

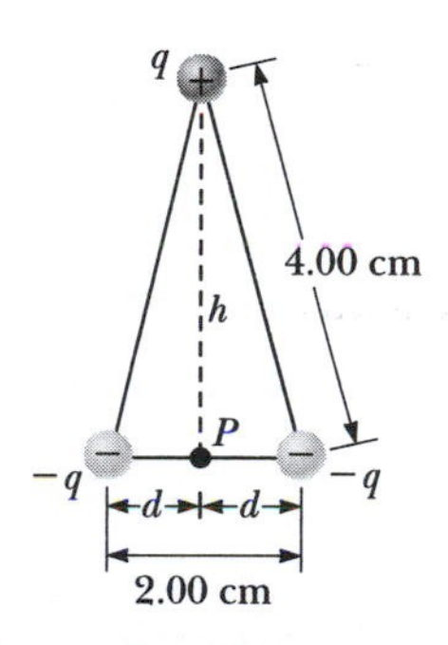

Figure P16.17 (modified)

$$V_P = \sum_i V_i = \sum_i \frac{k_e Q_i}{r_i} = \frac{k_e(-q)}{d} + \frac{k_e(-q)}{d} + \frac{k_e(+q)}{h} = k_e q\left(\frac{-2}{d} + \frac{1}{h}\right)$$

From the figure given above, we observe that $d = 1.00$ cm while the Pythagorean theorem gives $h = \sqrt{(4.00 \text{ cm})^2 - d^2} = \sqrt{(4.00 \text{ cm})^2 - (1.00 \text{ cm})^2} = \sqrt{15.0}$ cm. Thus, with $q = 7.00$ nC, the total electrical potential at point P due to the given charge distribution is

$$V_P = \left(8.99\times10^9 \frac{\text{N}\cdot\text{m}^2}{\text{C}^2}\right)\left(7.00\times10^{-9} \text{ C}\right)\left(\frac{-2}{1.00\times10^{-2} \text{ m}} + \frac{1}{\sqrt{15.0}\times10^{-2} \text{ m}}\right)$$

or $$V_P = -1.10\times10^4 \frac{\text{N}\cdot\text{m}}{\text{C}} = -11.0\times10^3 \frac{\text{J}}{\text{C}} = -11.0\times10^3 \text{ V} = -11.0 \text{ kV}$$ ◊

---

20. A proton and an alpha particle (charge $2e$, mass 6.64×10^{-27} kg) are initially at rest, separated by 4.00×10^{-15} m. (a) If they are both released simultaneously, explain why you can't find their velocities at infinity using only conservation of energy. (b) What other conservation law can be applied in this case? (c) Find the speeds of the proton and alpha particle, respectively, at infinity.

Solution

(a) The proton and the alpha particle are both positively charged and repel one another. Thus, when they are simultaneously released from rest, they accelerate in opposite directions, converting electrical potential energy into kinetic energy. The gain in kinetic energy is unequally divided between the two particles because of the difference in their masses, so the proton and the alpha particle will have two different final speeds. However, use of conservation of energy alone only provides one equation relating these two unknown final speeds and will not permit a full solution. ◊

(b) A second equation involving the two unknown final speeds may be obtained by applying the principle of conservation of linear momentum. Then, one will have two equations with two unknowns, and solving these equations simultaneously will yield values for both unknowns. ◊

(c) From conservation of energy: $KE_f + PE_f = KE_i + PE_i$, we have

$$\left(\frac{1}{2}m_p v_p^2 + \frac{1}{2}m_\alpha v_\alpha^2\right) + 0 = 0 + k_e\frac{q_p q_\alpha}{r_i} \quad \text{and} \quad m_p v_p^2 + m_\alpha v_\alpha^2 = 2k_e\frac{q_p q_\alpha}{r_i} \qquad \textbf{[1]}$$

From conservation of linear momentum:

$$m_p v_p + m_\alpha v_\alpha = 0 \quad \text{or} \quad v_\alpha = -\left(m_p/m_\alpha\right)v_p \qquad \textbf{[2]}$$

Substituting Equation [2] into [1] gives $m_p v_p^2 + m_\alpha\left(m_p/m_\alpha\right)^2 v_p^2 = 2k_e\dfrac{q_p q_\alpha}{r_i}$

and reduces to $\left(m_\alpha + m_p\right)\left(\dfrac{m_p}{m_\alpha}\right)v_p^2 = 2k_e\dfrac{q_p q_\alpha}{r_i}$ or $v_p = \sqrt{\left(\dfrac{m_\alpha}{m_p}\right)\dfrac{2k_e q_p q_\alpha}{\left(m_\alpha + m_p\right)r_i}}$

Thus, the final speed of the proton is

$$v_p = \sqrt{\left(\frac{6.64\times10^{-27}\text{ kg}}{1.67\times10^{-27}\text{ kg}}\right)\left[\frac{2\left(8.99\times10^9\ \text{N}\cdot\text{m}^2/\text{C}^2\right)\left(1.60\times10^{-19}\text{ C}\right)\left(3.20\times10^{-19}\text{ C}\right)}{\left(6.64\times10^{-27}\text{ kg} + 1.67\times10^{-27}\text{ kg}\right)\left(4.00\times10^{-15}\text{ m}\right)}\right]}$$

yielding $v_p = 1.05\times10^7$ m/s. ◊

Then, Equation [2] gives the final speed of the alpha particle as

$$\left|v_\alpha\right| = \left(\frac{m_p}{m_\alpha}\right)v_p = \left(\frac{1.67\times10^{-27}\text{ kg}}{6.64\times10^{-27}\text{ kg}}\right)\left(1.05\times10^7\text{ m/s}\right) = 2.64\times10^6\text{ m/s} \qquad ◊$$

27. An air-filled parallel-plate capacitor has plates of area 2.30 cm^2 separated by 1.50 mm. The capacitor is connected to a 12.0-V battery. (a) Find the value of its capacitance. (b) What is the charge on the capacitor? (c) What is the magnitude of the uniform electric field between the plates?

Solution

(a) A pair of parallel conducting plates, each having area A and separated by distance d, with an insulating material having dielectric constant κ filling the space between the two plates, has a capacitance of

$$C = \kappa \in_0 A/d$$

where the permittivity of free space is given by $\in_0 = 8.85 \times 10^{-12}\ \text{C}^2/\text{N}\cdot\text{m}^2$. When the insulator between the plates is air, the dielectric constant is $\kappa = 1.00$. Thus, if the area of one side of either plate is $A = 2.30\ \text{cm}^2 = 2.30 \times 10^{-4}\ \text{m}^2$ and the distance separating the parallel plates is $d = 1.50\ \text{mm} = 1.50 \times 10^{-3}\ \text{m}$, the capacitance is

$$C = \frac{(1.00)(8.85 \times 10^{-12}\ \text{C}^2/\text{N}\cdot\text{m}^2)(2.30 \times 10^{-4}\ \text{m}^2)}{1.50 \times 10^{-3}\ \text{m}} = 1.36 \times 10^{-12}\ \text{F} = 1.36\ \text{pF} \qquad \lozenge$$

(b) From the definition of capacitance, $C = Q/\Delta V$, the charge stored in a capacitor with capacitance C when a potential difference ΔV exists between the plates is $Q = C(\Delta V)$. If a battery having $\Delta V = 12.0$ V is applied to the parallel-plate capacitor described in part (a) above, the stored charge will be

$$Q = C(\Delta V) = (1.36 \times 10^{-12}\ \text{F})(12.0\ \text{V}) = 1.63 \times 10^{-11}\ \text{C} = 16.3 \times 10^{-12}\ \text{C} = 16.3\ \text{pC} \qquad \lozenge$$

(c) Since the electric field between the plates of a parallel-plate capacitor is uniform, the work required to carry charge q from one plate to the other has a magnitude $W = F_e d = (qE)d$. The work per unit charge, or the potential difference between the plates, is then $\Delta V = W/q = Ed$, giving the magnitude of the electric field between the plates as $E = \Delta V/d$. For the capacitor described in parts (a) and (b) above, the magnitude of this field is

$$E = \frac{\Delta V}{d} = \frac{12.0\ \text{V}}{1.50 \times 10^{-3}\ \text{m}} = 8.00 \times 10^3\ \text{V/m} = 8.00 \times 10^3\ \text{N/C} \qquad \lozenge$$

35. (a) Find the equivalent capacitance of the capacitors in Figure P16.35, (b) the charge on each capacitor, and (c) the potential difference across each capacitor.

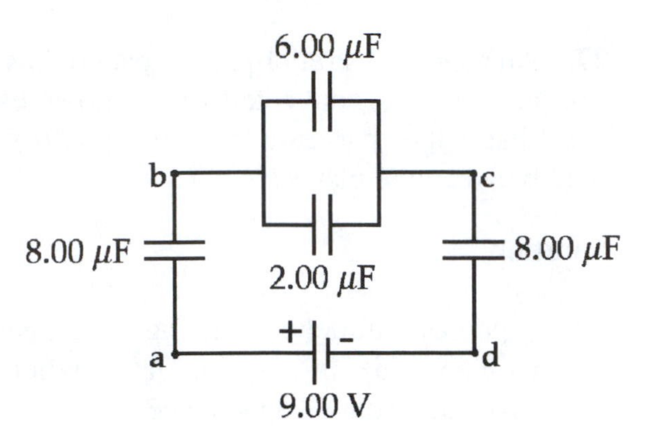

Figure P16.35 (modified)

Solution

(a) The equivalent capacitance of the parallel combination between points b and c is $C_{bc} = 6.00\ \mu\text{F} + 2.00\ \mu\text{F} = 8.00\ \mu\text{F}$.

Replacing this parallel combination by its equivalent capacitance gives us three 8.00-μF capacitors connected in series between points a and d. The equivalent capacitance for the full circuit is then

$$\frac{1}{C_{eq}} = \frac{1}{8.00\ \mu\text{F}} + \frac{1}{8.00\ \mu\text{F}} + \frac{1}{8.00\ \mu\text{F}} = \frac{3}{8.00\ \mu\text{F}} \quad \text{or } C_{eq} = \frac{8.00\ \mu\text{F}}{3} = 2.67\ \mu\text{F} \quad \lozenge$$

(b) Since capacitors in series all have the same charge, the charge on each capacitor in the series combination is

$$Q_{ab} = Q_{bc} = Q_{cd} = C_{eq}\left(\Delta V_{ad}\right) = (2.67\ \mu\text{F})(9.00\ \text{V}) = 24.0\ \mu\text{C}$$

Also, we find that $\Delta V_{bc} = \dfrac{Q_{bc}}{C_{bc}} = \dfrac{24.0\ \mu\text{C}}{8.00\ \mu\text{F}} = 3.00\ \text{V}$

The charge on each individual capacitor is:

For the 8.00-μF capacitor between points a and b: $Q_8 = Q_{ab} = 24.0\ \mu\text{C}$ ◊

For the 8.00-μF capacitor between points c and d: $Q_8 = Q_{cd} = 24.0\ \mu\text{C}$ ◊

For the 6.00-μF capacitor: $Q_6 = C_6\left(\Delta V_{bc}\right) = (6.00\ \mu\text{F})(3.00\ \text{V}) = 18.0\ \mu\text{C}$ ◊

And, for the 2.00-μF capacitor: $Q_2 = C_2\left(\Delta V_{bc}\right) = (2.00\ \mu\text{F})(3.00\ \text{V}) = 6.00\ \mu\text{C}$ ◊

Note that $Q_2 + Q_6 = Q_{bc}$ as it must.

(c) The potential difference across each capacitor in the circuit is:

For the 8.00-μF capacitor between points a and b: $\Delta V_8 = \dfrac{Q_8}{C_8} = \dfrac{24.0\ \mu\text{C}}{8.00\ \mu\text{F}} = 3.00\ \text{V}$ ◊

For the 6.00-μF and 2.00-μF capacitors in parallel: $\Delta V_6 = \Delta V_2 = \Delta V_{bc} = 3.00\ \text{V}$ ◊

For the 8.00-μF capacitor between points c and d: $\Delta V_8 = \dfrac{Q_8}{C_8} = \dfrac{24.0\ \mu\text{C}}{8.00\ \mu\text{F}} = 3.00\ \text{V}$ ◊

39. Find the charge on each of the capacitors in Figure P16.39.

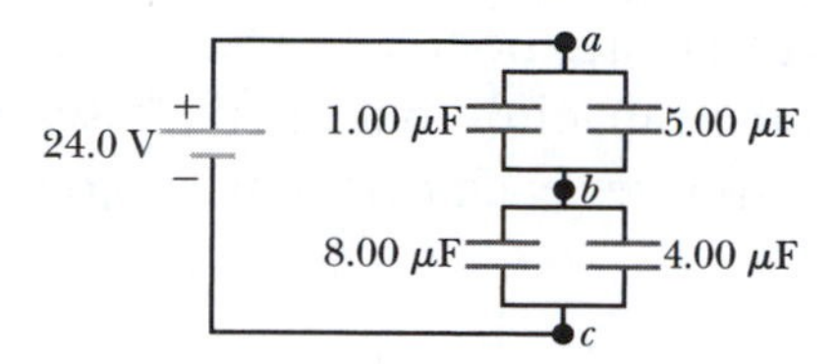

Figure P16.39 (modified)

Solution

The equivalent capacitance of the parallel combination between points a and b in the figure above is $C_{ab} = C_1 + C_5 = 1.00\ \mu\text{F} + 5.00\ \mu\text{F} = 6.00\ \mu\text{F}$, while that between points b and c is $C_{bc} = C_8 + C_4 = 8.00\ \mu\text{F} + 4.00\ \mu\text{F} = 12.0\ \mu\text{F}$. These capacitances are connected in series with each other and with the 24.0-V battery. The total capacitance, C_{ac}, for the circuit shown is given by

$$\frac{1}{C_{ac}} = \frac{1}{C_{ab}} + \frac{1}{C_{bc}} = \frac{C_{ab} + C_{bc}}{C_{ab}C_{bc}} \quad \text{or} \quad C_{ac} = \frac{C_{ab}C_{bc}}{C_{ab} + C_{bc}} = \frac{(6.00\ \mu\text{F})(12.0\ \mu\text{F})}{6.00\ \mu\text{F} + 12.0\ \mu\text{F}} = 4.00\ \mu\text{F}$$

The total stored charge is then $Q_{ac} = C_{ac}\Delta V_{ac} = (4.00\ \mu\text{F})(24.0\ \text{V}) = 96.0\ \mu\text{C}$, and since the change stored on each capacitor in a series combination equals the total charge stored by the combination, we have $Q_{ab} = Q_{bc} = Q_{ac} = 96.0\ \mu\text{C}$. The potential difference across the two parallel combinations in Figure P16.39 is then

$$\Delta V_{ab} = \frac{Q_{ab}}{C_{ab}} = \frac{96.0\ \mu\text{C}}{6.00\ \mu\text{F}} = 16.0\ \text{V} \quad \text{and} \quad \Delta V_{bc} = \frac{Q_{bc}}{C_{bc}} = \frac{96.0\ \mu\text{C}}{12.0\ \mu\text{F}} = 8.00\ \text{V}$$

The charge stored on each of the capacitors in these parallel combinations is given by

$$Q_1 = C_1\Delta V_{ab} = (1.00\ \mu\text{F})(16.0\ \text{V}) = 16.0\ \mu\text{C} \qquad \Diamond$$

$$Q_5 = C_5\Delta V_{ab} = (5.00\ \mu\text{F})(16.0\ \text{V}) = 80.0\ \mu\text{C} \qquad \Diamond$$

$$Q_8 = C_8\Delta V_{bc} = (8.00\ \mu\text{F})(8.00\ \text{V}) = 64.0\ \mu\text{C} \qquad \Diamond$$

and $$Q_4 = C_4\Delta V_{bc} = (4.00\ \mu\text{F})(8.00\ \text{V}) = 32.0\ \mu\text{C} \qquad \Diamond$$

Observe that $Q_1 + Q_5 = Q_{ab}$ and $Q_8 + Q_4 = Q_{bc}$, as they must if the equivalent capacitances have been calculated correctly.

47. A parallel-plate capacitor has capacitance 3.00 μF. (a) How much energy is stored in the capacitor if it is connected to a 6.00-V battery? (b) If the battery is disconnected and the distance between the charged plates doubled, what is the energy stored? (c) The battery is subsequently reattached to the capacitor, but the plate separation remains as in part (b). How much energy is stored? (Answer each part in microjoules.)

Solution

(a) The energy stored in a capacitor may be expressed in two equivalent ways:

$$\text{Energy Stored} = \frac{Q^2}{2C} \qquad \text{and} \qquad \text{Energy Stored} = \frac{1}{2}C(\Delta V)^2$$

In this case, the potential difference across the capacitor is known, and we use the latter form to compute

$$(\text{Energy Stored})_a = \frac{1}{2}(3.00\times 10^{-6}\ \text{F})(6.00\ \text{V})^2 = 54.0\times 10^{-6}\ \text{J} = 54.0\ \mu\text{J}$$ ◊

(b) The capacitance of an air-filled parallel-plate capacitor is $C = \epsilon_0\, A/d$, so it is seen that doubling the distance d between the plates reduces the capacitance by a factor of 2, $C_f = C_i/2$. Once the capacitor is disconnected from the battery, the charge on it has no way to leave and must remain constant. Thus, we use the first form for the energy stored to obtain

$$(\text{Energy Stored})_b = \frac{Q_f^2}{2C_f} = \frac{Q_i^2}{2(C_i/2)} = 2\left(\frac{Q_i^2}{2C_i}\right) = 2(\text{Energy Stored})_a$$

or $(\text{Energy Stored})_b = 2(54.0\ \mu\text{J}) = 108\ \mu\text{J}$ ◊

(c) Once the battery is reconnected, with the capacitance left at value $C_f = C_i/2$, the charge will readjust itself until the potential difference across the capacitor again matches that across the battery. Then, the second expression for the energy stored gives

$$(\text{Energy Stored})_c = \frac{1}{2}C_f(\Delta V)_f^2 = \frac{1}{2}\left(\frac{C_i}{2}\right)(\Delta V)_i^2 = \frac{1}{2}\left[\frac{1}{2}C_i(\Delta V)_i^2\right] = \frac{1}{2}(\text{Energy Stored})_a$$

or $(\text{Energy Stored})_c = \frac{1}{2}(54.0\ \mu\text{J}) = 27.0\ \mu\text{J}$ ◊

53. A model of a red blood cell portrays the cell as a spherical capacitor, a positively charged liquid sphere of surface area A separated from the surrounding negatively charged fluid by a membrane of thickness t. Tiny electrodes introduced into the interior of the cell show a potential difference of 100 mV across the membrane. The membrane's thickness is estimated to be 100 nm and has a dielectric constant of 5.00. (a) If an average red blood cell has a mass of 1.00×10^{-12} kg, estimate the volume of the cell and thus find its surface area. The density of blood is $1\,100\ \text{kg/m}^3$. (b) Estimate the capacitance of the cell by assuming the membrane surfaces act as parallel plates. (c) Calculate the charge on the surface of the membrane. How many electronic charges does the surface charge represent?

Solution

(a) Since the density of a substance is given by $\rho = m/V$, the volume of a blood cell having a mass of $m = 1.00\times10^{-12}$ kg and filled with blood of density $\rho = 1\,100\ \text{kg/m}^3$ is given by

$$V = \frac{m}{\rho} = \frac{1.00\times10^{-12}\ \text{kg}}{1100\ \text{kg/m}^3} = 9.09\times10^{-16}\ \text{m}^3 \qquad \Diamond$$

If the cell has a spherical shape, $V = \frac{4}{3}\pi r^3$ and the radius is $r = [3V/4\pi]^{\frac{1}{3}}$. The surface area is then $A = 4\pi r^2 = 4\pi(3V/4\pi)^{\frac{2}{3}} = \left[4\pi\cdot(3V)^2\right]^{1/3}$, or

$$A = \left[4\pi\cdot 9\cdot\left(9.09\times10^{-16}\ \text{m}^3\right)^2\right]^{1/3} = 4.54\times10^{-10}\ \text{m}^2 \qquad \Diamond$$

(b) If we consider the surfaces of the membrane to serve as the plates of a parallel-plate capacitor with plate separation $d = t$, where t is the membrane thickness, the capacitance is

$$C = \frac{\kappa \in_0 A}{t} = \frac{(5.00)\left(8.85\times10^{-12}\ \text{C}^2/\text{N}\cdot\text{m}^2\right)\left(4.54\times10^{-10}\ \text{m}^2\right)}{100\times10^{-9}\ \text{m}} = 2.01\times10^{-13}\ \text{F} \qquad \Diamond$$

(c) With a potential difference of $\Delta V = 100$ mV across the membrane (i.e., between the plates of the capacitor), the magnitude of the charge stored on each surface of the membrane is

$$Q = C\cdot\Delta V = \left(2.01\times10^{-13}\ \text{F}\right)\left(100\times10^{-3}\ \text{V}\right) = 2.01\times10^{-14}\ \text{C} \qquad \Diamond$$

The number of electronic charges required to yield this much charge is

$$n = \frac{Q}{|e|} = \frac{2.01\times10^{-14}\ \text{C}}{1.60\times10^{-19}\ \text{C/electron}} = 1.26\times10^5\ \text{electrons} \qquad \Diamond$$

57. A parallel-plate capacitor with a plate separation d has a capacitance C_0 in the absence of a dielectric. A slab of dielectric material of dielectric constant κ and thickness $d/3$ is then inserted between the plates as in Figure P16.57a. Show that the capacitance of this partially filled capacitor is given by

$$C = \left(\frac{3\kappa}{2\kappa + 1}\right) C_0$$

(*Hint:* Treat the system as two capacitors connected in series, one with dielectric in it and the other one empty.)

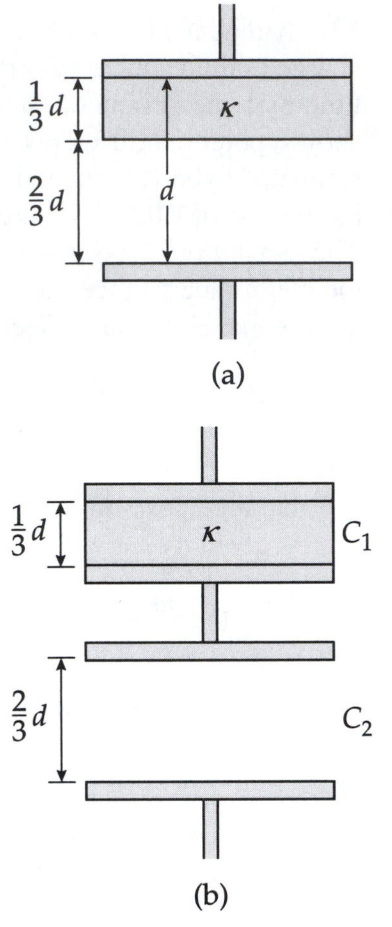

Solution

Without a dielectric present, the capacitance of the parallel-plate capacitor is

$$C_0 = \epsilon_0 A/d$$

Once the dielectric is inserted, it fills one-third of the gap between the plates as shown in part (a) of the sketch at the right. This situation can be thought of as having two capacitors, C_1 and C_2, in series as shown in part (b) of the sketch. In reality, the lower plate of C_1 and the upper plate of C_2 are the same surface, which is the lower surface of the dielectric shown in part (a) of the sketch.

The capacitances of the individual capacitors in our model are

$$C_1 = \frac{\kappa \epsilon_0 A}{(d/3)} = \frac{3\kappa \epsilon_0 A}{d} \quad \text{and} \quad C_2 = \frac{\epsilon_0 A}{(2d/3)} = \frac{3 \epsilon_0 A}{2d}$$

The capacitance of the parallel-plate capacitor after the dielectric is inserted is the equivalent capacitance of the series combination in our model of part (b) of the sketch. Thus,

$$\frac{1}{C} = \frac{1}{C_1} + \frac{1}{C_2} = \frac{d}{3\kappa \epsilon_0 A} + \frac{2d}{3 \epsilon_0 A} = \frac{d}{3 \epsilon_0 A}\left(\frac{1}{\kappa} + 2\right) = \frac{d}{3 \epsilon_0 A}\left(\frac{2\kappa + 1}{\kappa}\right)$$

which yields $$C = \frac{3 \epsilon_0 A}{d}\left(\frac{\kappa}{2\kappa + 1}\right) = \left(\frac{3\kappa}{2\kappa + 1}\right)\frac{\epsilon_0 A}{d} = \left(\frac{3\kappa}{2\kappa + 1}\right) C_0$$ ◊

60. Two charges of 1.0 μC and -2.0 μC are 0.50 m apart at two vertices of an equilateral triangle as in Figure P16.60. (a) What is the electric potential due to the 1.0-μC charge at the third vertex, point P? (b) What is the electric potential due to the -2.0-μC charge at P? (c) Find the total electric potential at P. (d) What is the work required to move a 3.0-μC charge from infinity to P?

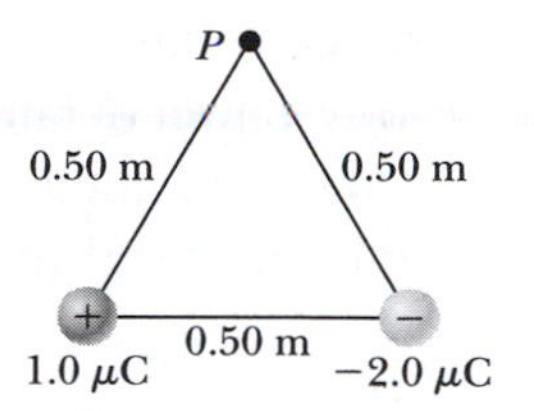

Figure P16.60

Solution

(a) The contribution to the electric potential at an observation position by a point charge Q located distance r away is given by $V = k_e Q/r$, where $k_e = 8.99\times10^9\ \text{N}\cdot\text{m}^2/\text{C}^2$. Thus, the electric potential at point P due to the charge $Q = 1.0\ \mu\text{C}$ that is a distance $r = 0.50$ m from point P is

$$V_1 = \frac{\left(8.99\times10^9\ \text{N}\cdot\text{m}^2/\text{C}^2\right)\left(+1.0\times10^{-6}\ \text{C}\right)}{0.50\ \text{m}} = +1.8\times10^4\ \text{V} = +18\ \text{kV} \quad \Diamond$$

(b) Similarly, the contribution to the electric potential at point P due to the presence of the charge $Q = -2.0\ \mu\text{C}$ located a distance of 0.50 m from point P is

$$V_2 = \frac{\left(8.99\times10^9\ \text{N}\cdot\text{m}^2/\text{C}^2\right)\left(-2.0\times10^{-6}\ \text{C}\right)}{0.50\ \text{m}} = -3.6\times10^4\ \text{V} = -36\ \text{kV} \quad \Diamond$$

(c) The total electric potential at point P due to the charges at the lower vertices of the equilateral triangle in Figure P16.60 is then

$$V_P = \sum_i V_i = V_1 + V_2 = +18\ \text{kV} - 36\ \text{kV} = -18\ \text{kV} \quad \Diamond$$

(d) The total electric potential at a point in space represents the work per unit charge required to move charge from a location of zero potential (of which a point at infinite distance from all charges is one) up to the point of interest. Hence, the work needed to move charge q from infinity to point P is $W = qV_P$. In particular, the work required to move a charge $q = +3.0\ \mu\text{C}$ from infinity to point P, where the potential is $V_P = -18$ kV, is

$$W = qV_P = \left(+3.0\times10^{-6}\ \text{C}\right)\left(-18\times10^3\ \text{V}\right) = -5.4\times10^{-2}\ \text{J} \quad \Diamond$$

The fact that the work computed above is negative means that the charge q would spontaneously move from infinity to point P if it is free to do so.

65. Capacitors $C_1 = 6.0\ \mu\text{F}$ and $C_2 = 2.0\ \mu\text{F}$ are charged as a parallel combination across a 250-V battery. The capacitors are disconnected from the battery and from each other. They are then connected positive plate to negative plate and negative plate to positive plate. Calculate the resulting charge on each capacitor.

Solution

When connected in parallel across the battery, the potential difference across each of the capacitors is $\Delta V_6 = \Delta V_2 = \Delta V_{\text{battery}} = 250\ \text{V}$. The charge initially stored on each of the capacitors by the battery is

$$Q_6 = C_6(\Delta V_6) = (6.0\ \mu\text{F})(250\ \text{V}) = 1.5 \times 10^3\ \mu\text{C}$$

and $$Q_2 = C_2(\Delta V_2) = (2.0\ \mu\text{F})(250\ \text{V}) = 5.0 \times 10^2\ \mu\text{C}$$

When the capacitors are reconnected, with the positive plate of one connected to the negative plate of the other and vise-versa, the magnitude of the net charge on each side of this new parallel combination is

$$Q_{\text{net}} = Q_1 - Q_2 = 1.5 \times 10^3\ \mu\text{C} - 5.0 \times 10^2\ \mu\text{C} = 1.0 \times 10^3\ \mu\text{C}$$

The equivalent capacitance of the parallel combination of these two capacitors is

$$C_{\text{eq}} = C_1 + C_1 = 6.0\ \mu\text{F} + 2.0\ \mu\text{F} = 8.0\ \mu\text{F}$$

so the potential difference across the new parallel combination is

$$\Delta V' = \frac{Q_{\text{net}}}{C_{\text{eq}}} = \frac{1.0 \times 10^3\ \mu\text{C}}{8.0\ \mu\text{F}} = 125\ \text{V}$$

and the final charge stored on each of the capacitors is

$$Q_6' = C_6(\Delta V') = (6.0\ \mu\text{F})(125\ \text{V}) = 7.5 \times 10^2\ \mu\text{C} = 0.75\ \text{mC}$$ ◊

and $$Q_2' = C_2(\Delta V') = (2.0\ \mu\text{F})(125\ \text{V}) = 2.5 \times 10^2\ \mu\text{C} = 0.25\ \text{mC}$$ ◊

17

Current and Resistance

NOTES FROM SELECTED CHAPTER SECTIONS

17.1 Electric Current

The direction of conventional current is designated as the direction in which positive charges would flow. *In an ordinary metal conductor, the direction of conventional current will be opposite the direction of flow of electrons (which are the charge carriers in this case).*

17.2 A Microscopic View: Current and Drift Speed

In the classical model of electronic conduction in a metal, electrons are treated like molecules in a gas. In the absence of an electric field, electrons have an average velocity of zero.

Under the influence of an electric field, the electrons move along a direction opposite the direction of the applied field with a drift velocity, v_d. The drift velocity of electrons in a conductor is proportional to the current and inversely proportional to the number of free electrons per unit volume, n. *The drift speed is much smaller than the average speed of electrons between collisions.*

17.3 Current and Voltage Measurements in Circuits

The figure at right illustrates a simple *circuit diagram.* The ammeter and voltmeter are arranged to measure the current through the bulb (a resistor) and the potential difference (voltage) across the filament of the bulb. *The ammeter is placed in series with the bulb and the voltmeter is in parallel with the bulb.*

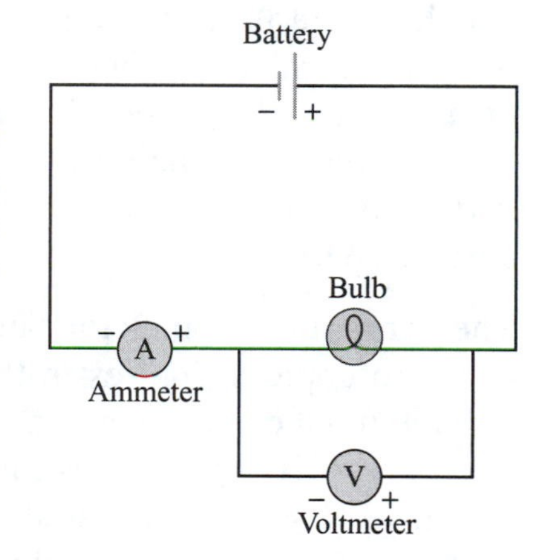

17.4 Resistance, Resistivity, and Ohm's Law

When a potential difference, ΔV, is applied across the ends of a metallic conductor, the current in the conductor is found to be proportional to the applied voltage. For many materials, including most metals, the resistance is constant over a wide range of applied voltages, and $\Delta V = IR$. This statement is known as Ohm's law, and R is called the resistance of the conductor.

Ohm's law is an empirical relationship that is valid only for certain materials. Materials which obey Ohm's law over a wide range of voltages are said to be ohmic.

Resistance has the SI units of volts per ampere, called ohms (Ω); if a potential difference of 1 V across a conductor produces a current of 1 A, the resistance of the conductor is 1 Ω.

A current can be established in a conductor by applying a potential difference between two points in the conductor (for example, between the ends of a wire). The average velocity of electrons moving through the conductor is limited by collisions with atoms of the conducting material. The overall effect of the conductor inhibiting electron flow is due to the **resistivity** (ρ) of the conducting material, which depends on the molecular structure and temperature of the conductor. *Resistivity is characteristic of a particular* **type** *of material (for example, copper), and the value of ρ for a material does not depend on the size or shape of the material.* Electrical resistance, R, is associated with a particular **sample** of material (for example, a cylinder of copper of specific length and cross-sectional area). Good conductors have low values of ρ.

17.5 Temperature Variation of Resistance

The manner in which resistivity of a material (and resistance of a conductor) changes with temperature depends on a parameter called the temperature coefficient of resistance, α. For most metals, α is positive and the resistivity changes approximately linearly with temperature over a limited temperature range; α has a negative value for most semiconductors.

EQUATIONS AND CONCEPTS

Average current is defined as the rate at which charge flows through a cross section of a conductor. *Under the influence of an electric field, electric charges will move through conducting gases, liquids, and solids.* The SI unit of current is the ampere (A).

$$I_{av} \equiv \frac{\Delta Q}{\Delta t} \qquad (17.1a)$$

$$1\text{ A} = 1\text{ C/s}$$

The **current** in a conductor can be related to microscopic quantities in the conductor: the number of charge carriers (positive or negative) per unit volume (n), the quantity of charge (q) associated with each carrier, and the drift velocity (v_d) of the carriers.

$$I = nqv_dA \qquad (17.2)$$

n = the number of mobile charge carriers per unit volume of conductor

In **Ohm's law**, as expressed in Equation 17.4, R is understood to be independent of ΔV. This form of Ohm's law is valid only for ohmic materials which have a linear current-voltage relationship over a wide range of applied voltages.

$$\Delta V = IR \qquad (17.4)$$

The **resistance, *R*, of a given conductor** made of a homogeneous material of uniform cross section, A, can be expressed in terms of the dimensions of the conductor, and an intrinsic property of the material of which the conductor is made called its **resistivity**. *The value of the resistivity of a given material depends on the electronic structure of the material and on the temperature.*

$$R = \rho \frac{\ell}{A} \tag{17.5}$$

R = resistance

ρ = resistivity

The **ohm** is the SI unit of resistance. If a potential difference of 1 V across a conductor produces a current of 1 A, the resistance of the conductor is 1 ohm. Resistivity is expressed in SI units of ohm-meters (Ω·m).

$$1\ \Omega = 1\ \text{V/A}$$

The **temperature coefficient of resistivity** determines the rate at which the resistivity of a material (and therefore the resistance of a conductor) varies with temperature. Over a limited range of temperatures, this variation is approximately linear. *Semiconductors (for example, carbon) are characterized by a negative temperature coefficient of resistivity, and in these materials the resistance decreases as the temperature increases.*

$$\rho = \rho_0[1 + \alpha(T - T_0)] \tag{17.6}$$

$$R = R_0[1 + \alpha(T - T_0)] \tag{17.7}$$

α = temperature coefficient of resistivity

T_0 is a reference temperature (usually taken to be 20.0 °C)

Power delivered to a resistor or other current-carrying device can be expressed in alternative forms as shown in Equations 17.8 and 17.9.

$$P = I\Delta V \tag{17.8}$$

$$P = I^2R = \frac{\Delta V^2}{R} \tag{17.9}$$

The **SI unit of power** is the watt (W).

$$1\ \text{W} = 1\ \text{J/s} = 1\ \text{V} \cdot \text{A}$$

The **kilowatt-hour** is the quantity of energy consumed in one hour at a constant use rate (or power) of 1 kW. It is the unit used by electric companies to bill for energy consumption.

$$1\ \text{kW h} = 3.60 \times 10^6\ \text{J} \tag{17.10}$$

REVIEW CHECKLIST

- Define the term "electric current" in terms of rate of charge flow and its corresponding unit of measure, the ampere. Calculate electron drift velocity in a conductor of specified characteristics carrying a current, I. (Sections 17.1–17.2)
- Sketch a simple single-loop circuit to illustrate the use of basic circuit element symbols and direction of conventional current. (Section 17.3)
- Determine the resistance of a conductor using Ohm's law. Also, calculate the resistance based on the physical characteristics of a conductor. Distinguish between ohmic and nonohmic conductors. (Section 17.4)
- Make calculations of the variation of resistance with temperature, which involves the concept of the temperature coefficient of resistivity. (Section 17.5)
- Calculate the power dissipated in a resistor. (Section 17.6)

SOLUTIONS TO SELECTED END-OF-CHAPTER PROBLEMS

3. In the Bohr model of the hydrogen atom, an electron in the lowest energy state moves at a speed of 2.19×10^{6} m/s in a circular path having a radius of 5.29×10^{-11} m. What is the effective current associated with this orbiting electron?

Solution

The current at a point in space is defined as the rate at which charge flows past that point. Thus, to determine the magnitude of the effective current associated with the orbiting electron, we consider a fixed point in the orbit and count the total charge passing that point each second.

The time required for the electron to complete one revolution around its orbit (i.e., the period of its motion) is given by

$$T=\frac{circumference}{speed}=\frac{2\pi r}{v}$$

and the orbital frequency (or the number of times the electron passes our observation point each second) is

$$f=\frac{1}{T}=\frac{v}{2\pi r}$$

Each time the electron completes a revolution around its orbit, charge of magnitude $|-e|$ passes our observation point. Therefore, the total charge passing this point in one second (i.e., the effective current) is

$$I = |-e| \cdot f = ef = \frac{ev}{2\pi r}$$

and if $v = 2.19 \times 10^6$ m/s while $r = 5.29 \times 10^{-11}$ m, the effective current is

$$I = \frac{\left(1.60 \times 10^{-19}\ \text{C}\right)\left(2.19 \times 10^6\ \text{m/s}\right)}{2\pi\left(5.29 \times 10^{-11}\ \text{m}\right)} = 1.05 \times 10^{-3}\ \text{C/s} = 1.05 \times 10^{-3}\ \text{A}$$

$$= 1.05\ \text{mA}$$

◊

---

8. An aluminum wire having a cross-sectional area of 4.00×10^{-6} m^2 carries a current of 5.00 A. The density of aluminum is 2.70 g/cm^3. Assume each aluminum atom supplies one conduction electron per atom. Find the drift speed of the electrons in the wire.

Solution

The current in a conductor may be expressed in terms of the drift speed of the charge carriers, v_d and the density of free charge carriers, n, by $I = nqv_dA$, where q is the charge of each charge carrier and A is the cross-sectional area of the conductor. Thus, the drift speed is $v_d = I/nqA$.

In a metallic conductor, such as an aluminum wire, the charge carriers are electrons and $q = |-e| = 1.60 \times 10^{-19}$ C. If each aluminum atom supplies one conduction electron, the density of charge carriers is the same as the number of atoms per unit volume, which may be computed from

$$n = \frac{\text{mass per unit volume}}{\text{mass per atom}} = \frac{\text{density}}{(\text{mass per mole})/(\text{number of atoms per mole})}$$

or $n = \rho/\left(M/N_A\right) = \rho N_A/M$, where M is the molecular weight of the conducting material and N_A is Avogadro's number. For the aluminum wire, the density of charge carriers is then

$$n = \frac{\left(2.70 \times 10^3\ \text{kg/m}^3\right)\left(6.02 \times 10^{23}\ \text{atoms/mol}\right)}{\left(26.982\ \text{g/mol}\right)\left(10^{-3}\ \text{kg/g}\right)} = 6.02 \times 10^{28}\ \text{electrons/m}^3$$

The drift speed of the charge carriers in an aluminum wire with cross-sectional area $A = 4.00 \times 10^{-6}\ \text{m}^2$ and carrying current $I = 5.00\ \text{A}$ is given by

$$v_d = \frac{5.00\ \text{C/s}}{(6.02 \times 10^{28}\ \text{electrons/m}^3)(1.60 \times 10^{-19}\ \text{C/electron})(4.00 \times 10^{-6}\ \text{m}^2)}$$

which yields

$$v_d = 1.30 \times 10^{-4}\ \text{m/s} = 0.130\ \text{mm/s} \qquad \Diamond$$

---

15. A potential difference of 12 V is found to produce a current of 0.40 A in a 3.2-m length of wire with a uniform radius of 0.40 cm. What is (a) the resistance of the wire? (b) the resistivity of the wire?

Solution

(a) According to Ohm's law, the potential difference across a conductor is directly proportional to the current flowing through that conductor, or $\Delta V = IR$. The proportionality constant is the resistance of the conductor. Thus, the resistance of this wire is

$$R = \frac{\Delta V}{I} = \frac{12\ \text{V}}{0.40\ \text{A}} = 30\ \Omega \qquad \Diamond$$

(b) The resistance of a conductor is directly proportional to the conductor's length L and inversely proportional to its cross-sectional area A, where the proportionality constant, ρ, is called the resistivity of the material. That is, $R = \rho L/A$. The resistivity of the wire is then

$$\rho = \frac{RA}{L} = \frac{R(\pi r^2)}{L} = \frac{(30\ \Omega)\pi(0.40 \times 10^{-2}\ \text{m})^2}{3.2\ \text{m}} = 4.7 \times 10^{-4}\ \Omega \cdot \text{m} \qquad \Diamond$$

---

19. A wire of initial length L_0 and radius r_0 has a measured resistance of 1.0 Ω. The wire is drawn under tensile stress to a new uniform radius of $r = 0.25 r_0$. What is the new resistance of the wire?

Solution

The volume of material making up the wire is constant as the wire is placed under stress and drawn to have a new length with a new cross-sectional area. This constant volume is

$$V = (\pi r^2)L = (\pi r_0^2)L_0$$

Thus, if the final radius of the wire is $r = 0.25r_0 = r_0/4$, the final length of the wire must be

$$L = L_0\left(\frac{r_0^2}{r^2}\right) = L_0\left(\frac{r_0}{r}\right)^2 = L_0\left(\frac{r_0}{0.25r_0}\right)^2 = L_0(4)^2 = 16L_0$$

The resistance of the cylindrical wire is given by

$$R = \rho\frac{L}{A} = \rho\frac{L}{(\pi r^2)}$$

so the new resistance of the wire is

$$R = \rho\frac{L}{(\pi r^2)} = \rho\frac{16L_0}{\left[\pi(r_0/4)^2\right]} = \rho\frac{256L_0}{(\pi r_0^2)} = 256\left[\rho\frac{L_0}{(\pi r_0^2)}\right] = 256R_0$$

or $\quad R = 256(1.0\ \Omega) = 256\ \Omega$ ◊

---

27. At 20°C, the carbon resistor in an electric circuit connected to a 5.0-V battery has a resistance of $2.0\times10^2\ \Omega$. What is the current in the circuit when the temperature of the carbon rises to 80°C?

Solution

Carbon is an interesting conducting material because its resistance decreases as the temperature increases. The temperature coefficient of resistivity for carbon is given by $\alpha = -0.50\times10^{-3}\ (°\text{C})^{-1}$.

At the reference temperature of $T_0 = 20°\text{C}$, the resistance of our carbon resistor is $R_0 = 2.0\times10^2\ \Omega$. When the temperature of the carbon rises to T, the resistance of this resistor will be given by $R = R_0\left[1+\alpha(T-T_0)\right]$. Thus, if this resistor is connected to a 5.0-V battery when the temperature is $T = 80°C$, Ohm's law gives the expected current through it as $I = \Delta V/R$, or

$$I = \frac{\Delta V}{R_0\left[1+\alpha(T-T_0)\right]} = \frac{5.00\ \text{V}}{(2.0\times10^2\ \Omega)\left[1+(-0.50\times10^{-3}\ °\text{C}^{-1})(80°\text{C}-20°\text{C})\right]}$$

which gives

$$I = 2.6\times10^{-2}\ \text{A} = 26\times10^{-3}\ \text{A} = 26\ \text{mA}$$ ◊

---

32. A platinum resistance thermometer has resistances of 200.0 Ω when placed in a 0°C ice bath and 253.8 Ω when immersed in a crucible containing melting potassium. What is the melting point of potassium? [*Hint:* First determine the resistance of the platinum resistance thermometer at room temperature, 20°C.]

Solution

The resistance of a conductor varies with temperature according to the relation $R = R_0\left[1+\alpha\left(T-T_0\right)\right]$, where R_0 is the resistance at the reference temperature T_0 (normally 20.0°C), and α is a property of the conducting material called the temperature coefficient of resistivity. For platinum, Table 17.1 in the textbook gives $\alpha = 3.92\times10^{-3}\ (°\text{C})^{-1}$.

Thus, if the thermometer has resistance $R = 200.0\ \Omega$ at $T = 0°\text{C}$, its resistance at the reference temperature of $T_0 = 20.0°\text{C}$ is

$$R_0 = \frac{R}{1+\alpha\left(T-T_0\right)} = \frac{200.0\ \Omega}{1+\left[3.92\times10^{-3}\ (°\text{C})^{-1}\right](0°\text{C}-20.0°\text{C})} = 217\ \Omega$$

and the temperature at which it has a resistance of $R = 253.8\ \Omega$ (and hence, the temperature of the melting potassium) is

$$T = T_0 + \frac{\left(R/R_0\right)-1}{\alpha} = 20.0°\text{C} + \frac{\left(253.8\ \Omega/217\ \Omega\right)-1}{3.92\times10^{-3}\ (°\text{C})^{-1}} = 63.3°\text{C}$$ ◊

---

35. Residential building codes typically require the use of 12-gauge copper wire (diameter 0.205 cm) for wiring receptacles. Such circuits carry currents as large as 20.0 A. If a wire of smaller diameter (with a higher gauge number) carried that much current, the wire could rise to a high temperature and cause a fire. (a) Calculate the rate at which internal energy is produced in 1.00 m of 12-gauge copper wire carrying 20.0 A. (b) Repeat the calculation for a 12-gauge aluminum wire. (c) Explain whether a 12-gauge aluminum wire would be as safe as a copper wire.

Solution

(a) From Table 17.1 of the textbook, the resistivity of copper at room temperature is seen to be $\rho_{\text{Cu}} = 1.7\times10^{-8}\ \Omega\cdot\text{m}$. Therefore, the resistance of a 1.00-m length of 12-gauge copper wire is

$$R = \frac{\rho_{\text{Cu}}L}{A} = \frac{\rho_{\text{Cu}}L}{\left(\pi d^2/4\right)} = \frac{4\rho_{\text{Cu}}L}{\pi d^2} = \frac{4\left(1.7\times10^{-8}\ \Omega\cdot\text{m}\right)(1.00\ \text{m})}{\pi\left(0.205\times10^{-2}\ \text{m}\right)^2} = 5.2\times10^{-3}\ \Omega$$

The rate at which internal energy is produced by the 1.00-m copper wire equals the electrical power it consumes or dissipates. When the wire carries a current of 20.0 A, this power is given by $P_{Cu} = (\Delta V)I = (IR)I = I^2 R$ as

$$P_{Cu} = (20.0\ \text{A})^2 (5.2 \times 10^{-3}\ \Omega) = 2.1\ \text{J/s} = 2.1\ \text{W} \qquad \Diamond$$

(b) The resistivity of aluminum is $\rho_{Al} = 2.82 \times 10^{-8}\ \Omega \cdot \text{m}$, so a 1.00-m length of 12-gauge aluminum wire has a resistance of

$$R = \frac{\rho_{Al} L}{A} = \frac{4\rho_{Al} L}{\pi d^2} = \frac{4(2.82 \times 10^{-8}\ \Omega \cdot \text{m})(1.00\ \text{m})}{\pi (0.205 \times 10^{-2}\ \text{m})^2} = 8.54 \times 10^{-3}\ \Omega$$

The rate this wire generates internal energy when carrying a 20.0-A current is

$$P_{Al} = (20.0\ \text{A})^2 (8.54 \times 10^{-3}\ \Omega) = 3.42\ \text{W} \qquad \Diamond$$

(c) When the two wires carry the same current, the 1.00-m length of aluminum wire produces internal energy at a rate that is approximately 160% the production rate of an equal length of copper wire. Hence, the aluminum wire is not as safe as the copper wire. If both wires were surrounded by insulating materials, the aluminum wire would get much hotter than the copper wire and could present a fire hazard. ◊

---

39. Lightbulb A is marked "25 W 120 V," and lightbulb B is marked "100 W 120 V." These labels mean that each lightbulb has its respective power delivered to it when it is connected to a constant 120-V source. (a) Find the resistance of each lightbulb. (b) During what time interval does 1.00 C pass into lightbulb A? (c) Is this charge different upon its exit versus its entry into the lightbulb? Explain. (d) In what time interval does 1.00 J pass into lightbulb A? (e) By what mechanisms does this energy enter and exit the lightbulb? Explain. (f) Find the cost of running lightbulb A continuously for 30.0 days, assuming the electric company sells its product at $0.110 per kWh.

Solution

(a) The power consumed by a resistive element that obeys Ohm's law may be expressed as $P = (\Delta V)I = (\Delta V)(\Delta V/R) = (\Delta V)^2/R$, where ΔV is the potential difference across that element and R is its resistance. The resistance of each bulb is then

Bulb A: $$R_A = \frac{(\Delta V)^2}{P_A} = \frac{(120\ \text{V})^2}{25.0\ \text{W}} = 576\ \Omega \qquad \Diamond$$

Bulb B: $$R_B = \frac{(\Delta V)^2}{P_B} = \frac{(120\ \text{V})^2}{100\ \text{W}} = 144\ \Omega \qquad \Diamond$$

(b) The current passing through Bulb A is given by $I = P_A/\Delta V$, and the charge passing through this bulb in time Δt is $Q = I\Delta t = (P_A/\Delta V)\Delta t$. Thus, if a charge $Q = 1.00$ C is to pass through the bulb, the time required is

$$\Delta t = \frac{Q(\Delta V)}{P_A} = \frac{(1.00\text{ C})(120\text{ V})}{25.0\text{ W}} = \frac{(1.00\text{ C})(120\text{ J/C})}{25.0\text{ J/s}} = 4.80\text{ s}$$ ◊

(c) The charge emerging from the bulb is the same as that which entered the bulb. However, it has given up electrical potential energy as it passed through the resistance of the bulb and emerges at an electrical potential that is 120 V lower that the potential at which it entered. ◊

(d) Energy is delivered to the bulb at the rate of $P_A = 25.0$ W, or at a rate of 25 joules each second. Therefore, the time during which $W = 1.00$ J enters the bulb is

$$\Delta t = \frac{W}{P_A} = \frac{1.00\text{ J}}{25.0\text{ J/s}} = 4.00 \times 10^{-2}\text{ s} = 0.040\,0\text{ s}$$ ◊

(e) The quantity of energy leaving the bulb each second equals the energy entering each second; however, it leaves by different mechanisms (and in different forms) than was the case as it entered. The energy enters the bulb by transmission as electrical potential energy and emerges by heat and electromagnetic radiation. ◊

(f) Lightbulb A has a power rating of $P_A = 25.0\text{ W} = 25.0 \times 10^{-3}\text{ kW}$. If the utility company charges \$0.110 per kilowatt-hour for electrical energy, the cost of operating this bulb continuously for 30.0 days would be

$$cost = (\text{energy used})(\text{rate}) = (P_A \cdot \Delta t)(\text{rate})$$

or $$cost = \left[(25.0 \times 10^{-3}\text{ kW}) \cdot (30.0\text{ }\cancel{\text{days}})\left(\frac{24.0\text{ h}}{1\text{ }\cancel{\text{day}}}\right)\right](\$0.110/\text{kWh}) = \underline{\underline{\$1.98}}$$ ◊

46. An office worker uses an immersion heater to warm 250 g of water in a light, covered, insulated cup from 20°C to 100°C in 4.00 minutes. The heater is a Nichrome resistance wire connected to a 120-V power supply. Assume the wire is at 100°C throughout the 4.00-min time interval. (a) Calculate the average power required to warm the water to 100°C in 4.00 min. (b) Calculate the required resistance in the heating element at 100°C. (c) Calculate the resistance of the heating element at 20.0°C. (d) Derive a relationship between the diameter of the wire, the resistivity at 20.0°C, ρ_0, the resistance at 20.0°C, R_0, and the length L. (e) If $L = 3.00$ m, what is the diameter of the wire?

Solution

(a) The thermal energy required to raise the temperature of the liquid water by ΔT is $E = mc\Delta T$, and the average power required to deliver this energy in time Δt is $P = E/\Delta t = mc\Delta T/\Delta t$. If $m = 0.250$ kg, $\Delta T = 100°\text{C} - 20°\text{C} = 80°\text{C}$, and $\Delta t = 4.00 \text{ min} = 240$ s, the needed power is

$$P = \frac{mc(\Delta T)}{\Delta t} = \frac{(0.250 \text{ kg})(4\,186 \text{ J/kg}\cdot°\text{C})(80°\text{C})}{240 \text{ s}} = 3.5\times10^2 \text{ W} \qquad \Diamond$$

(b) Since the power dissipation by a resistor may be expressed as $P = (\Delta V)^2/R$, the needed resistance of the heating element at the operating temperature of 100°C is

$$R = \frac{(\Delta V)^2}{P} = \frac{(120 \text{ V})^2}{3.5\times10^2 \text{ W}} = 41\ \Omega \qquad \Diamond$$

(c) Using $R = R_0\left[1+\alpha\left(T-T_0\right)\right]$, and observing in Table 17.1 of the textbook that $\alpha = 0.40\times10^{-3}\ (°\text{C})^{-1}$ for Nichrome, the resistance of the heating element at $T_0 = 20°\text{C}$ is found to be

$$R_0 = \frac{R}{1+\alpha\left(T-T_0\right)} = \frac{41\ \Omega}{1+\left[0.40\times10^{-3}\ (°\text{C})^{-1}\right](100°\text{C}-20°\text{C})} = 4.0\times10^1\ \Omega \qquad \Diamond$$

(d) With diameter d, cross-sectional area $A = \pi d^2/4$, and length L, the resistance at 20°C is given by $R_0 = \rho_0 L/A = 4\rho_0 L/\pi d^2$. Thus, the wire diameter, with a length L, needed to create the desired heating element is given by $d^2 = 4\rho_0 L/\pi R_0$, or

$$d = \sqrt{4\rho_0 L/\pi R_0} \qquad \Diamond$$

(e) With $\rho = 150\times10^{-8}\ \Omega\cdot\text{m}$ for Nichrome (see Table 17.1 in the textbook), and $L = 3.00$ m, the diameter wire needed to produce the desired heating element is

$$d = \sqrt{\frac{4\rho_0 L}{\pi R_0}} = \sqrt{\frac{4\left(150\times10^{-8}\ \Omega\cdot\text{m}\right)(3.00\text{ m})}{\pi\left(4.0\times10^1\ \Omega\right)}} = 3.8\times10^{-4}\text{ m} = 0.38\text{ mm} \qquad \Diamond$$

50. A car owner forgets to turn off the headlights of his car while it is parked in his garage. If the 12.0-V battery in his car is rated at 90.0 A · h and each headlight requires 36.0 W of power, how long will it take the battery to completely discharge?

Solution

The ampere-hour rating of a battery is a measure of the total amount of charge the battery can provide before it will be completely discharged. It is the product of the current the battery supplies and the length of time the battery could supply a constant current of that magnitude, or it is $Q = I \cdot \Delta t$. For example, a fully charged battery that is rated at 90.0 A · h could supply a current of 90 amperes for 1 hour, or could supply a steady current of 9 amperes for 10 hours.

Assuming two identical headlights on the car, the total power used to operate the headlights is $P = 2(36.0 \text{ W}) = 72.0 \text{ W}$. If the battery maintains a potential difference of $\Delta V = 12.0 \text{ V}$, the current it must supply for the headlights is

$$I = \frac{P}{\Delta V} = \frac{72.0 \text{ W}}{12.0 \text{ V}} = 6.00 \text{ A}$$

If the battery is rated 90.0 A · h, and was fully charged when the headlights were left on, the time it will take to totally discharge the battery is

$$\Delta t = \frac{Q}{I} = \frac{90.0 \text{ A} \cdot \text{h}}{6.00 \text{ A}} = 15.0 \text{ h}$$ ◊

---

57. You are cooking breakfast for yourself and a friend using a 1 200-W waffle iron and a 500-W coffeepot. Usually, you operate these appliances from a 110-V outlet for 0.500 h each day. (a) At 12 cents per kWh, how much do you spend to cook breakfast during a 30.0 day period? (b) You find yourself addicted to waffles and would like to upgrade to a 2 400-W waffle iron that will enable you to cook twice as many waffles during a half-hour period, but you know that the circuit breaker in your kitchen is a 20-A breaker. Can you do the upgrade?

Solution

(a) The total power used by the two appliances is

$$P_{\text{total}} = P_{\substack{\text{waffle} \\ \text{iron}}} + P_{\substack{\text{coffee} \\ \text{maker}}} = 1\,200 \text{ W} + 500 \text{ W} = 1\,700 \text{ W} = 1.70 \text{ kW}$$

The total energy used making breakfast during a 30.0-day period is then

$$E = P \cdot t = (1.70 \text{ kW})\left[(0.500 \text{ h/d})(30.0 \text{ d})\right] = 25.5 \text{ kWh}$$

and the cost is

$$cost = E \cdot rate = (25.5 \text{ kWh})(\$0.120/\text{kWh}) = \$3.06$$ ◊

(b) With the upgraded waffle maker, the total power requirement would be

$$P_{\text{total}} = P_{\substack{\text{waffle}\\ \text{iron}}} + P_{\substack{\text{coffee}\\ \text{maker}}} = 2\,400\text{ W} + 500\text{ W} = 2\,900\text{ W} = 2.90\text{ kW}$$

At $\Delta V = 110$ V, the current needed to supply this much power is

$$I = \frac{P_{\text{total}}}{\Delta V} = \frac{2\,900\text{ W}}{110\text{ V}} = 26.4\text{ A}$$

which exceeds the 20-A limit of your circuit breaker. Thus, the present electrical service in your kitchen will not permit the upgrade. ◊

---

65. An x-ray tube used for cancer therapy operates at 4.0 MV, with a beam current of 25 mA striking the metal target. Nearly all the power in the beam is transferred to a stream of water flowing through holes drilled in the target. What rate of flow, in kilograms per second, is needed if the rise in temperature (ΔT) of the water is not to exceed 50°C?

Solution

The power used by the x-ray tube and transferred to the target by the beam is

$$P = (\Delta V)I = \left(4.0\times10^{6}\text{ V}\right)\left(25\times10^{-3}\text{ A}\right) = 1.0\times10^{5}\text{ W} = 1.0\times10^{5}\text{ J/s}$$

The energy deposited in the target each second is

$$E = P\cdot t = \left(1.0\times10^{5}\text{ J/s}\right)(1.0\text{ s}) = 1.0\times10^{5}\text{ J}$$

When this energy is transferred as thermal energy to a mass m of water, the water will experience a temperature increase described by the relation $E = mc_{\text{water}}(\Delta T)$. If the increase in temperature is to be $\Delta T < 50°\text{C}$, the minimum mass of water needed to flow through the cooling channels in the target each second is

$$m_{\text{min}} = \frac{E}{c_{\text{water}}(\Delta T)_{\text{max}}} = \frac{1.0\times10^{5}\text{ J}}{(4\,186\text{ J/kg}\cdot°\text{C})(50°\text{C})} = 0.48\text{ kg}$$

Thus, the minimum flow rate needed is $(\Delta m/\Delta t)_{\text{min}} = 0.48\text{ kg/s}$. ◊

18

Direct-Current Circuits

NOTES FROM SELECTED CHAPTER SECTIONS

18.1 Sources of EMF

A source of emf (for example, a battery or generator) maintains the current in a closed circuit. The emf of the source is the work done per unit charge in increasing the electric potential energy of the circulating charges. *One joule of work is required to move one coulomb of charge between two points which differ in potential by one volt.*

The terminal voltage (ΔV between the positive and negative terminals of a battery or other source) is equal to the emf of the battery when the current in the circuit is zero. In this case the terminal voltage is called the **open circuit voltage**. When the battery is delivering current to a circuit, the terminal voltage **(closed circuit voltage)** is less than the emf. This is due to the internal resistance of the battery.

18.2 Resistors in Series

For a group of resistors connected in series:

- The equivalent resistance of the series combination is the algebraic sum of the individual resistances. See Equation 18.4.
- The current is the same for each resistor in the group.
- The total potential difference across the group of resistors equals the sum of the potential differences across the individual resistors.

18.3 Resistors in Parallel

For a group of resistors connected in parallel:

- The reciprocal of the equivalent resistance of the group is the algebraic sum of the reciprocals of the individual resistors. See Equation 18.6.
- The potential differences across the individual resistors have the same value.
- The total current associated with the parallel group is the sum of the currents in the individual resistors.

18.4 Kirchhoff's Rules and Complex DC Circuits

When all resistors in a circuit are connected in series and (or) parallel combinations, Ohm's law and the properties stated above in Sections 18.2 and 18.3 are adequate to analyze the circuit to determine the current in each resistor and the potential difference between any two points in the circuit.

When circuits are formed so that they cannot be reduced to a single equivalent resistor, it is necessary to use Kirchhoff's rules to analyze the circuit.

1. **Junction rule (conservation of charge)**: The sum of the currents entering any junction must equal the sum of the currents leaving that junction. (A junction is any point in the circuit where the current can split.)

2. **Loop rule (conservation of energy)**: The algebraic sum of the changes in potential around any closed circuit loop must be zero.

18.5 *RC* Circuits

When the switch in a dc circuit consisting only of a battery and a resistor is moved from the "open" to the "closed" position, the current in the circuit will have a constant value. Consider the case of a circuit which includes a battery, resistor, and capacitor. When the switch in this circuit is moved from the "open" to the "closed" position (the capacitor will begin charging), the charge on the capacitor increases with time, the current decreases exponentially, and the charge on the capacitor approaches its maximum value of $Q = C\mathcal{E}$ as the time, t, is allowed to go to infinity.

When the switch in a circuit containing an initially charged capacitor and a resistor is moved from the "open" to the "closed" position (the capacitor will begin discharging), the charge on the capacitor, the current in the circuit, and the voltage across the resistor will all decrease exponentially with time.

In the process of charging or discharging a capacitor, the potential difference between the plates changes as charges are transferred from one plate of the capacitor to the other. *The transfer of charge produces a current in the circuit; the charges do not move across the gap between the plates of the capacitor.*

EQUATIONS AND CONCEPTS

The **terminal voltage of a battery** will be less than the emf when the battery is providing a current to an external circuit. This is due to **internal resistance** of the battery. The terminal voltage equals the emf when the current is zero. *Often the internal resistance, r, of the source is very small compared to the external load resistance, R, of the circuit, and r can be neglected.*

$$\Delta V = \mathcal{E} - Ir \tag{18.1}$$

$$I = \frac{\mathcal{E}}{R + r}$$

Battery with internal resistance

For resistors connected in series:

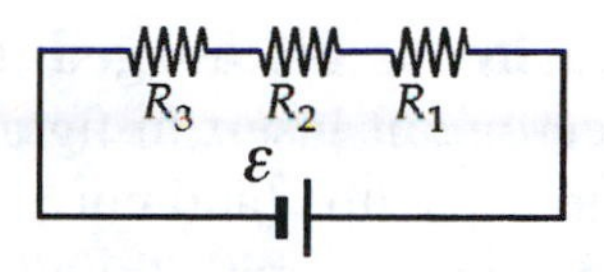

- There is only one common circuit point per pair.
- The total or equivalent resistance of a series combination of resistors is equal to the sum of the resistances of the individual resistors.

$$R_{eq} = R_1 + R_2 + R_3 + \ldots \quad (18.4)$$

- Each resistor has the same current.

$$I_1 = I_2 = I_3 = \ldots$$

- The total potential difference equals the sum of the individual potential differences.

$$\Delta V = \Delta V_1 + \Delta V_2 + \Delta V_3 + \ldots$$

For resistors connected in parallel:

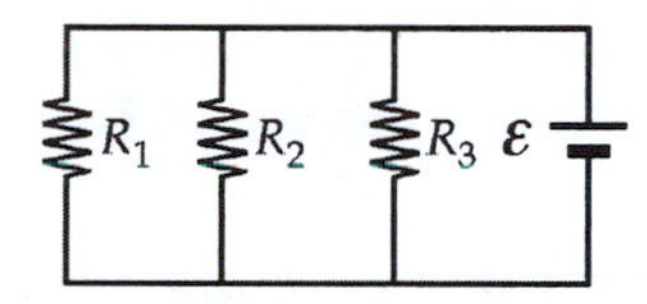

- Each resistor has two circuit points in common with each of the other resistors.
- The reciprocal of the equivalent resistance is the algebraic sum of the reciprocals of the individual resistances.

$$\frac{1}{R_{eq}} = \frac{1}{R_1} + \frac{1}{R_2} + \frac{1}{R_3} \cdots \quad (18.6)$$

- There is a common potential difference across the group of resistors.

$$\Delta V = \Delta V_1 = \Delta V_2 = \Delta V_3 = \ldots$$

- The total current equals the sum of the currents in the individual resistors.

$$I_{total} = I_1 + I_2 + I_3 + \ldots$$

When a **capacitor is charging in series with a resistor** (see circuit insert in figure), the charge on the capacitor increases as given by Equation 18.7 and shown in the figure.

$$q = Q\left(1 - e^{-t/RC}\right) \quad (18.7)$$

$$\tau = RC \quad (18.8)$$

The **time constant** of the circuit, τ, represents the time required for the charge on the capacitor to reach 63.2% of maximum value. *Also, during one time constant, the charging current decreases from its initial maximum value of $I_i = \mathcal{E}/R$ to 36.8% of I_i.*

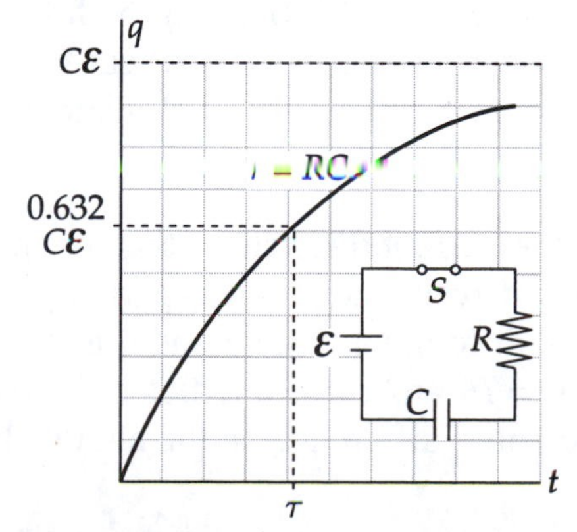

Capacitor charging in an RC circuit

When a **capacitor is discharged** through a resistor (see circuit insert in figure), the charge on the capacitor (and current) will decrease exponentially with time from an initial value of Q_0/RC to 0.368 of the initial value in a time $t = \tau$.

$$q = Qe^{-t/RC} \tag{18.9}$$

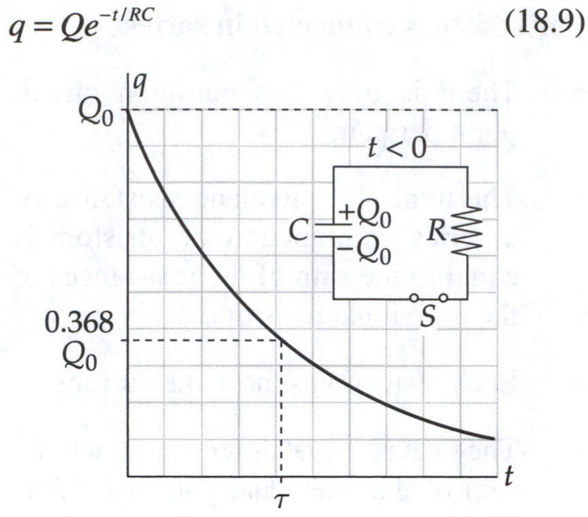

Capacitor discharging through a resistor

SUGGESTIONS, SKILLS, AND STRATEGIES

PROBLEM-SOLVING STRATEGY FOR RESISTORS

1. When two or more unequal resistors are connected in series, they carry the same current, but the potential differences across them are not the same. The resistances add directly to give the equivalent resistance of the series combination.

2. When two or more resistors (whether or not of equal value) are connected in parallel, the potential differences across them are the same. Since the current is inversely proportional to the resistance, the currents through them are not the same (except when the resistors are of equal value). *The equivalent resistance of a parallel combination of resistors is found through reciprocal addition, and the equivalent resistance is always less than the smallest individual resistance.*

3. A circuit consisting of several resistors, in series-parallel combination, can often be reduced to a simple circuit containing only one equivalent resistor. To do so, examine the initial circuit and replace any resistors in series or any in parallel using Equations 18.4 and 18.6, along with the properties outlined in Steps 1 and 2. Sketch the new circuit after these changes have been made. Examine the new circuit and replace any series or parallel combinations. **Continue this process until a single equivalent resistance is found.**

4. If the current through, or the potential difference across, a resistor in a circuit consisting of resistors in series-parallel combination is to be identified, start with the final circuit found in Step 3 and gradually work your way back through the circuits, using $\Delta V = IR$ and the properties of Steps 1 and 2. Record calculated values of current and potential difference on the circuit sketches as you proceed.

STRATEGY FOR USING KIRCHHOFF'S RULES

1. First, draw the circuit diagram and assign labels and symbols to all the known and unknown quantities. **You must assign directions to the currents in each branch of the circuit.** Do not be alarmed if you guess the direction of a current incorrectly; the resulting value will be negative, but its magnitude will be correct. Although the assignment of current directions is arbitrary, you must stick with your choice throughout as you apply Kirchhoff's rules.

2. Apply the junction rule to any junction in the circuit. The junction rule may be applied as many times as a new current (one not used in a previous application) appears in the resulting equation. **In general, the number of times the junction rule can be used is one fewer than the number of junction points in the circuit.**

3. Now apply Kirchhoff's loop rule to as many loops in the circuit as are needed to solve for the unknowns. Remember you must have as many equations as there are unknowns (I's, R's, and $\mathcal{E}$'s). **For each loop, you must first choose a direction to sum the voltage changes, clockwise or counterclockwise.** Then, you must correctly identify the increase or decrease in potential as you cross each resistor or source of emf in the closed loop.

4. Finally, you must solve the equations simultaneously for the unknown quantities. Be careful in your algebraic steps, and check your numerical answers for consistency.

Rules which should be used in determining the increase or decrease in potential difference as resistors or sources of emf are crossed in a circuit loop are stated below. *In the accompanying figures, assume that the direction of travel on the section of circuit shown is from point "a" to point "b" (left to right).*

a. The potential decreases (changes by $-IR$) when a resistor is traversed in the direction of the current.

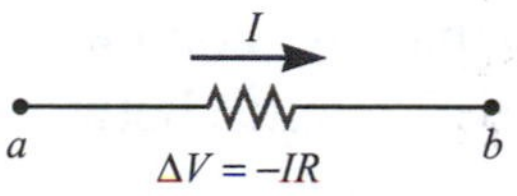

b. The potential increases (changes by $+IR$) when a resistor is traversed in the direction opposite the direction of the current.

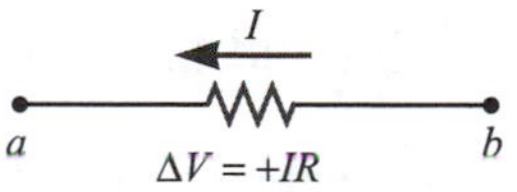

c. The potential increases by $+\mathcal{E}$ when a source of emf is traversed in the direction of the emf (from – to +).

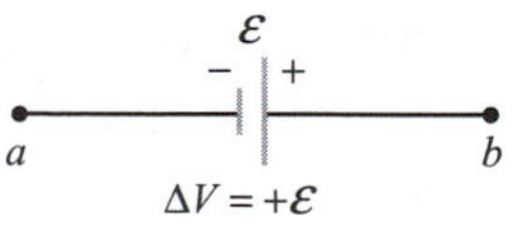

d. The potential decreases by $-\mathcal{E}$ when a source of emf is traversed opposite to the direction of the emf (from + to –).

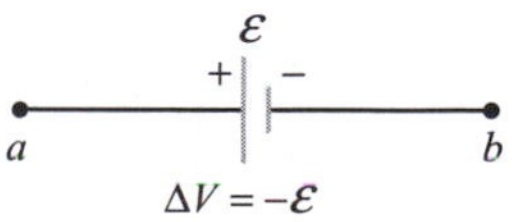

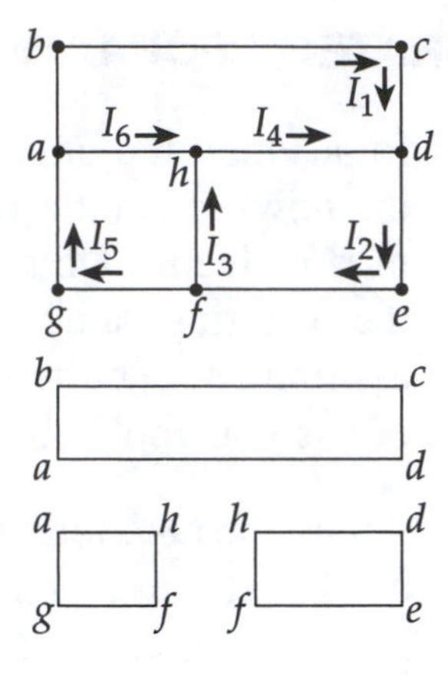

An illustration of the use of Kirchhoff's rules for a three-loop circuit. In this illustration, the actual circuit elements, R's and $\mathcal{E}$'s, are not shown but assumed to be known. There are six possible different values of I in the circuit; therefore, you will need six independent equations to solve for the six values of I. There are four junction points in the circuit (at points a, d, f, and h). The first rule applied at **any three of these points** will yield three equations. The circuit can be thought of as a group of three "blocks" as shown in the figure. Kirchhoff's second law, when applied to each of these loops (*abcda*, *ahfga*, and *defhd*), will yield three additional equations. You can then solve the total of six equations simultaneously for the six values of I_1, I_2, I_3, I_4, I_5, and I_6 (or some combination of currents, resistances, and emf's totaling six quantities).

The sum of the changes in potential difference around any other closed loop in the circuit will be zero (for example, *abcdefga* or *ahfedcba*); however, equations found by applying Kirchhoff's second rule to these additional loops will not be independent of the six equations found previously.

REVIEW CHECKLIST

- Calculate the terminal voltage, internal resistance, and current in a single loop. (Section 18.1)
- Determine the equivalent resistance of a group of resistors in parallel, series, or series-parallel combination and find the current at any point in the circuit and the potential difference between any two points. (Sections 18.2 and 18.3)
- Apply Kirchhoff's rules to solve multiloop circuits; that is, find the currents at any point and the potential difference between any two points. (Section 18.4)
- Describe in qualitative terms the manner in which charge accumulates on a capacitor or current flow changes through a resistor with time in a series circuit with battery, capacitor, resistor, and switch. Use Equations 18.7 and 18.9 to calculate q at any time t and also to find the time t for which the ratio q/Q_0 will have a specified value. (Section 18.5)

SOLUTIONS TO SELECTED END-OF-CHAPTER PROBLEMS

3. A lightbulb marked "75 W [at] 120 V" is screwed into a socket at one end of a long extension cord in which each of the two conductors has a resistance of 0.800 Ω. The other end of the extension cord is plugged into a 120-V outlet. (a) Draw a circuit diagram, and (b) find the actual power of the bulb in the circuit described.

Solution

(a) If a lightbulb consumes 75.0 W of power when connected to a 120-V power source, the resistance of the filament in the bulb is given by $P = (\Delta V)^2/R$ as

$$R_{\text{filament}} = \frac{(\Delta V)^2}{P} = \frac{(120\text{ V})^2}{75.0\text{ W}} = 192\ \Omega$$

The 192-Ω resistance of the filament is in series with each of the 0.800-Ω resistances of the conductors in the extension cord. This series combination of resistors is then connected to the 120-V source as shown in the circuit diagram given below.

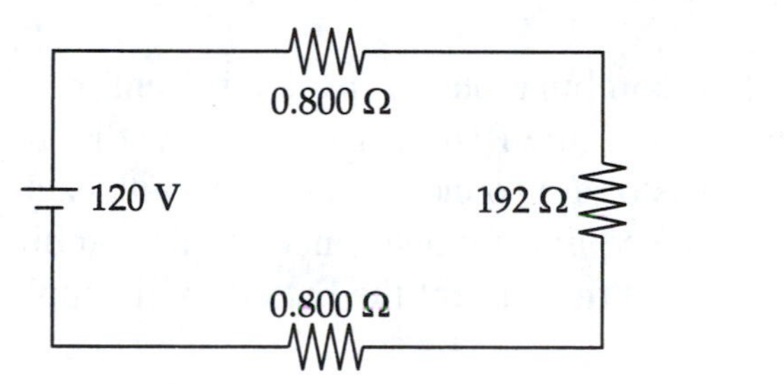

◊

(b) The equivalent resistance of this series combination is

$$R_{\text{eq}} = R_{\text{line}} + R_{\text{filament}} + R_{\text{line}} = 0.800\ \Omega + 192\ \Omega + 0.800\ \Omega$$

and the current which will flow through each resistance in the series combination is

$$I = \frac{\Delta V}{R_{\text{eq}}} = \frac{120\text{ V}}{0.800\ \Omega + 192\ \Omega + 0.800\ \Omega} = 0.620\text{ A}$$

The power actually delivered to the "75-W bulb" is then

$$P_{\text{bulb}} = I^2 R_{\text{filament}} = (0.620\text{ A})^2 (192\ \Omega) = 73.8\text{ W}$$

◊

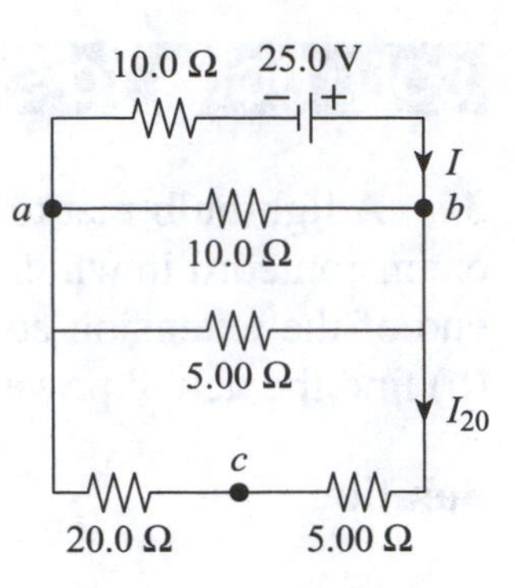

Figure P18.9a

9. Consider the circuit shown in Figure P18.9. Find (a) the potential difference between points *a* and *b*, and (b) the current in the 20.0-Ω resistor.

Solution

In Figure P18.9a at the right, the 20.0-Ω and the 5.00-Ω resistors in the lowermost branch are in series between points *a* and *b*. Replacing this series combination by its equivalent resistance of $R_{eq1} = 20.0\ \Omega + 5.00\ \Omega = 25.0\ \Omega$ reduces the circuit to that of Figure P18.9b. Then, it is seen that R_{eq1}, the lower 10.0-Ω resistor, and the remaining 5.00-Ω resistor are in parallel between points *a* and *b*. The equivalent resistance of this combination is

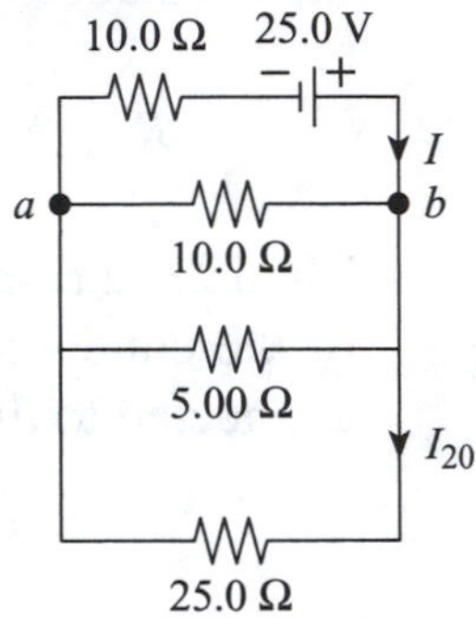

Figure P18.9b

$$\frac{1}{R_{eq2}} = \frac{1}{10.0\ \Omega} + \frac{1}{5.00\ \Omega} + \frac{1}{25.0\ \Omega} = \frac{5+10+2}{50.0\ \Omega}$$

$$\text{or } R_{eq2} = \frac{50.0\ \Omega}{17} = 2.94\ \Omega$$

Replacing this parallel combination by its equivalent resistance yields the circuit shown in Figure P18.9c. Here, we see that R_{eq2} and the remaining 10.0-Ω resistor are connected in series with each other and across the battery. The equivalent resistance of this combination is $R_{total} = 10.0\ \Omega + 2.94\ \Omega$, and the current the battery will supply to this series combination is

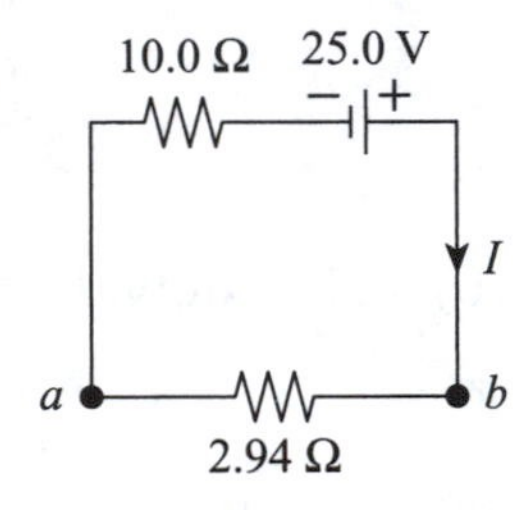

Figure P18.9c

$$I = \frac{\Delta V}{R_{total}} = \frac{25.0\ \text{V}}{10.0\ \Omega + 2.94\ \Omega} = 1.93\ \text{A}$$

(a) The potential difference between points *a* and *b* is then

$$\Delta V_{ab} = IR_{eq2} = (1.93\ \text{A})(50/17\ \Omega) = 5.68\ \text{V}$$ ◊

(b) From Figures P18.9b and P18.9a, we see that the current through the lower branch of the circuit (i.e., the current through $R_{eq1} = 25.0\ \Omega$ in Figure P18.9b and also the current through each resistor of the series combination of the 20.0-Ω and rightmost 5.00-Ω resistor in Figure P18.9a) will be

$$I_{eq1} = I_{20} = I_5 = \frac{\Delta V_{ab}}{R_{eq1}} = \frac{5.68\ \text{V}}{25.0\ \Omega} = 0.227\ \text{A}$$ ◊

12. A battery with $\mathcal{E} = 6.00$ V and no internal resistance supplies current to the circuit shown in Figure P18.12. When the double-throw switch S is open as shown in the figure, the current in the battery is 1.00 mA. When the switch is closed in position *a*, the current in the battery is 1.20 mA. When the switch is closed in position *b*, the current in the battery is 2.00 mA. Find the resistances (a) R_1, (b) R_2, and (c) R_3.

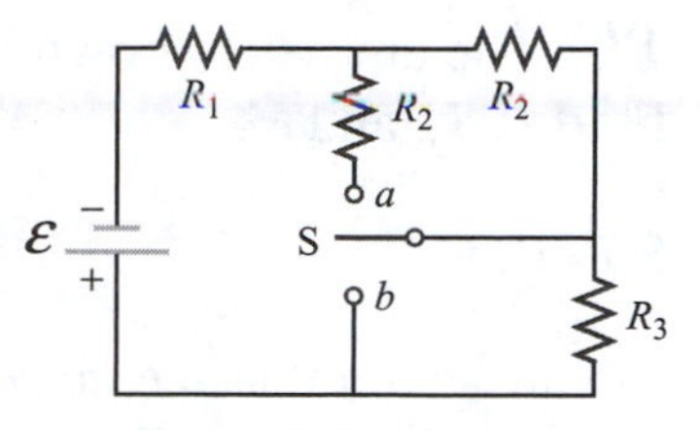

Figure P18.12

Solution

With the switch open as shown in Figure P18.12, no current can flow through the vertical resistor labeled R_2. The only current in the circuit, I_O, flows around the outer perimeter, through a series connection of three resistors R_1, R_2, and R_3. The equivalent resistance in this case is $R_O = R_1 + R_2 + R_3$, and if the current is $I_O = 1.00$ mA, Ohm's law gives $R_O = \mathcal{E}/I_O = 6.00\text{ V}/1.00\text{ mA} = 6.00 \times 10^3\ \Omega$, or

$$R_1 + R_2 + R_3 = 6.00\text{ k}\Omega \qquad [1]$$

When the switch is closed to position *a*, the two resistors labeled R_2 are connected in parallel with each other, giving an equivalent resistance of $R_p = \frac{1}{2}R_2$ for that combination. This parallel combination is then in series with the other two resistors, R_1 and R_3, so the total resistance seen by the battery with the switch in this position is $R_a = R_1 + \frac{1}{2}R_2 + R_3$. If the current supplied by the battery in this case is $I_a = 1.20$ mA, Ohm's law gives $R_a = \mathcal{E}/I_a = 6.00\text{ V}/1.20 \times 10^{-3}\text{ A} = 5.00 \times 10^3\ \Omega$, or

$$R_1 + \tfrac{1}{2}R_2 + R_3 = 5.00\text{ k}\Omega \qquad [2]$$

When in position *b*, the switch forms a short-circuit path around R_3, allowing all of the current to bypass R_3. Again, no current will be able to pass through the vertical resistor labeled R_2, so the only resistance limiting the current is a series combination of the two horizontal resistors labeled R_1 and R_2. If the current in this case is $I_b = 2.00$ mA, Ohm's law gives $R_b = \mathcal{E}/I_b = 3.00 \times 10^3\ \Omega$, or

$$R_1 + R_2 = 3.00\text{ k}\Omega \qquad [3]$$

Subtracting Equation [3] from Equation [1] gives $R_3 = 3.00\text{ k}\Omega$. Subtracting Equation [2] from Equation [1] yields $R_2 = 2.00\text{ k}\Omega$, and substituting this result into Equation [3] gives $R_1 = 1.00\text{ k}\Omega$. In summary, we have:

(a) $R_1 = 1.00\text{ k}\Omega$ (b) $R_2 = 2.00\text{ k}\Omega$ (c) $R_3 = 3.00\text{ k}\Omega$ ◊

17. The ammeter shown in Figure P18.17 reads 2.00 A. Find I_1, I_2, and $\mathcal{E}$.

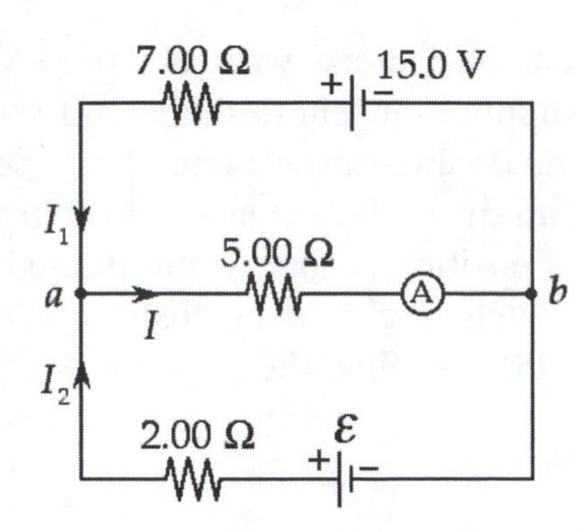

Figure P18.17

Solution

We are given that the current in the middle branch of the circuit is $I = 2.00$ A. Thus, the potential difference between points a and b in the circuit diagram at the right is

$$\Delta V_{ab} = (2.00\text{ A})(5.00\ \Omega) = 10.0\text{ V}$$

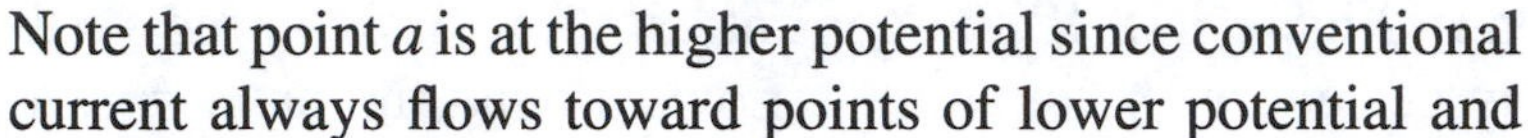

Note that point a is at the higher potential since conventional current always flows toward points of lower potential and the current I flows from a toward b. Next, we start at point a and go counterclockwise around the upper loop recording changes in potential. It is essential to realize that as we go through the 7.00-Ω resistor, in the direction the conventional current I_1 flows, the change in potential is negative. Kirchhoff's loop rule gives

$$-10.0\text{ V} + 15.0\text{ V} - (7.00\ \Omega)I_1 = 0 \qquad \text{or} \qquad I_1 = \frac{+5.00\text{ V}}{7.00\ \Omega} = 0.714\text{ A} \qquad \Diamond$$

Then, applying Kirchhoff's junction rule at point a yields $I = I_1 + I_2$, or

$$I_2 = I - I_1 = 2.00\text{ A} - 0.714\text{ A} = 1.29\text{ A} \qquad \Diamond$$

Finally, we start at point a and go clockwise around the loop forming the outer perimeter of the circuit. Note that as we pass through the 7.00-Ω resistor this time, we are going opposite to the flow of I_1, and the resulting change in potential will be positive. However, as we pass through the 2.00-Ω resistor, we will be going in the direction of the current I_2, and the change in potential will be negative. For this loop, Kirchhoff's loop rule gives

$$+(7.00\ \Omega)I_1 - 15.0\text{ V} + \mathcal{E} - (2.00\ \Omega)I_2 = 0$$

so the emf of the battery in the lower branch is

$$\mathcal{E} = 15.0\text{ V} + (2.00\ \Omega)(1.29\text{ A}) - (7.00\ \Omega)(0.714\text{ A}) = 12.6\text{ V} \qquad \Diamond$$

---

21. In the circuit of Figure P18.21, determine (a) the current in each resistor and (b) the potential difference across the 200-Ω resistor.

Solution

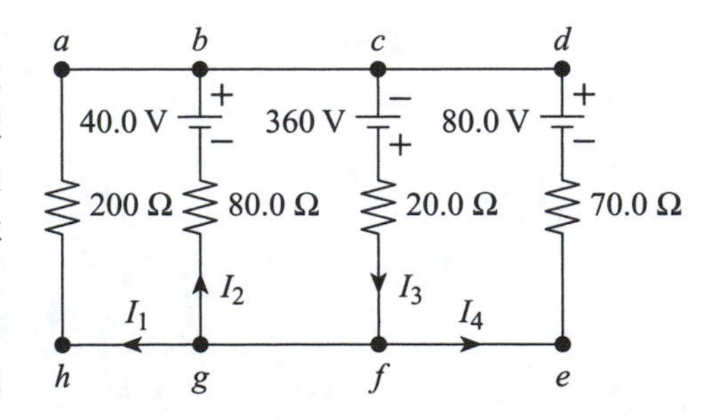

The assumed direction of the current in each branch of the circuit in Figure P18.21 is shown in the circuit diagram at the right. If the correct direction has been assumed for a given current, the solution for that current will have a positive value when calculated. The circuit has already been simplified as far as possible by replacing series and parallel combinations of resistors with equivalent resistances. Kirchhoff's rules must now be used to complete the circuit analysis.

Applying the loop rule to the loop *abcfgha* gives $+360\text{ V}-(20.0\ \Omega)I_3-(200\ \Omega)I_1=0$,

or $$I_1 = 1.80\text{ A} - 0.100 I_3 \qquad \textbf{[1]}$$

Applying the loop rule around the loop *bcfgb* yields

$$+360\text{ V}-(20.0\ \Omega)I_3-(80.0\ \Omega)I_2+40.0\text{ V}=0$$

or $$I_2 = 5.00\text{ A} - 0.250 I_3 \qquad \textbf{[2]}$$

Next, we apply the loop rule to loop *cfedc* to obtain

$$+360\text{ V}-(20.0\ \Omega)I_3-(70.0\ \Omega)I_4+80.0\text{ V}=0$$

or $$I_4 = \left(440\text{ A} - 20.0 I_3\right)/70.0 \qquad \textbf{[3]}$$

Finally, the junction rule tells us that the total current flowing into the conductor between points *f* and *g* must be equal to the total current flowing out of it. This gives

$$I_3 = I_1 + I_2 + I_4 \qquad \textbf{[4]}$$

(a) Substituting Equations [1], [2], and [3] into Equation [4] and multiplying by 70.0 to clear the fraction yields

$$70.0 I_3 = 126\text{ A} - 7.00 I_3 + 350\text{ A} - 17.5 I_3 + 440\text{ A} - 20.0 I_3$$

and $$I_3 = \frac{(126+350+440)\text{ A}}{70.0+7.00+17.5+20.0} = +8.00\text{ A} \qquad \Diamond$$

Substituting this result into Equations [1], [2], and [3] gives the other currents as

$$I_1 = +1.00\text{ A}, \qquad I_2 = 3.00\text{ A}, \qquad \text{and} \qquad I_4 = 4.00\text{ A} \qquad \Diamond$$

Note that all currents were found to have positive values, meaning that the direction shown by the arrows in the circuit diagram above is the correct direction for each current.

(b) Now that the current I_1 has a known magnitude and direction, Ohm's law gives the potential difference across the 200-Ω resistor in the leftmost branch of the circuit as

$$\Delta V_{ha} = I_1 R_{ha} = (1.00\text{ A})(200\ \Omega) = 200\text{ V} \qquad \Diamond$$

with point *a* at a lower potential that point *h*, since conventional current always flows from points of high potential toward points of lower potential.

27. (a) Can the circuit shown in Figure P18.27 be reduced to a single resistor connected to the batteries? Explain. (b) Find the magnitude of the current and its direction in each resistor.

Solution

(a) A modified version of Figure P18.27, with currents assumed to flow in the directions shown, is given below. Note that no two of the resistors in this circuit are connected in series since the current in any one of the resistors may split up, with a portion of it going through each of the other resistors. Also, neither do any two of these resistors necessarily have the same potential difference across them, so the circuit does not have any resistors connected in parallel with each other. Therefore, the circuit cannot be further simplified to reduce it to a single resistor connected to the batteries. ◊

(b) To determine the magnitudes and directions of the currents in this circuit, we use Kirchhoff's rules. Applying the junction rule at point *a* gives

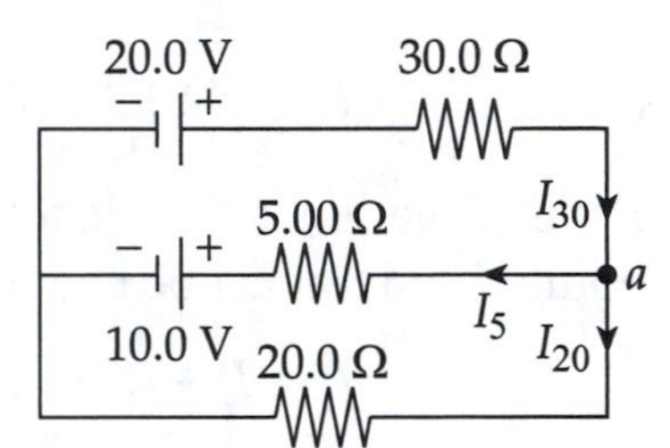

$$I_{30} = I_5 + I_{20} \quad [1]$$

Starting at point *a*, and applying Kirchhoff's loop rule as we go clockwise around the upper loop of this circuit, gives

$$-(5.00\ \Omega)I_5 - 10.0\ \text{V} + 20.0\ \text{V} - (30.0\ \Omega)I_{30} = 0 \quad \text{or} \quad I_5 = 2.00\ \text{A} - 6.00 I_{30} \quad [2]$$

Finally, when we start at point *a* and go clockwise around the outer perimeter of the circuit, the loop rule yields $-(20.0\ \Omega)I_{20} + 20.0\ \text{V} - (30.0\ \Omega)I_{30} = 0$, or

$$I_{20} = 1.00\ \text{A} - 1.50 I_{30} \quad [3]$$

Substituting Equations [2] and [3] into Equation [1] and simplifying gives

$$8.50 I_{30} = 3.00\ \text{A} \quad \text{or} \quad I_{30} = +0.353\ \text{A}$$ ◊

The positive value for I_{30} means that this current has a magnitude of 0.353 A and flows in the direction assumed in the circuit diagram given above. ◊

Using this value for I_{30} in Equation [2], we obtain $I_5 = -0.118$ A. This result tells us that a current of magnitude $|I_5| = 0.118$ A actually flows from left to right (i.e., opposite the direction assumed in the above circuit diagram) through the center branch of this circuit. ◊

Substituting I_{30} into Equation [3] yields $I_{20} = +0.471$ A. Thus, this current flows in the direction given in the above diagram and has magnitude $I_{20} = 0.471$ A. ◊

35. A charged capacitor is connected to a resistor and a switch as in Figure P18.35. The circuit has a time constant of 1.50 s. Soon after the switch is closed, the charge on the capacitor is 75.0% of its initial charge. (a) Find the time interval required for the capacitor to reach this charge. (b) If $R = 250\ \text{k}\Omega$, what is the value of C?

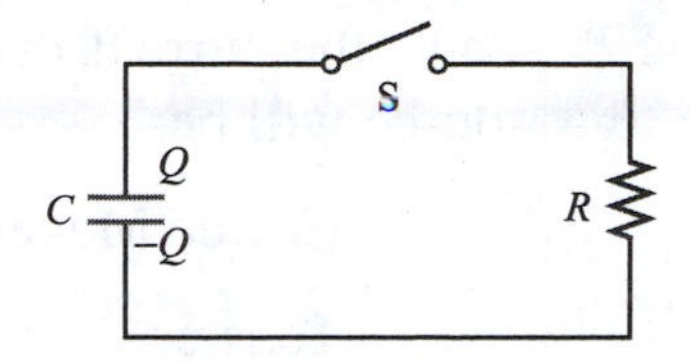

Figure P18.35

Solution

(a) We take $t = 0$ at the instant the switch in Figure P18.35 is closed. The charge Q_0 stored on the capacitor at this time will then begin to leak off, discharging the capacitor, with a conventional current flowing from the positive plate, through the resistance R, to the negative plate. The charge remaining of the capacitor at elapsed time t after the switch is closed is given by $q = Q_0 e^{-t/\tau}$, where e is the base of the natural logarithmic system, and $\tau = RC$ is known as the time constant of the circuit. When the capacitor has 75.0% of its initial charge (i.e., when $q = 0.750Q_0$), we have

$$0.750\,\cancel{Q_0} = \cancel{Q_0}\,e^{-t/\tau} \quad \text{or} \quad e^{-t/\tau} = 0.750$$

Taking the natural logarithm of each side of this result gives

$$-t/\tau = \ln(0.750) \quad \text{or} \quad t = -\tau \ln(0.750)$$

If the time constant of this circuit is $\tau = 1.50$ s, the elapsed time when $q = 0.750Q_0$ will be

$$t = -(1.50\ \text{s})\ln(0.750) = 0.432\ \text{s}$$ ◊

(b) With a time constant of $\tau = RC = 1.50$ s, and a total resistance of $R = 250\ \text{k}\Omega$ in the circuit, the capacitance of the circuit is given by

$$C = \frac{\tau}{R} = \frac{1.50\ \text{s}}{250 \times 10^3\ \Omega} = 6.00 \times 10^{-6}\ \text{F} = 6.00\ \mu\text{F}$$ ◊

40. A coffee maker is rated at 1200 W, a toaster at 1100 W, and a waffle maker at 1400 W. The three appliances are connected in parallel to a common 120-V household circuit. (a) What is the current in each appliance when operating independently? (b) What total current is delivered to the appliances when all are operating simultaneously? (c) Is a 15-A circuit breaker sufficient in this situation? Explain.

Solution

(a) Considering units, we see that the power consumed by a device is given by the product of the potential difference, ΔV, across that device (measured in units of volts or work per unit charge) and the current, I, in that device (measured in amperes or charge per unit time). That is,

$$\left(\frac{\text{Joules}}{\cancel{\text{Coulomb}}}\right)\left(\frac{\cancel{\text{Coulomb}}}{\text{second}}\right) = \frac{\text{Joules}}{\text{second}} = \text{Watts} \quad \text{or} \quad (\Delta V)(I) = P$$

Thus, the current in each of the given appliances when connected to a 120-V power source will be

Coffee Maker: $I = P/\Delta V = 1\,200 \text{ W}/120 \text{ V} = 10 \text{ A}$ ◊

Toaster: $I = P/\Delta V = 1100 \text{ W}/120 \text{ V} = 9.2 \text{ A}$ ◊

Waffle Maker: $I = P/\Delta V = 1\,400 \text{ W}/120 \text{ V} = 12 \text{ A}$ ◊

(b) When these three appliances are connected in parallel, as in a household circuit, and operated simultaneously, the total current delivered by the power source is $I_{\text{total}} = I_{\substack{\text{coffee}\\\text{maker}}} + I_{\text{toaster}} + I_{\substack{\text{waffle}\\\text{maker}}} = (10 + 9.2 + 12) \text{ A} = 31 \text{ A}$. ◊

(c) Since the three appliances require a total current $I_{\text{total}} = 31$ A when operated simultaneously, a 15-A circuit breaker will be insufficient to permit simultaneous operation of these devices. In fact, with a 15-A breaker, no two of the devices could be on at the same time without exceeding the capacity of the circuit breaker and causing it to trip. ◊

---

46. For the circuit shown in Figure P18.46, the voltmeter reads 6.0 V and the ammeter reads 3.0 mA. Find (a) the value of R, (b) the emf of the battery, and (c) the voltage across the 3.0-kΩ resistor. (d) What assumptions did you have to make to solve this problem?

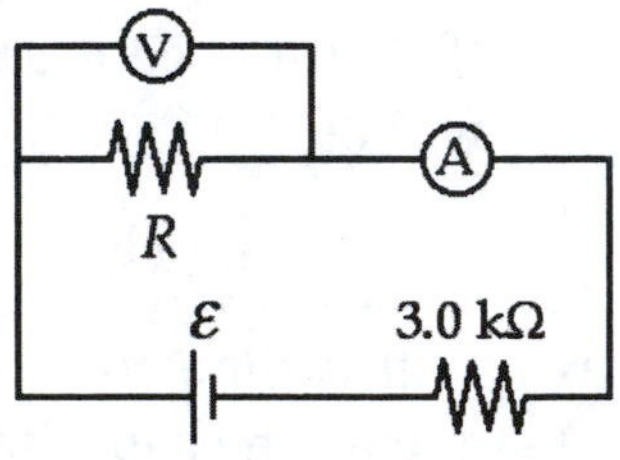

Figure P18.46

Solution

(a) The voltmeter gives a measure of the potential difference across the resistance R, $\Delta V = 6.0$ V. If we neglect any current passing through the voltmeter, the current measured by the ammeter is the same as the current through R, $I = 3.0 \text{ mA} = 3.0 \times 10^{-3}$ A. Then, Ohm's law gives the resistance as

$$R = \frac{\Delta V}{I} = \frac{6.0 \text{ V}}{3.0 \times 10^{-3} \text{ }\Omega} = 2.0 \times 10^{3} \text{ }\Omega = 2.0 \text{ k}\Omega$$ ◊

(b) Neglecting any current through the voltmeter, the current $I = 3.0 \times 10^{-3}$ A flows clockwise around the lower loop of the circuit (i.e., through R, through the ammeter, on through the 3.0 kΩ resistor, and back to the battery. We start at the lower left corner of the circuit, go clockwise around the lower loop, and neglect any change in potential across the ammeter. Kirchhoff's loop rule gives $-RI - (3.0 \text{ k}\Omega)I + \mathcal{E} = 0$, or

$$\mathcal{E} = (R + 3.0 \text{ k}\Omega)I = (5.0 \times 10^{3} \text{ }\Omega)(3.0 \times 10^{-3} \text{ A}) = 15 \text{ V}$$ ◊

(c) The potential difference, or voltage, across the 3.0-kΩ resistor is

$$\Delta V_3 = (3.0 \text{ k}\Omega)I = (3.0 \times 10^{3} \text{ }\Omega)(3.0 \times 10^{-3} \text{ A}) = 9.0 \text{ V}$$ ◊

(d) We have made three basic assumptions in this solution. First, we have assumed that the voltmeter and the ammeter are ideal devices. That is, the voltmeter has extremely high resistance and allows a negligible amount of current through it, while the ammeter has extremely low resistance and a negligible change in potential across it. Finally, we have assumed the battery is an ideal device, having a constant emf and negligible internal resistance. ◊

---

55. The student engineer of a campus radio station wishes to verify the effectiveness of the lightning rod on the antenna mast (Fig. P18.55). The unknown resistance R_x is between points C and E. Point E is a "true ground" but is inaccessible for direct measurement, since the stratum in which it is located is several meters below the Earth's surface. Two identical rods are driven into the ground at A and B, introducing an unknown resistance R_y. The procedure for finding the unknown resistance R_x is as follows. Measure resistance R_1 between points A and B. Then connect A and B with a heavy conducting wire, and measure resistance R_2 between points A and C. (a) Derive a formula for R_x in terms of the observable resistances R_1 and R_2. (b) A satisfactory ground resistance would be $R_x < 2.0\ \Omega$. Is the grounding of the station adequate if measurements give $R_1 = 13\ \Omega$ and $R_2 = 6.0\ \Omega$?

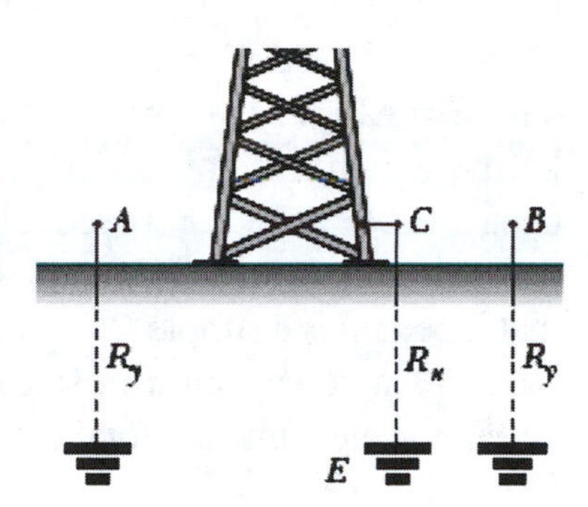

Figure P18.55

Solution

(a) For the first measurement (the resistance between points A and B), the equivalent circuit is as shown in Figure 1 at the right. Note that the lower ends of the resistances at A and B are connected by the Earth and that point C is not involved in this measurement. The measured resistance R_1 is the total resistance of a series combination

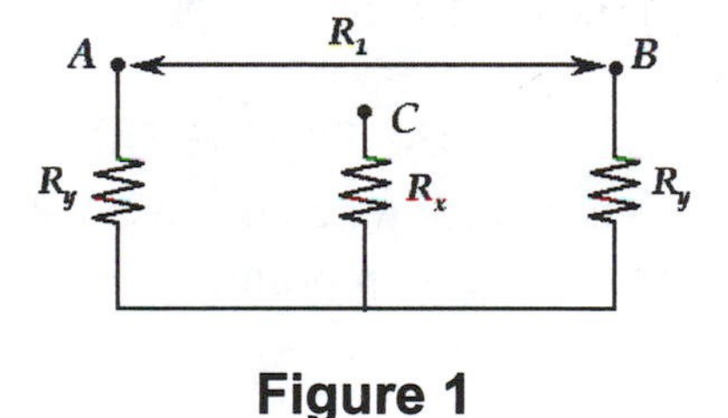

Figure 1

$$R_1 = R_y + R_y = 2R_y \qquad \text{so} \qquad R_y = R_1/2$$

Connecting points A and B makes a parallel combination of the two resistances labeled R_y as shown in Figure 2. This parallel combination has an equivalent resistance

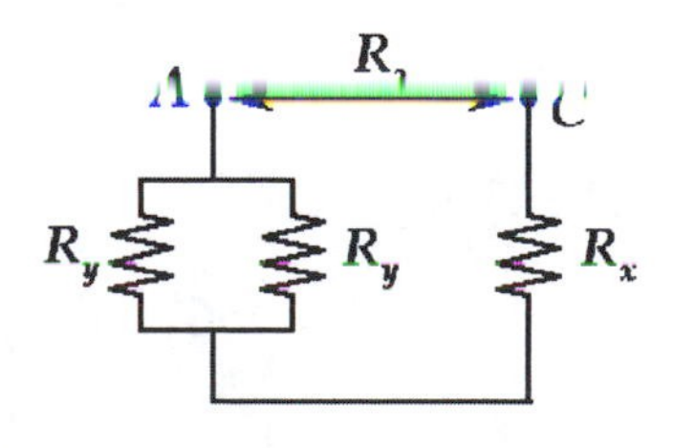

Figure 2

$$\frac{1}{R_p} = \frac{1}{R_y} + \frac{1}{R_y} = \frac{2}{R_y} \qquad \text{or} \qquad R_p = \frac{R_y}{2} = \frac{R_1}{4}$$

The measured R_2 is the total resistance of a series combination of R_p and R_x.

$$R_2 = R_p + R_x = \frac{R_1}{4} + R_x \qquad \text{so} \qquad R_x = R_2 - \frac{R_1}{4}$$ ◊

(b) If $R_1 = 13\ \Omega$ and $R_2 = 6.0\ \Omega$, then $R_x = 6.0\ \Omega - \dfrac{13\ \Omega}{4} = 2.8\ \Omega > 2.0\ \Omega$ ◊

so the antenna is inadequately grounded. ◊

---

60. The circuit in Figure P18.60 contains two resistors, $R_1 = 2.0\ \text{k}\Omega$ and $R_2 = 3.0\ \text{k}\Omega$, and two capacitors, $C_1 = 2.0\ \mu\text{F}$ and $C_2 = 3.0\ \mu\text{F}$, connected to a battery with emf $\mathcal{E} = 120$ V. If there are no charges on the capacitors before switch *S* is closed, determine the charges q_1 and q_2 on capacitors C_1 and C_2, respectively, as functions of time after the switch is closed. (*Hint*: First reconstruct the circuit so that it becomes a simple *RC* circuit containing a single resistor and single capacitor in series, connected to the battery, and then determine the total charge *q* stored in the circuit.)

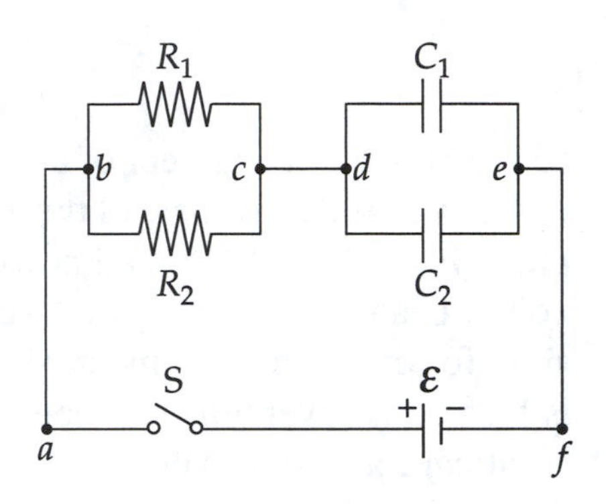

Figure P18.60

Solution

The equivalent resistance of the parallel combination of resistors between *b* and *c* is

$$\frac{1}{R_{bc}} = \frac{1}{2.0\ \text{k}\Omega} + \frac{1}{3.0\ \text{k}\Omega} = \frac{5.0}{6.0\ \text{k}\Omega} \qquad \text{or} \qquad R_{bc} = \frac{6.0\ \text{k}\Omega}{5.0} = 1.2\ \text{k}\Omega$$

Also, the total capacitance between points *d* and *e* is $C_{de} = C_1 + C_2 = 5.0\ \mu\text{F}$

so the time constant for this circuit is

$$\tau = R_{bc}C_{de} = \left(1.2\times10^3\ \Omega\right)\left(5.0\times10^{-6}\ \text{F}\right) = 6.0\times10^{-3}\ \text{s} = 6.0\ \text{ms}$$

Thus, if the switch is closed at $t = 0$, with the capacitors initially uncharged, the total charge stored between points *d* and *e* at time $t > 0$ is

$$q_{de} = Q\left(1 - e^{-t/\tau}\right) = C_{de}\mathcal{E}\left(1 - e^{-t/6.0\ \text{ms}}\right)$$

and the potential difference between *d* and *e* as a function of time is

$$\Delta V_{de} = \frac{q_{de}}{C_{de}} = \mathcal{E}\left(1 - e^{-t/6.0\ \text{ms}}\right) = (120\ \text{V})\left(1 - e^{-t/6.0\ \text{ms}}\right)$$

The charges on the capacitors C_1 and C_2 as functions of time are then

$$q_1 = C_1\left(\Delta V_{de}\right) = (2.0\ \mu\text{F})\left[(120\ \text{V})\left(1 - e^{-t/6.0\ \text{ms}}\right)\right] = (240\ \mu\text{C})\left(1 - e^{-t/6.0\ \text{ms}}\right)$$ ◊

and

$$q_2 = C_2\left(\Delta V_{de}\right) = (3.0\ \mu\text{F})\left[(120\ \text{V})\left(1 - e^{-t/6.0\ \text{ms}}\right)\right] = (360\ \mu\text{C})\left(1 - e^{-t/6.0\ \text{ms}}\right)$$ ◊

---

65. The given pair of capacitors in Figure P18.65 is fully charged by a 12.0-V battery. The battery is disconnected and the circuit closed. After 1.00 ms, how much charge remains on (a) the 3.00-μF capacitor? (b) The 2.00-μF capacitor? (c) What is the current in the resistor?

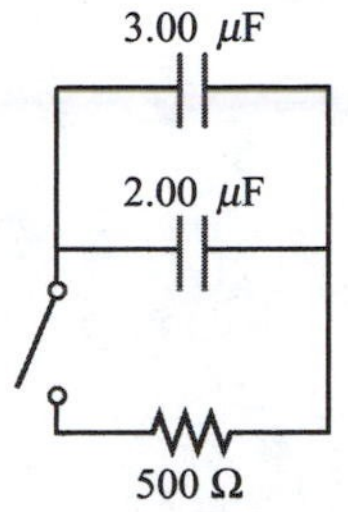

3.00 μF

2.00 μF

500 Ω

Figure P18.65

Solution

The parallel combination of the two capacitors has a total capacitance of $C_{eq} = 3.00\ \mu\text{F} + 2.00\ \mu\text{F} = 5.00\ \mu\text{F}$. Thus, when connected to a 12.0-V battery and charged, the total charge stored is

$$Q = C_{eq}(\Delta V) = (5.00\ \mu\text{F})(12.0\ \text{V}) = 60.0\ \mu\text{C}$$

If the switch shown in Figure P18.65 is closed at time $t = 0$, with the capacitors initially fully charged, the total charge still stored in the parallel combination of capacitors later at time t is $q = Qe^{-t/\tau}$, where the time constant is

$$\tau = RC_{eq} = (5.00\times 10^{2}\ \Omega)(5.00\times 10^{-6}\ \text{F}) = 2.50\times 10^{-3}\ \text{s} = 2.50\ \text{ms}$$

Hence, the potential difference across the combination of capacitors and the resistor, all in parallel with each other, at time t will be

$$\Delta V = \frac{q}{C_{eq}} = \frac{Qe^{-t/\tau}}{C_{eq}} = \frac{(60.0\ \mu\text{C})e^{-t/\tau}}{5.00\ \mu\text{F}} = (12.0\ \text{V})e^{-t/\tau}$$

(a) At $t = 1.00$ ms, the charge q_3 on the 3.00-μF capacitor will be

$$q_3 = C_3(\Delta V) = (3.00\ \mu\text{F})(12.0\ \text{V})e^{-1.00\ \text{ms}/2.50\ \text{ms}} = (36.0\ \mu\text{C})e^{-0.400} = 24.1\ \mu\text{C} \quad \Diamond$$

(b) Also at this time, the charge q_2 on the 2.00-μF capacitor will be

$$q_2 = C_2(\Delta V) = (2.00\ \mu\text{F})(12.0\ \text{V})e^{-1.00\ \text{ms}/2.50\ \text{ms}} = (24.0\ \mu\text{C})e^{-0.400} = 16.1\ \mu\text{C} \quad \Diamond$$

(c) The current in the 500-Ω resistor at $t = 1.00$ ms will be

$$i = \frac{\Delta V}{R} = \frac{(12.0\ \text{V})e^{-1.00\ \text{ms}/2.50\ \text{ms}}}{5.00\times 10^{2}\ \Omega} = (2.40\times 10^{-2}\ \text{A})e^{-0.400}$$

yielding $i = 16.1\times 10^{-3}\ \text{A} = 16.1\ \text{mA}$. $\Diamond$

19

Magnetism

NOTES FROM SELECTED CHAPTER SECTIONS

19.1 Magnets
19.2 Earth's Magnetic Field

Like magnetic poles (north-north or south-south) repel each other, and unlike poles (north-south) attract each other. Magnetic poles cannot be isolated and always occur in north-south pairs. **Soft magnetic materials**, such as iron, are easily magnetized but also tend to lose their magnetism easily; **hard magnetic materials**, such as cobalt and nickel, are difficult to magnetize but tend to retain their magnetism.

The region of space surrounding any magnetic material can be characterized by a magnetic field; the magnetic field of the Earth can be pictured as having a pattern of magnetic field lines similar to the field lines surrounding a bar magnet. The magnetic pole of the Earth in the vicinity of the Earth's geographic North Pole corresponds to a magnetic south pole.

At any given location on the surface of the Earth, a compass needle will indicate two angles related to the Earth's magnetic field:

> **Dip angle:** the angle between the horizontal, relative to the surface of the Earth, and the direction of the magnetic field at that point.
>
> **Magnetic declination:** the angle between true north (geographic North Pole) and magnetic north (indicated by a compass needle).

19.3 Magnetic Fields

The magnetic field is a vector quantity designated by the symbol $\vec{\mathbf{B}}$ and has the SI unit tesla (T).

By convention, the direction of the magnetic field vector is shown graphically as illustrated here.

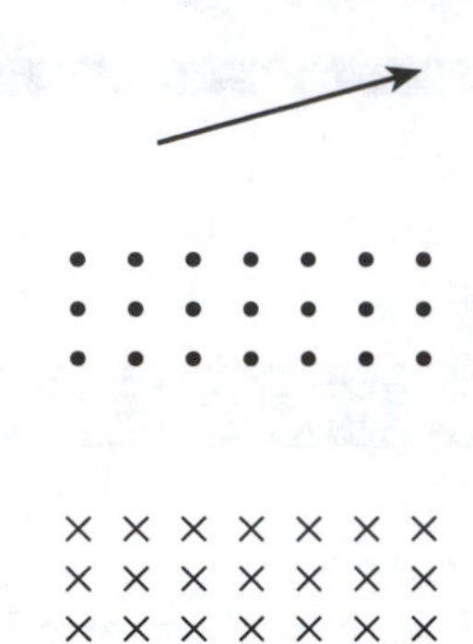

A magnetic field vector lying in the plane of the page is shown by an arrow in the plane of the page.

Magnetic field vectors directed out of the page are shown by dots representing tips of arrows coming outward.

Magnetic field vectors directed into the page are shown by crosses representing the feathers of arrows going inward.

A particle with charge q, moving with velocity $\vec{\mathbf{v}}$ in a magnetic field $\vec{\mathbf{B}}$, will experience a magnetic force $\vec{\mathbf{F}}$ with magnitude $F = qvB\sin\theta$. Properties of the magnetic force are:

- The magnetic force is proportional to the charge q and speed v of the particle.
- The magnitude of the magnetic force depends on the angle between the velocity vector of the particle and the magnetic field vector.
- When a charged particle moves in a direction parallel to the magnetic field vector, the magnetic force $\vec{\mathbf{F}}$ on the charge is zero.
- The magnetic force acts in a direction perpendicular to both $\vec{\mathbf{v}}$ and $\vec{\mathbf{B}}$; that is, $\vec{\mathbf{F}}$ is perpendicular to the plane formed by $\vec{\mathbf{v}}$ and $\vec{\mathbf{B}}$.
- The magnetic force on a positive charge is in the direction opposite to the force on a negative charge moving in the same direction.
- If the velocity vector makes an angle θ with the magnetic field, the magnitude of the magnetic force is proportional to $\sin\theta$.

The direction of the magnetic force on a **positive charge** can be determined by using **right-hand rule number 1**:

1. With your thumb extended, point the fingers of your right hand in the direction of the velocity, $\vec{\mathbf{v}}$.
2. Curl the fingers to point in the direction of the magnetic field, $\vec{\mathbf{B}}$, moving through the smaller of the two possible angles.
3. Your thumb is now pointing in the direction of the magnetic force $\vec{\mathbf{F}}$ exerted on a positive charge.

Another version of right-hand-rule number 1 is described in Suggestions, Skills, and Strategies.

19.4 Magnetic Force on a Current-Carrying Conductor

A magnetic force will be exerted on a current-carrying conductor when placed in a magnetic field. For a straight conductor, the magnitude of the force will be maximum when the conductor is perpendicular to the magnetic field and zero when the conductor is parallel to the magnetic field. *The net magnetic force on a closed current loop in a uniform magnetic field is zero.*

The direction of the force on a straight conductor can be determined by using **right-hand rule number 1**.

In this case, point the fingers on your right hand in the direction of conventional current (I) and curl your fingers to point in the direction of the magnetic field. Your thumb will now point in the direction of the magnetic force on the conductor.

Can you use the right-hand rule to determine the direction of the magnetic field that exerts a force as shown on the conductor at the right? (Answer: into the plane of the page)

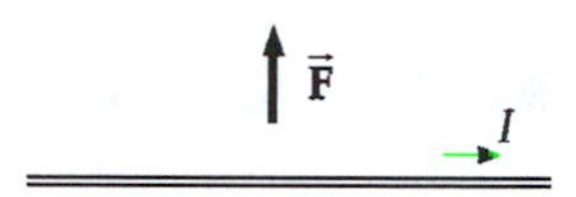

19.5 Torque on a Current Loop and Electric Motors

As stated above, zero net force is exerted on a closed current loop (coil) in a uniform magnetic field; however, there is a **net torque** exerted on a current loop in a magnetic field. The magnitude of the torque is proportional to the area of the loop, magnetic field strength, loop current, and the sine of the angle between the perpendicular to the plane of the loop and the direction of the magnetic field. The torque on the loop will be maximum when the field is parallel to the plane of the loop, and the torque will be zero when the field is perpendicular to the plane of the loop. *The torque will cause the loop to rotate so that the normal to the plane of the loop will turn toward the direction parallel to the magnetic field.*

19.6 Motion of a Charged Particle in a Magnetic Field

A charged particle entering a uniform magnetic field with its velocity vector initially along a direction perpendicular to the field will move in a circular path in a plane perpendicular to the magnetic field. The magnetic force on the moving charge will be directed toward the center of the circular path and produce a centripetal acceleration. *The magnetic force changes the direction of the velocity vector but does not change its magnitude; the work done by the magnetic force on the particle is zero.*

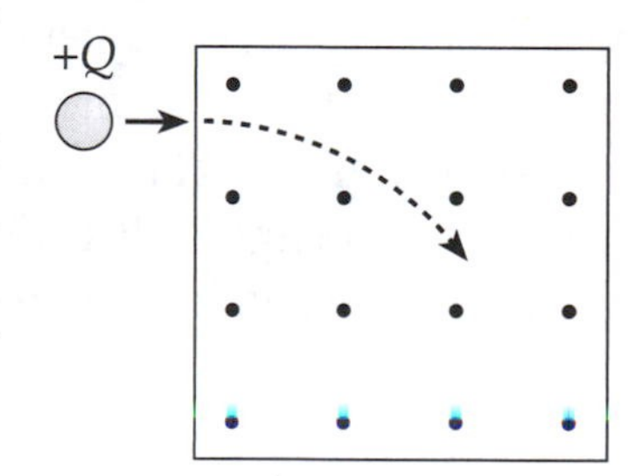

19.7 Magnetic Field of a Long, Straight Wire and Ampère's Law

A magnetic field will exist in the vicinity of a current-carrying wire. For a current in a long wire (where the length of the wire is much greater than the distance from the wire) the magnetic field is inversely proportional to the distance from the conductor. The direction of the magnetic field due to a current in a long straight wire is given by **right-hand rule number 2**:

> **If the wire is grasped in the right hand with the thumb in the direction of the positive current, the fingers will wrap (or curl) in the direction of the circular magnetic field lines.**

As illustrated in the figure, the magnetic field lines are circular in a plane perpendicular to the length of the conductor.

At any point in space surrounding the conductor, the magnetic field vector, $\vec{\mathbf{B}}$, is directed along the tangent to the circular field line through that point.

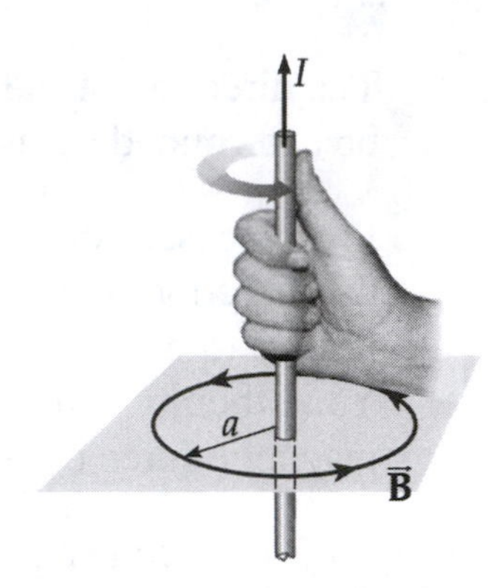

Ampère's circuital law states a method for calculating the magnetic field in the vicinity of a current-carrying conductor. *This technique is valid only for steady currents and can be easily used only in those cases where the current configuration has a high degree of symmetry.*

19.8 Magnetic Force Between Two Parallel Conductors

Parallel conductors carrying currents in the same direction attract each other; if the currents are in opposite directions, the conductors repel each other. *The forces on the two conductors are equal in magnitude regardless of the relative value of the two currents.*

The force between two parallel wires each carrying a current is used to define the ampere. If two long, parallel wires 1 m apart in a vacuum carry the same current and the force per unit length on each wire is 2×10^{-7} N/m, then the current is defined to be 1 A.

19.9 Magnetic Fields of Current Loops and Solenoids

The direction of the magnetic field at the center of a current loop is perpendicular to the plane of the loop and directed in the sense given by right-hand rule number 2. For the situation illustrated in the figure, $\vec{\mathbf{B}}$ is directed out of the page within the area of the loop and into the page outside the loop.

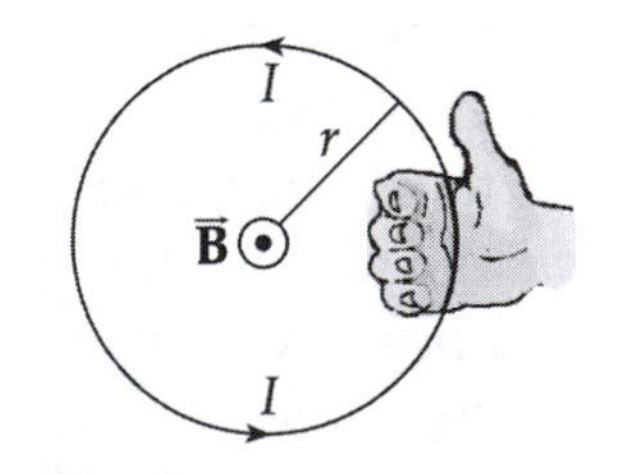

Within a solenoid, the magnetic field is parallel to the axis of the solenoid and pointing in a sense determined by applying right-hand rule number 2 to one of the coils in the winding. *The magnetic field is uniform within the volume of the solenoid (except near the ends).*

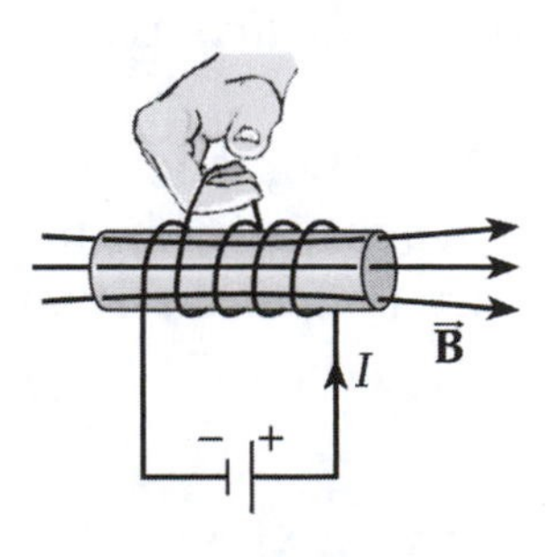

EQUATIONS AND CONCEPTS

The **magnitude of the magnetic force on a charge moving in a magnetic field** can have values ranging from zero to a maximum value, depending on the angle between the direction of the magnetic field and the direction of the velocity of the charge. The **direction** of the force exerted on a positive charge moving in a magnetic field is given by using right-hand rule number 1. (See Suggestions, Skills, and Strategies.)

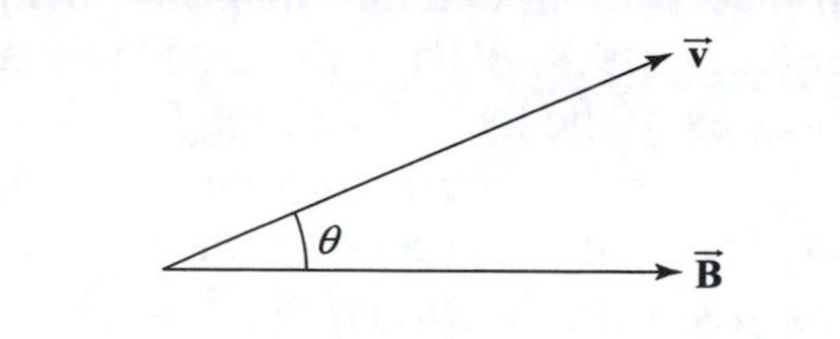

$$F = qvB\sin\theta \qquad (19.1)$$

θ is the smaller of the two angles between the directions of $\vec{\mathbf{v}}$ and $\vec{\mathbf{B}}$.

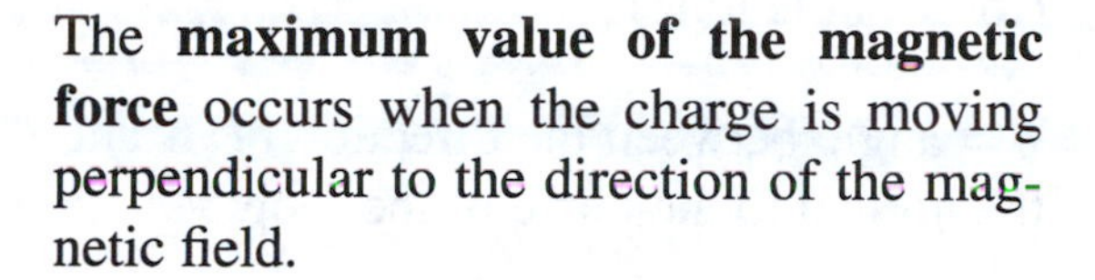

The **maximum value of the magnetic force** occurs when the charge is moving perpendicular to the direction of the magnetic field.

$$F_{\text{max}} = qvB \qquad (19.4)$$

The **magnitude of the magnetic field intensity** at some point in space is defined in terms of the magnetic force exerted on a moving positive electric charge at that point. Note in Equation 19.1 above that when $\theta = 0$, the force $F = 0$. *Therefore, do not misinterpret Equation 19.2 to mean that B will be infinite when $\theta = 0$.*

$$B \equiv \frac{F}{qv\sin\theta} \qquad (19.2)$$

The **SI unit of magnetic field intensity** is the tesla (T).

$$[B] = \text{T} = \text{Wb/m}^2 = \text{N/A}\cdot\text{m} \qquad (19.3)$$

The **magnetic force exerted on a current-carrying conductor** placed in a magnetic field depends on the angle between the direction of the current in the conductor and the direction of the field. The magnetic force will be maximum when the conductor is directed perpendicular to the magnetic field. *Equation 19.6 applies in the case of a straight conductor.* What is the direction of force on the conductor in the figure?

$$F = BI\ell\sin\theta \qquad (19.6)$$

$$F_{\text{max}} = BI\ell \qquad (19.5)$$

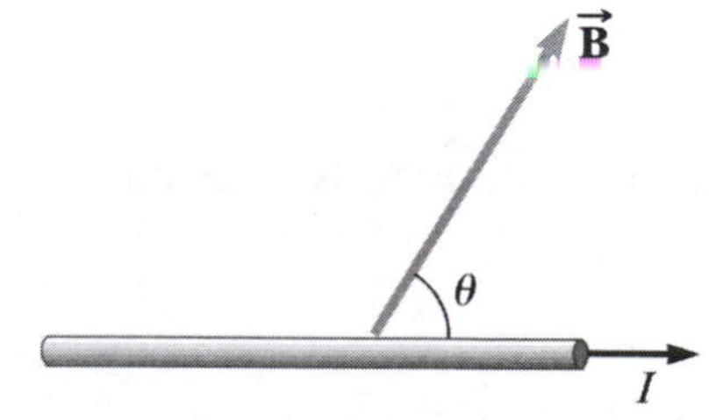

Current-carrying conductor in a magnetic field

A **net torque** will be exerted on a current loop placed in an external magnetic field. In Equation 19.8, $\vec{\mathbf{A}}$ has a magnitude equal to the area of the loop. *When the fingers of the right hand are curled around the loop in the direction of the current, the extended thumb points in the direction of* $\vec{\mathbf{A}}$.

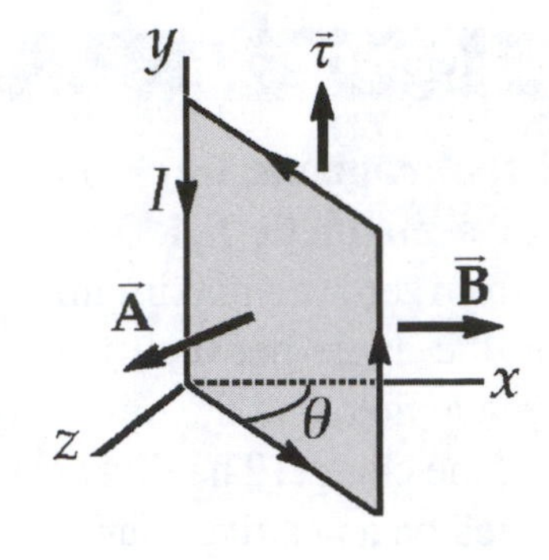

For a multi-loop coil, the torque is proportional to the number of turns, N.

- The **magnitude of the net torque** depends on the area of the loop, the angle between the direction of the magnetic field, and the direction of the normal (or perpendicular) to the plane of the loop.

$$\tau = BIA\sin\theta \quad (19.8)$$

(for a single loop)

θ = angle between the direction of $\vec{\mathbf{B}}$ and the normal to the plane of the loop

- **Maximum magnitude of the torque** will occur when the magnetic field is parallel to the plane of the loop (i.e., $\theta = 90°$).

$$\tau_{max} = BIA \quad (19.7)$$

- The **direction of rotation of a current loop** in a magnetic field decreases the angle between the normal to the loop and the magnetic field; *the loop turns toward a position of minimum torque*. The loop shown in the figure will rotate counterclockwise as seen from above.

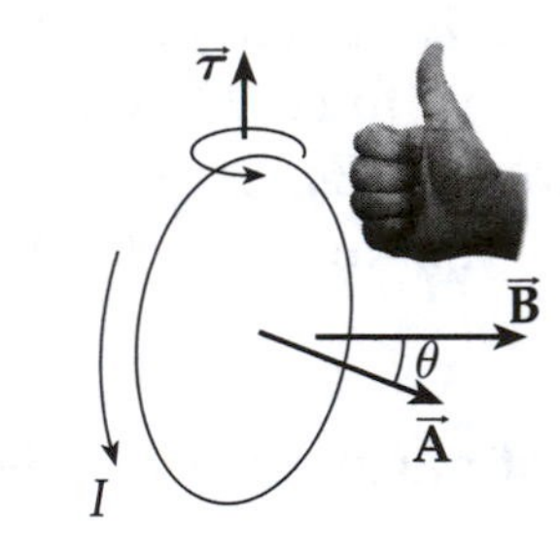

The **magnetic dipole moment of a current loop**, $\vec{\mu}$, can be used to express the torque on the loop when placed in a magnetic field. The direction of $\vec{\mu}$ is perpendicular to the plane of the loop, and θ is the angle between the directions of $\vec{\mu}$ and $\vec{\mathbf{B}}$.

$$\mu = IAN$$

$$\tau = \mu B\sin\theta \quad (19.9b)$$

Motion of a charged particle entering a uniform magnetic field with the velocity vector initially perpendicular to the field has the following characteristics:

- The **path** of the particle will be circular and in a plane perpendicular to the direction of the field. The direction of rotation of the particle will be as determined by the right-hand rule.

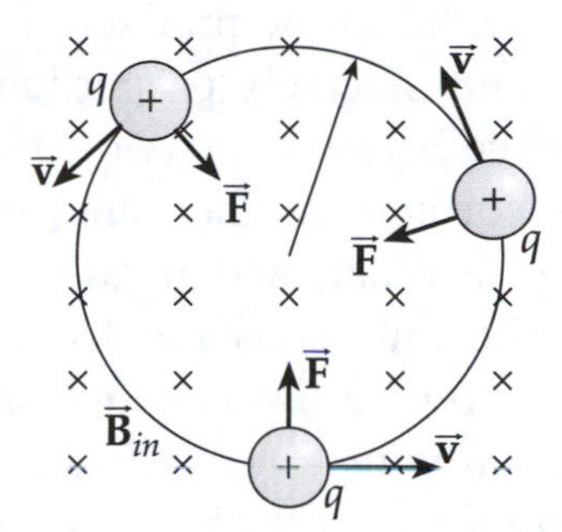

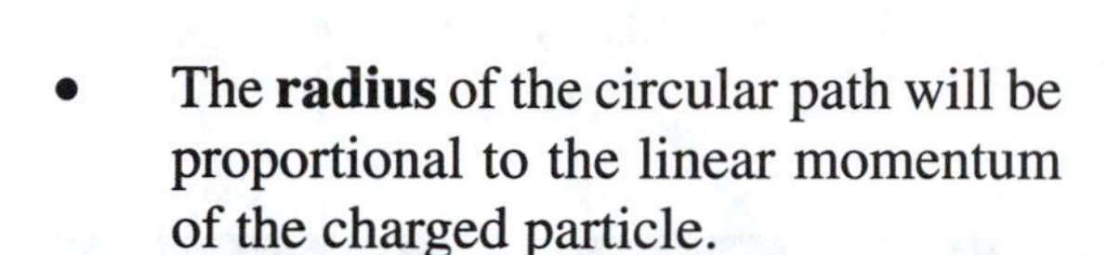

- The **radius** of the circular path will be proportional to the linear momentum of the charged particle.

$$r = \frac{mv}{qB} \quad (19.10)$$

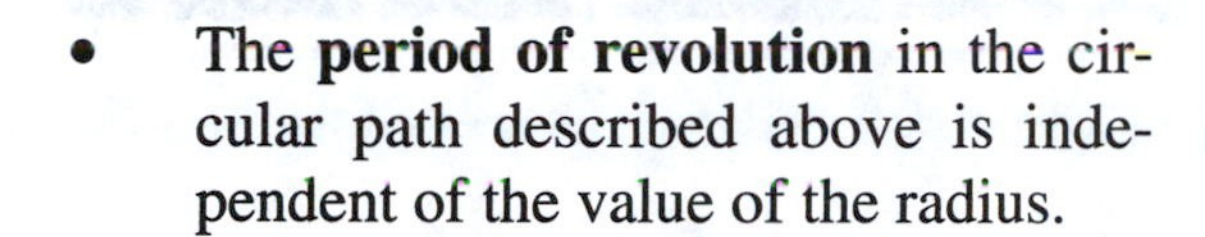

- The **period of revolution** in the circular path described above is independent of the value of the radius.

$$T = \frac{2\pi r}{v} = \frac{2\pi m}{qB}$$

- A **helical path** will result if the initial velocity is not perpendicular to the magnetic field. The axis of the helix will be parallel to the magnetic field and the "pitch" (distance between adjacent coils) will depend on the component of velocity parallel to the field.

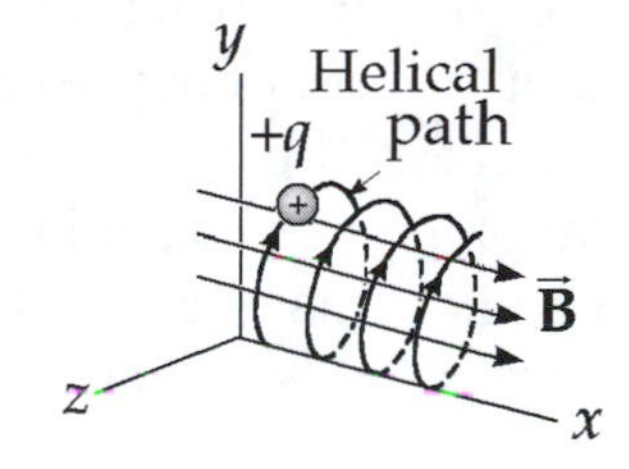

Equations for calculating the magnitude of the magnetic field due several current geometries with important applications are given below. The value of the permeability of free space constant appearing in the equations is $\mu_0 = 4\pi \times 10^{-7}\ \text{T}\cdot\text{m/A}$.

The magnitude of the magnetic field:

At a distance r from a **long straight conductor**

$$B = \frac{\mu_0 I}{2\pi r} \quad (19.15)$$

At the **center of a current loop** of N turns and radius R

$$B = N\frac{\mu_0 I}{2R}$$

Inside a solenoid of n turns of conductor per unit length

$$B = \mu_0 nI \quad (19.16)$$

N = total turns
n = turns per meter

Ampère's circuital law can be used to find the magnetic field around certain simple current-carrying conductors. The product ($B_{\parallel}\,\Delta\ell$) must be summed around a closed path containing the current.

$$\sum B_{\parallel}\,\Delta\ell = \mu_0 I \quad (19.13)$$

$B_{\parallel}$ is the component of $\vec{\mathbf{B}}$ tangent to a small displacement of length $\Delta\ell$.

The **magnitude of the force, per unit length, between parallel conductors** is proportional to the product of the two currents and inversely proportional to d, the distance between the two conductors. Parallel currents in the same direction attract each other, and parallel currents in opposite directions repel each other. *The forces on the two conductors will be equal in magnitude regardless of the relative magnitude of the two currents.*

$$\frac{F_1}{\ell} = \frac{\mu_0 I_1 I_2}{2\pi d} \quad (19.14)$$

SUGGESTIONS, SKILLS, AND STRATEGIES

Using the right-hand rule

Application of the right-hand rule as illustrated below applies to *positive charges;* results are reversed in the case of negative charges. The right-hand rule is used to find the direction of force on a charged particle moving in a magnetic field. There are two versions of the right-hand rule; in order to avoid confusion, you should pick the version that suits you. In either version, note that the stated order of $\vec{\mathbf{v}}$ and $\vec{\mathbf{B}}$ is very important.

Version 1

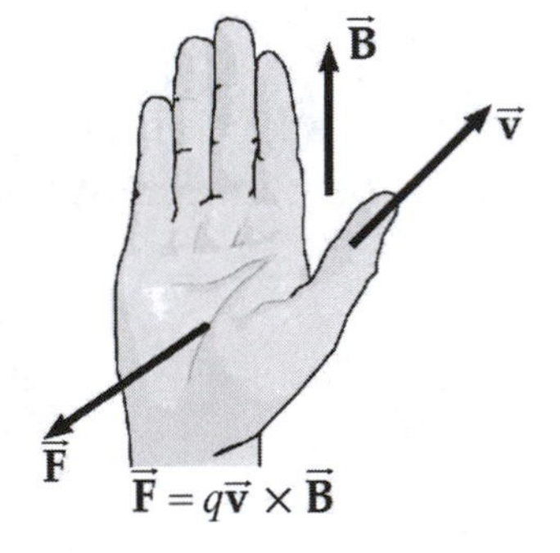

Hold your open right hand with your thumb pointing in the direction of $\vec{\mathbf{v}}$ and your fingers pointing in the direction of $\vec{\mathbf{B}}$. The force is now directed out of the palm of your hand. If the angle between $\vec{\mathbf{v}}$ and $\vec{\mathbf{B}}$ is zero, your thumb and fingers will be aligned and there is no force.

Version 2

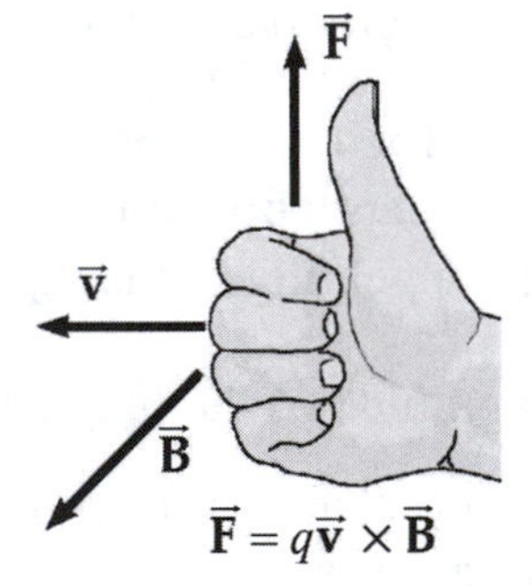

Orient your right hand so that your fingers point in the direction of $\vec{\mathbf{v}}$ and then "curl" your fingers to point in the direction of $\vec{\mathbf{B}}$. Your thumb now points in the direction of force. Since your fingers cannot bend more than 180°, you may have to flip your hand upside down to accomplish the action described. If in a given situation, you do not need to curl your fingers because they already point along $\vec{\mathbf{v}}$, the angle between $\vec{\mathbf{v}}$ and $\vec{\mathbf{B}}$ is zero and there is no force.

A different, but useful, right-hand rule applies in some specific situations. An example is when one of the quantities involves circulation (e.g., a current) and another a vector. The rule looks similar to version 2 described above; curl your fingers around in the direction of the circulation, and your thumb will point in the direction of the vector. This rule applies, for example, when finding the direction of the area vector for a current loop.

REVIEW CHECKLIST

- Use right-hand rule number 1 and appropriate equations to determine the direction and the magnitude of the magnetic force exerted on a moving electric charge or current-carrying conductor in a region where there is a magnetic field. Practice the right-hand rule for situations in which different quantities (velocity, force, and magnetic field) represent the unknown direction. (Sections 19.3 and 19.4)
- Determine the magnitude of the torque exerted on a closed current loop in an external magnetic field and state the direction of rotation. (Section 19.5)
- Calculate the period and radius of the circular orbit of a charged particle moving in a uniform magnetic field. (Section 19.6)
- Calculate the magnitude and determine the direction (using right-hand rule number 2) of the magnetic field for the following cases: (i) a distance r from a long, straight current-carrying conductor, (ii) at the center of a current loop of radius R, and (iii) at interior points of a solenoid with n turns per unit length. (Sections 19.7 and 19.9)
- Determine the magnetic force between parallel conductors. (Section 19.8)

SOLUTIONS TO SELECTED END-OF-CHAPTER PROBLEMS

1. Consider an electron near the Earth's equator. In which direction does it tend to deflect if its velocity is (a) directed downward? (b) Directed northward? (c) Directed westward? (d) Directed southeastward?

Solution

The direction of the magnetic force exerted on a charged particle moving through a magnetic field (and hence the direction of the deflection of that particle) is given by **right-hand rule number 1**, introduced in Section 19.3 of the textbook and illustrated by the drawing at the right. To employ this rule, hold your right hand with the fingers extended in the direction of the velocity of the particle, $\vec{\mathbf{v}}$, and so your fingers will move toward the direction of the magnetic field vector, $\vec{\mathbf{B}}$, as you begin to close your hand. If the particle of interest has a positive charge, the thumb of your right hand will now be pointing in the direction of the magnetic force experienced by the particle.

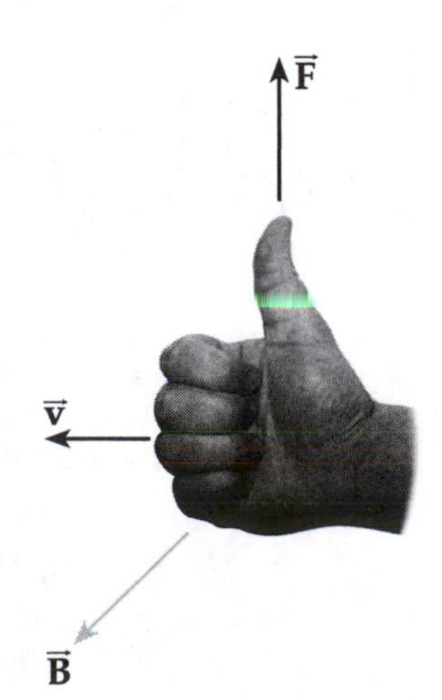

If the particle has a negative charge, as does the electron of this problem, the magnetic force will be in the direction opposite to the direction the thumb is pointing.

Near the equator, the Earth's magnetic field points northward in a horizontal plane.

(a) If the electron is moving downward toward Earth's surface, hold your right hand with the extended fingers pointing downward and turned so your fingers will move northward (the direction of $\vec{\mathbf{B}}$) as you start closing the hand. Your thumb is now pointing eastward. However, since the electron is negative, the magnetic force acting on it is opposite to the direction the thumb points, or the force is westward in this case. ◊

(b) When the electron moves northward, it is moving parallel to the magnetic field, and right-hand rule number 1 cannot be employed. However, this is not a concern since the magnitude of the magnetic force acting on a charged particle is $F_m = qvB\sin\theta$, where θ is the angle between the directions of $\vec{\mathbf{v}}$ and $\vec{\mathbf{B}}$. Thus, when $\theta = 0°$ (as it does in this case), and when $\theta = 180°$ (as it would if the particle moved anti-parallel to the field), the particle experiences zero magnetic force. ◊

(c) When the electron moves westward, application of right-hand rule number 1 leaves your thumb pointing downward. However, since the electron is negative, the magnetic force acting on the electron is directed upward. ◊

(d) If the electron moves southeastward, use of the rule described above leaves the thumb of the right hand pointing upward. Due to its negative charge, the electron will experience a downward force in this case. ◊

---

5. A laboratory electromagnet produces a magnetic field of magnitude 1.50 T. A proton moves through this field with a speed of 6.00×10^6 m/s. (a) Find the magnitude of the maximum magnetic force that could be exerted on the proton. (b) What is the magnitude of the maximum acceleration of the proton? (c) Would the field exert the same magnetic force on an electron moving through the field with the same speed? (d) Would the electron undergo the same acceleration? Explain.

Solution

(a) When a particle having charge q moves at speed v through a magnetic field of magnitude B, that particle experiences a magnetic force whose magnitude is given by $F_m = qvB\sin\theta$, where θ is the angle the particle's velocity makes with the direction of the magnetic field. Thus, the particle will experience maximum force when it moves perpendicular to the magnetic field, so $\theta = 90°$ and $\sin\theta = 1$. For a proton, $q = +e = 1.60 \times 10^{-19}$ C, so the maximum force in this case is

$$(F_m)_{\max} = qvB = \left(1.60\times10^{-19}\text{ C}\right)\left(6.00\times10^{6}\text{ m/s}\right)(1.50\text{ T}) = 1.44\times10^{-12}\text{ N}$$ ◊

(b) From Newton's second law, the maximum acceleration this force would give the proton, a particle with mass $m_p = 1.67\times10^{-27}$ kg, is

$$a_{\max} = \frac{(F_m)_{\max}}{m_p} = \frac{1.44\times10^{-12}\ \text{N}}{1.67\times10^{-27}\ \text{kg}} = 8.62\times10^{14}\ \text{m/s}^2$$ ◊

(c) The magnitude of the charge of an electron is the same as that of a proton, $|q| = e = 1.60\times10^{-19}$ C. Thus, if it moves at the same speed as the proton through the given magnetic field, it will experience the same magnitude maximum force as that found for the proton in part (a). However, since the electron has negative charge, the force it experiences will have the direction opposite to that of the force acting on the proton. ◊

(d) While the electron and the proton will experience the same magnitude maximum force, their accelerations will be very different due to the difference in the masses of these two particles. The magnitude of the maximum acceleration experienced by the electron under these conditions will be

$$a_{\max} = \frac{(F_m)_{\max}}{m_e} = \frac{1.44\times10^{-12}\ \text{N}}{9.11\times10^{-31}\ \text{kg}} = 1.58\times10^{18}\ \text{m/s}$$ ◊

---

16. A wire having a mass per unit length of 0.500 g/cm carries a 2.00-A current horizontally to the south. What are the direction and magnitude of the minimum magnetic field needed to lift this wire vertically upward?

Solution

In order to lift the wire using a minimum magnetic field (and hence, minimum magnetic force), the magnetic force should be directed vertically upward and have a magnitude equal to the weight of the wire. If the wire has length L and carries current I through a magnetic field with magnitude B, the magnitude of the magnetic force acting on it is

$$F = BIL\sin\theta$$

where θ is the angle the direction of the current makes with the direction of the magnetic field. If this force has magnitude equal to the weight of the wire, then

$$BIL\sin\theta = mg \qquad \text{or} \qquad B = \frac{(m/L)g}{I\sin\theta}$$

Thus, for a minimum magnitude field, one should have $\theta = 90°$, or the field should be perpendicular to the current. The minimum acceptable magnitude for the field is

$$B_{\min} = \left(\frac{m}{L}\right)\frac{g}{I\sin 90°} = \left[\left(0.500\ \frac{\text{g}}{\text{cm}}\right)\left(\frac{1\ \text{kg}}{10^3\ \text{g}}\right)\left(\frac{10^2\ \text{cm}}{1\ \text{m}}\right)\right]\frac{(9.80\ \text{m/s}^2)}{(2.00\ \text{A})(1)} = 0.245\ \text{T}$$ ◊

To determine the direction of this minimum magnitude magnetic field, use a variation of right-hand rule number 1. With your right hand fully open, hold it so your extended fingers point in the direction of the current (horizontally southward) and the thumb points in the required direction of the force (vertically upward). Then, after moving the fingers 90° as you close your hand, the fingers point in the required direction of the field. You should find that this is horizontal and due east. ◊

---

25. A wire is formed into a circle having a diameter of 10.0 cm and is placed in a uniform magnetic field of 3.00 mT. The wire carries a current of 5.00 A. Find the maximum torque on the wire.

Solution

When the wire is bent into the form of a circular loop having a diameter of $d = 10.0 \text{ cm} = 0.100 \text{ m}$, it encloses an area of

$$A = \frac{\pi d^2}{4} = \frac{\pi (0.100 \text{ m})^2}{4} = 7.85 \times 10^{-3} \text{ m}^2$$

If this loop carries current I while in a uniform magnetic field having a magnitude B, the loop will experience a torque that attempts to rotate the loop until the line normal to the plane of the loop aligns with the direction of the magnetic field. The magnitude of this torque is given by

$$\tau = BIA \sin\theta$$

where θ is the angle the normal line makes with the direction of the field. Therefore, the loop experiences maximum torque when the plane of the loop is parallel to the direction of the field, so the normal line makes an angle $\theta = 90°$ with the field direction, and $\sin\theta = 1$.

In the present case, $I = 5.00 \text{ A}$, $B = 3.00 \text{ mT} = 3.00 \times 10^{-3} \text{ T}$, and $A = 7.85 \times 10^{-3} \text{ m}^2$ as computed above. Thus, the maximum torque exerted on the current loop formed by the wire is

$$\tau_{\max} = BIA = \left(3.00 \times 10^{-3} \text{ T}\right)(5.00 \text{ A})\left(7.85 \times 10^{-3} \text{ m}^2\right) = 1.18 \times 10^{-4} \text{ N} \cdot \text{m}$$

or $\tau_{\max} = 118 \times 10^{-6} \text{ N} \cdot \text{m} = 118 \ \mu\text{N} \cdot \text{m}$ ◊

---

29. A 200-turn rectangular coil having dimensions of 3.0 cm by 5.0 cm is placed in a uniform magnetic field of magnitude 0.90 T. (a) Find the current in the coil if the maximum torque exerted on it by the magnetic field is $0.15\ \text{N}\cdot\text{m}$. (b) Find the magnitude of the torque on the coil when the magnetic field makes an angle of 25° with the normal to the plane of the coil.

Solution

If a coil having N turns of wire wound on it carries current I while in a magnetic field of magnitude B, that coil will experience a torque of magnitude $\tau = \mu B \sin\theta$, where $\mu = IAN$ is the magnitude of the magnetic moment of the coil. The direction of the magnetic moment $\vec{\mu}$ is defined as the direction the thumb of the right hand points when the fingers of the right hand curl around the coil in the direction of the current. Note that this means the magnetic moment of a coil is directed perpendicular to the plane of the coil. The angle θ is the angle between the directions of the magnetic moment $\vec{\mu}$ and the magnetic field $\vec{\mathbf{B}}$.

(a) The coil experiences maximum torque when $\theta = 90°$ and $\sin\theta = 1$, or when the magnetic moment is perpendicular to the field. The magnitude of the maximum torque exerted on the coil is $\tau_{max} = \mu B = IANB$, and if $\tau_{max} = 0.15\ \text{N}\cdot\text{m}$, the current in the coil must be

$$I = \frac{\tau_{max}}{ANB} = \frac{0.15\ \text{N}\cdot\text{m}}{\left[\left(3.0\times10^{-2}\ \text{m}\right)\left(5.0\times10^{-2}\ \text{m}\right)\right](200)(0.90\ \text{T})} = 0.56\ \text{A} \qquad \Diamond$$

(b) If the magnetic field makes an angle of $\theta = 25°$ with the normal to the plane of the coil (and hence with the direction of the magnetic moment), the torque exerted on the coil will be $\tau = \mu B \sin 25° = IANB \sin 25°$, or

$$\tau = (0.56\ \text{A})\left[\left(3.0\times10^{-2}\ \text{m}\right)\left(5.0\times10^{-2}\ \text{m}\right)\right](200)(0.90\ \text{T})\sin 25°$$

This yields $\tau = 6.4\times10^{-2}\ \text{N}\cdot\text{m} = 0.064\ \text{N}\cdot\text{m}$. $\Diamond$

37. A singly charged positive ion has a mass of 2.50×10^{-26} kg. After being accelerated through a potential difference of 250 V, the ion enters a magnetic field of 0.500 T, in a direction perpendicular to the field. Calculate the radius of the path of the ion in the field.

Solution

The singly charged $(q = 1e)$ ion starts from rest and accelerates through a 250-V potential difference before entering the magnetic field. Since electric fields are conservative force fields, the speed of the ion as it enters the field may be found using conservation of energy, $KE_f + PE_f = KE_i + PE_i$. The electrical potential energy of the ion at a location having potential V is $PE = qV$. Thus, we have

$$\frac{1}{2}mv^2 + qV_f = 0 + qV_i \qquad \text{or} \qquad \frac{1}{2}mv^2 = q\left(V_i - V_f\right) = q(\Delta V)$$

giving $v = \sqrt{\dfrac{2q|\Delta V|}{m}} = \sqrt{\dfrac{2\left(1.60\times10^{-19}\text{ C}\right)(250\text{ V})}{2.50\times10^{-26}\text{ kg}}} = 5.66\times10^{4}\text{ m/s}$

When a charged particle moves perpendicularly to a magnetic field, the magnetic force exerted, always perpendicular to the motion, deflects the particle into a circular path and supplies the needed centripetal acceleration. That is,

$$F = qvB = m\frac{v^2}{r}$$

The radius of the path the particle follows is then $r = \dfrac{mv^2}{qvB} = \dfrac{mv}{qB}$. In this case, the radius is

$$r = \frac{\left(2.50\times10^{-26}\text{ kg}\right)\left(5.66\times10^{4}\text{ m/s}\right)}{\left(1.60\times10^{-19}\text{ C}\right)(0.500\text{ T})} = 1.77\times10^{-2}\text{ m} = 1.77\text{ cm}$$ ◊

41. A particle passes through a mass spectrometer as illustrated in Figure P19.36. The electric field between the plates of the velocity selector has a magnitude of 8 250 V/m, and the magnetic fields in both the velocity selector and the deflection chamber have magnitudes of 0.093 1 T. In the deflection chamber the particle strikes a photographic plate 39.6 cm removed from its exit point after traveling in a semicircle. (a) What is the mass-to-charge ratio of the particle? (b) What is the mass of the particle if it is doubly ionized? (c) What is its identity, assuming it's an element?

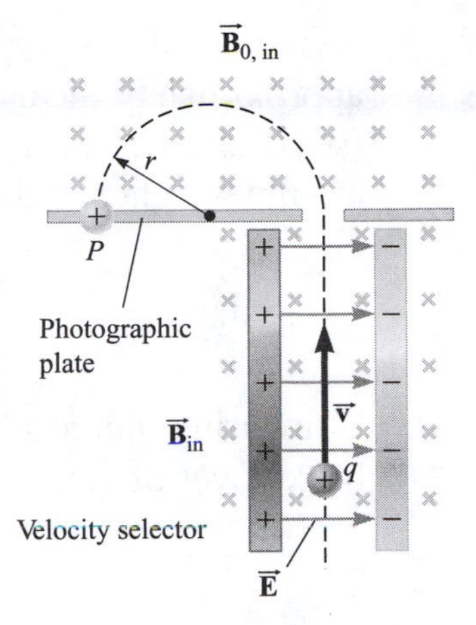

Figure P19.36

Solution

The charged particle shown moving through the electric and magnetic fields of the velocity selector in Figure P19.36 experiences an electrical force of magnitude $F_e = qE$ directed to the right, and a magnetic force of magnitude $F_m = qvB$ directed to the left. If the speed of the particle is $v = E/B$, then these oppositely-directed forces have equal magnitudes and the particle passes undeviated through the velocity selector and into the deflection chamber. Particles having any other speed will be deflected either to the right or the left and not pass through the narrow slit at the end of the velocity selector.

(a) Once a positive ion with speed $v = E/B$ enters the deflection chamber, it experiences only the magnetic force, which is always perpendicular to the velocity of the particle. This force deflects the particle into a circular path of radius r and provides the needed centripetal acceleration. Thus, $F_m = qvB = mv^2/r$, and the radius of the circular path is

$$r = \frac{mv^2}{qvB} = \left(\frac{m}{q}\right)\frac{v}{B} = \left(\frac{m}{q}\right)\frac{E/B}{B} = \left(\frac{m}{q}\right)\frac{E}{B^2}$$

The mass-to-charge ratio for the particle is then $m/q = rB^2/E$. If the particle strikes the photographic plate at distance $d = 2r = 39.6$ cm from the slit entrance into the deflection chamber when $B = 0.093\,1$ T and $E = 8\,250$ V/m, the mass-to-charge ratio for these particles is

$$\frac{m}{q} = \frac{\left[\frac{1}{2}\left(39.6\times10^{-2}\text{ m}\right)\right]\left(0.093\,1\text{ T}\right)^2}{8\,250\text{ V/m}} = 2.08\times10^{-7}\text{ kg/C} \quad \Diamond$$

(b) If the particle is doubly ionized (i.e., two electrons have been removed from the neutral atom), its charge is $q = +2e$. Thus, the mass of these particles must be

$$m = q\left(\frac{m}{q}\right) = (2e)\left(\frac{m}{q}\right) = 2\left(1.60\times10^{-19}\text{ C}\right)\left(2.08\times10^{-7}\text{ kg/C}\right)$$

or $\quad m = 6.66\times10^{-26}$ kg $\quad \Diamond$

(c) Assuming these particles are ionized atoms of some element, their mass should be approximately the atomic weight of that element multiplied by the atomic mass unit (see Table C.5 in Appendix C of the textbook). That is, $m = (\text{At. wt.})(1\text{ u})$, and the atomic weight should be

$$\text{At. wt.} = \frac{m}{1\text{ u}} = \frac{6.66\times10^{-26}\text{ kg}}{1.66\times10^{-27}\text{ kg}} = 40.1$$

Comparing this result to the atomic weights listed in the periodic table on the inside back cover of your textbook suggests that this element is very likely calcium. ◊

---

46. In 1962, measurements of the magnetic field of a large tornado were made at the Geophysical Observatory in Tulsa, Oklahoma. If the magnitude of the tornado's field was $B = 1.50\times10^{-8}$ T pointing north when the tornado was 9.00 km east of the observatory, what current was carried up or down the funnel of the tornado? Model the vortex as a long, straight wire carrying a current.

Solution

Ionized gases, moving at fairly high speeds in a tornado, constitute electrical currents which can produce measurable magnetic fields. Here, we model the vortex as a long, straight, current-carrying conductor. The magnetic field lines produced by this current are circular, centered on the funnel of the tornado, and the magnitude of the field at distance *r* from the vortex is

$$B = \frac{\mu_0 I}{2\pi r}$$

Thus, if the measured field is $B = 1.50\times10^{-8}$ T at a distance $r = 9.00\text{ km} = 9.00\times10^{3}\text{ m}$ from the tornado, the currents moving up or down the funnel must be

$$I = \frac{2\pi r B}{\mu_0} = \frac{2\pi\left(9.00\times10^{3}\text{ m}\right)\left(1.50\times10^{-8}\text{ T}\right)}{4\pi\times10^{-7}\text{ T}\cdot\text{m/A}} = 675\text{ A}$$ ◊

In the cited case, the tornado was located east of the observatory, or the observers were measuring the magnetic field on the *west side of the funnel*. To determine the direction of the current producing the observed northward directed field, imagine gripping the vertical funnel with your right hand so as your fingers curl around the vortex, they point northward (in the direction of the observed field) on the west side of the funnel. Then you should find that the thumb of your right hand is pointing downward, toward the ground. Therefore, the conventional current in the funnel was directed downward. ◊

Most likely, the current consisted of a mix of positive ions moving downward (in the direction of the conventional current) and free electrons (negative charges) moving upward, opposite the direction of the conventional current. Such movements of both types of charge would contribute to a northward-directed magnetic field on the west side of the funnel.

---

57. A wire with a weight per unit length of 0.080 N/m is suspended directly above a second wire. The top wire carries a current of 30.0 A, and the bottom wire carries a current of 60.0 A. Find the distance of separation between the wires so that the top wire will be held in place by magnetic repulsion.

Solution

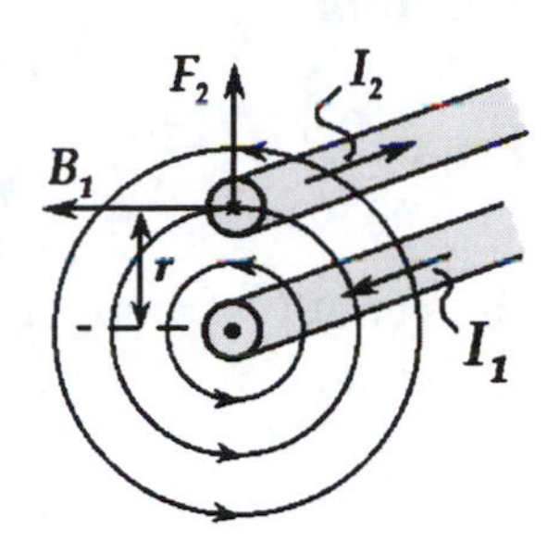

The sketch at the right shows the lower wire (wire 1) carrying a current I_1 directed out of the page. This current will produce a magnetic field whose field lines are circular and centered on this wire as shown. According to **right-hand rule number 2** (see Section 19.7 of the textbook), the direction of this field is counterclockwise around wire 1, meaning that, at the location of the upper wire (wire 2), the field is directed to the left as shown. At the location of wire 2, distance r above wire 1, the magnitude of the field due to the current in wire 1 is $B_1 = \mu_0 I_1/2\pi r$.

It is desired that wire 2 experience an upward force which will counteract the weight of this wire and cause it to float above wire 1. This force is the result of wire 2 carrying a current I_2 perpendicularly through the magnetic field generated by current in wire 1. Note that according to right-hand rule number 1, the current in wire 2 must be directed into the page (opposite the direction of the current in wire 1) if the two wires are to repel each other.

The magnitude of the force experienced by wire 2, of length L_2, as it carries current I_2 through the field B_1 is

$$F_2 = B_1 I_2 L_2 = \left(\frac{\mu_0 I_1}{2\pi r}\right) I_2 L_2 = \frac{\mu_0 I_1 I_2 L_2}{2\pi r}$$

If this force is to support the weight of wire 2 and cause it to float above wire 1, it is necessary that $F_2 = m_2 g$, or

$$\frac{\mu_0 I_1 I_2 L_2}{2\pi r} = m_2 g \qquad \text{and} \qquad r = \frac{\mu_0 I_1 I_2}{2\pi\left(m_2 g/L_2\right)}$$

The distance separating the centers of the two wires should be

$$r = \frac{\left(4\pi\times10^{-7}\ \mathrm{T\cdot m/A}\right)(60.0\ \mathrm{A})(30.0\ \mathrm{A})}{2\pi\left(0.080\ \mathrm{N/m}\right)} = 4.5\times10^{-3}\ \mathrm{m} = 4.5\ \mathrm{mm} \qquad \Diamond$$

61. It is desired to construct a solenoid that will have a resistance of 5.00 Ω (at 20°C) and produce a magnetic field of 4.00×10^{-2} T at its center when it carries a current of 4.00 A. The solenoid is to be constructed from copper wire having a diameter of 0.500 mm. If the radius of the solenoid is to be 1.00 cm, determine (a) the number of turns of wire needed and (b) the length the solenoid should have.

Solution

From $R = \dfrac{\rho L_{\text{wire}}}{A} = \dfrac{\rho L_{\text{wire}}}{\pi d^2/4} = \dfrac{4\rho L_{\text{wire}}}{\pi d^2}$,

the length of copper wire with diameter $d = 0.500$ mm needed to provide the desired resistance is

$$L_{\text{wire}} = \frac{\pi d^2 R}{4\rho_{\text{Cu}}} = \frac{\pi\left(0.500 \times 10^{-3}\ \text{m}\right)^2 (5.00\ \Omega)}{4\left(1.7 \times 10^{-8}\ \Omega\cdot\text{m}\right)} = 58\ \text{m}$$

where the resistivity of copper was obtained from Table 17.1 in the textbook.

The number of turns this length of wire will make on the cylindrical form of the solenoid having a 1.00-cm radius is

$$N = \frac{L_{\text{wire}}}{2\pi r} = \frac{58\ \text{m}}{2\pi\left(1.00 \times 10^{-2}\ \text{m}\right)} = 9.2 \times 10^2 = 920 \qquad \Diamond$$

The magnetic field inside a solenoid is given by $B = \mu_0 n I$, where $n = N/L_{\text{solenoid}}$ is the number of turns of wire per unit length on the solenoid. If it is desired to have $B = 4.00 \times 10^{-2}$ T when the current in the solenoid is $I = 4.00$ A, then

$$n = \frac{B}{\mu_0 I} = \frac{4.00 \times 10^{-2}\ \text{T}}{\left(4\pi \times 10^{-7}\ \text{T}\cdot\text{m/A}\right)(4.00\ \text{A})} = 7.96 \times 10^3\ \text{turns/m}$$

and the required length for the solenoid is

$$L_{\text{solenoid}} = \frac{N}{n} = \frac{9.2 \times 10^2\ \text{turns}}{7.96 \times 10^3\ \text{turns/m}} = 0.12\ \text{m} = 12\ \text{cm} \qquad \Diamond$$

69. Using an electromagnetic flowmeter (Fig. P19.69), a heart surgeon monitors the flow rate of blood through an artery. Electrodes A and B make contact with the outer surface of the blood vessel, which has interior diameter 3.00 mm. (a) For a magnetic field magnitude of 0.040 0 T, a potential difference of 160 μV appears between the electrodes. Calculate the speed of the blood. (b) Verify that electrode A is positive, as shown. Does the sign of the emf depend on whether the mobile ions in the blood are predominantly positively or negatively charged? Explain.

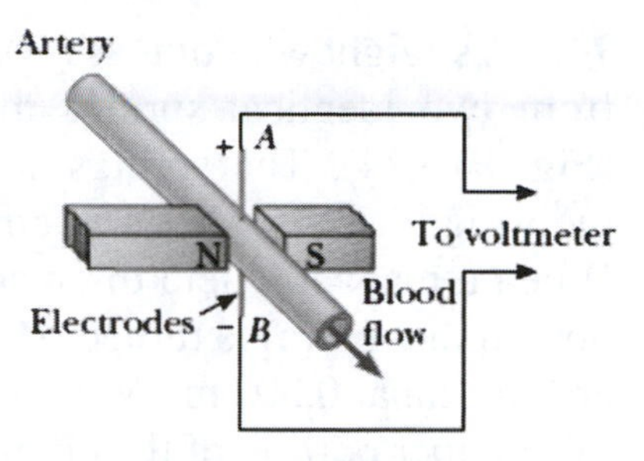

Figure P19.69

Solution

(a) The magnetic field inside the artery is directed from left to right (from the north (N) pole toward the south (S) pole of the magnet). According to right-hand rule number 1, positive ions moving with the blood flow experience an upward magnetic force, deflecting them toward electrode A. Similarly, negative ions will experience a downward magnetic force and deflect toward electrode B. The accumulation of positive charge at A and negative charge at B creates an electric field directed from A toward B. The electric field exerts a downward force on positive ions and upward force on negative ions (opposite to the directions of the magnetic forces acting on the ions). At equilibrium, the magnitudes of the electric and magnetic forces are equal, leaving zero net force acting on the ions as they pass through the meter. Then

$$qE = qvB \qquad \text{and the speed of the blood is} \qquad v = E/B$$

Since the magnitude of the electric field between the electrodes is $E = \Delta V/d$, where ΔV is the potential difference (i.e., the voltmeter reading) between A and B, and d is the distance between the electrodes (diameter of the artery), the speed of the blood is given by

$$v = \frac{\Delta V}{Bd} = \frac{160 \times 10^{-6}\ \text{V}}{(0.040\,0\ \text{T})(3.00 \times 10^{-3}\ \text{m})} = 1.33\ \text{m/s} \qquad \Diamond$$

(b) The polarity of the emf does not depend on the sign of the charge carriers. Positive charge carriers deflect upward to electrode A, making it positive relative to B. Likewise, negative charge carriers would deflect downward to electrode B, making it negative relative to A. These two situations are equivalent, yielding the same polarity emf (A at a higher potential than B). $\Diamond$

74. A straight wire of mass 10.0 g and length 5.0 cm is suspended from two identical springs that, in turn, form a closed circuit (Fig. P19.74). The springs stretch a distance of 0.50 cm under the weight of the wire. The circuit has a total resistance of 12 Ω. When a magnetic field directed out of the page (indicated by the dots in the figure) is turned on, the springs are observed to stretch an additional 0.30 cm. What is the strength of the magnetic field? (The upper portion of the circuit is fixed.)

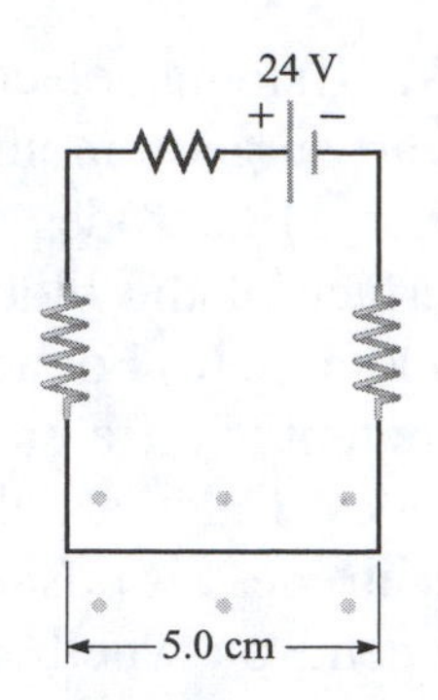

Figure P19.74

Solution

Before the magnetic field is turned on, the only force tending to stretch the springs is the weight of the wire. Thus, $F_s = kx_1 = mg$, where the mass of the wire is $m = 10.0$ g, and $x_1 = 0.50$ cm is the distance the springs are stretched at this time. Therefore, the force constant of the set of springs is given by $k = mg/x_1$.

The current in the straight wire at the lower end of the circuit flows from left to right and has magnitude of $I = \Delta V/R = 24\text{ V}/12\ \Omega = 2.0$ A. Once the magnetic field, of magnitude B and directed out of the page, is turned on, the wire will experience a downward magnetic force of magnitude $F_m = BIL$, where $L = 5.0$ cm is the length of the wire. The total force stretching the springs is now $F_s = F_m + mg$, and the total elongation of the springs is $x_{\text{total}} = x_1 + 0.30\text{ cm} = 0.80$ cm. Thus,

$$BIL + mg = kx_{\text{total}} = \left(\frac{mg}{x_1}\right)x_{\text{total}}$$

and the strength of the magnetic field is given by

$$B = \frac{mg}{IL}\left(\frac{x_{\text{total}}}{x_1} - 1\right) = \frac{(10.0\times10^{-3}\text{ kg})(9.80\text{ m/s}^2)}{(2.0\text{ A})(5.0\times10^{-2}\text{ m})}\left(\frac{0.80\text{ cm}}{0.50\text{ cm}} - 1\right) = 0.59\text{ T} \qquad \Diamond$$

20

Induced Voltages and Inductance

NOTES FROM SELECTED CHAPTER SECTIONS

20.1 Induced EMF and Magnetic Flux

An electric current can be produced by a changing magnetic field. *An induced emf is produced in a secondary circuit by the changing magnetic field in a primary circuit.* The emf is induced by a change in a quantity called the magnetic flux rather than simply by a change in the magnetic field.

As described in Chapter 19, magnetic field lines can be drawn to illustrate the direction and relative strength of a magnetic field throughout a region of space. Consider a plane area, A, immersed in a magnetic field with field lines intersecting the area; the magnetic flux is the product of the magnetic field, the area, and the cosine of the angle between the direction of the magnetic field and the perpendicular (normal) to the area. Magnetic flux, Φ_B, is proportional to the number of magnetic field lines passing through the area and has SI units of webers ($1\ \text{Wb} = 1\ \text{T}\cdot\text{m}^2$).

20.2 Faraday's Law of Induction and Lenz's Law

The emf induced in a circuit (or conducting loop) is proportional to the time rate of change of magnetic flux through the circuit; the resulting current is called an induced current. The polarity of the induced emf can be predicted by **Lenz's law:**

> **The polarity of an induced emf is such that it produces a current whose magnetic field opposes the change in magnetic flux through the loop. The induced current will be in the direction that will tend to maintain the original flux through the loop.**

20.3 Motional emf

A motional emf (potential difference) will be maintained across a conductor moving in a magnetic field as long as the direction of motion through the field is not parallel to the field direction. If the motion is reversed, the polarity of the potential difference will also be reversed.

20.4 Generators

An **alternating current (AC) generator** consists of a wire loop rotated by some external means in a magnetic field. The ends of the loop are connected to **slip rings** that rotate with the loop; connections to the external circuit are made by stationary brushes in contact with the slip rings. The induced emf varies sinusoidally with time with a frequency of 60 Hz (in the US and Canada.)

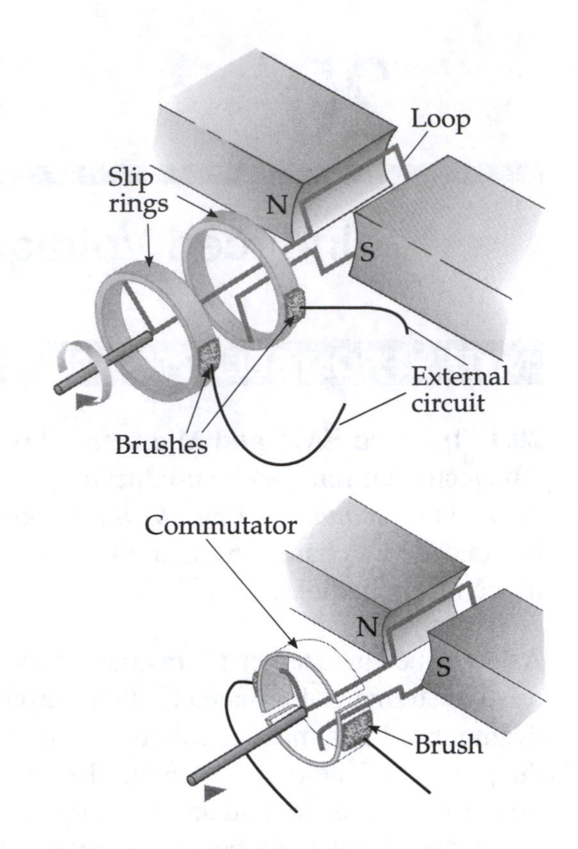

In a **direct current (DC) generator** the contacts to the rotating loop are made by a split ring, or **commutator**. In this design, the output voltage always has the same polarity (always in the same direction) and varies in magnitude from zero to some maximum value.

In the **operation of a motor**, a current is supplied to the loop by an external source of emf, and the magnetic torque on the current-carrying loop causes it to rotate, thereby doing mechanical work.

20.5 Self-Inductance

20.6 *RL* Circuits

When the switch is closed in a series circuit consisting of a resistor, a coil, and a battery, the current does not instantly go from zero to its maximum value. As the current increases in the coil, an induced emf is produced which opposes the emf of the battery. This effect, called **self-induction**, can be considered in the following steps:

- Current, initially at zero, increases through the coil resulting in an increasing magnetic flux through the coil.
- The changing flux induces an emf in the coil (Faraday's law).
- The induced emf is proportional to the rate at which the current is changing.
- The induced emf opposes the emf of the battery (Lenz's law).
- The net potential difference across the resistor is the emf of the battery minus the induced emf.
- The current continues to increase to a maximum value. *However, the rate at which the current increases becomes smaller.*

The property of a coil, solenoid, or other device (acting as an inductor) to limit the rate of change of current is called inductance (denoted by the symbol L) and has the SI unit of the henry (H). The inductance of a device (an inductor) depends on geometric factors (cross-sectional area, length, and number of turns of wire).

Remember, resistance (R) limits the current; inductance (L) limits the rate at which the current changes.

20.7 Energy Stored in a Magnetic Field

In an RL circuit, the rate at which energy is supplied by the battery equals the sum of the rate at which energy is dissipated in the resistor plus the rate at which energy is stored in the magnetic field of the inductor. *The energy stored in the magnetic field of an inductor is proportional to the square of the current in that inductor.*

EQUATIONS AND CONCEPTS

The **total magnetic flux** (Φ_B) through a plane area placed in a uniform magnetic field depends on the angle between the direction of the magnetic field and the direction perpendicular (normal) to the surface area. The maximum flux through the area occurs when the magnetic field is perpendicular to the surface area (i.e., $\vec{\mathbf{B}}$ is along the direction of the normal and $\theta = 0°$). The unit of magnet flux is the weber (Wb). 1 Wb = 1 T · m²

$$\Phi_B \equiv B_{\perp} A = BA\cos\theta \qquad (20.1)$$

$$\Phi_{B,\max} = BA$$

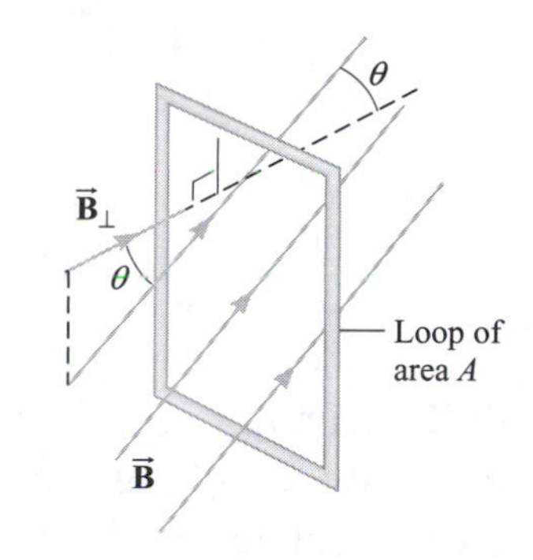

Faraday's law of induction states that the average emf induced in a circuit during a time interval Δt is proportional to the rate of change of magnetic flux through the circuit. *The minus sign is included to indicate the polarity of the induced emf, which can be found by use of Lenz's law.*

$$\mathcal{E} = -N\frac{\Delta\Phi_B}{\Delta t} \qquad (20.2)$$

Lenz's law states that the polarity of the induced emf (and the direction of the associated current in a closed circuit) produces a current whose magnetic field opposes the *change in the flux through the loop. That is, the induced current tends to maintain the original flux through the circuit.* Note: If the circuit contains a source of emf (i.e., a battery), the current in the circuit may not be in the direction of the induced emf.

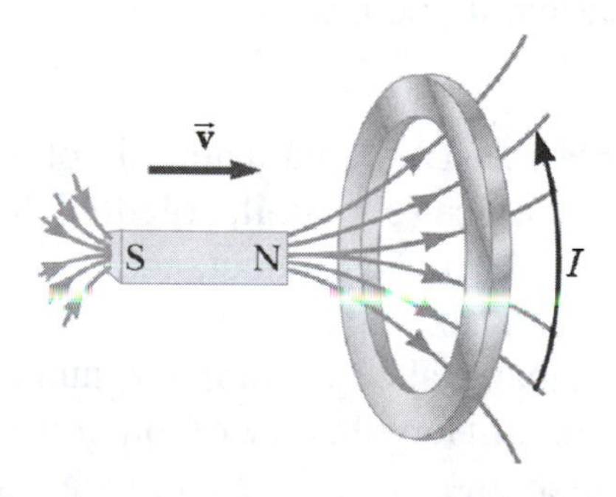

As the magnet moves toward the right, positive flux increases through the coil. This produces an induced current as shown (counterclockwise as seen from the direction of the approaching magnet).

A **motional emf** is induced in a conductor of length, ℓ, moving with speed, v, perpendicular to a magnetic field. *Remember the significance of the minus sign as required by Lenz's law.*

$$|\mathcal{E}| = \frac{\Delta\Phi_B}{\Delta t} = B\ell v \qquad (20.4)$$

An **induced current** will be present when a conductor moving in a magnetic field is part of a complete circuit of resistance, R. Equation 20.5 applies when the conductor moves perpendicular to the field. As seen in the figure, the motion of the conductor toward the right increases the area of the closed circuit and results in an increase in flux directed into the page. *Use the right-hand-rule and Lenz's law to confirm the direction of current as shown in the figure.*

$$I = \frac{|\mathcal{E}|}{R} = \frac{B\ell v}{R} \qquad (20.5)$$

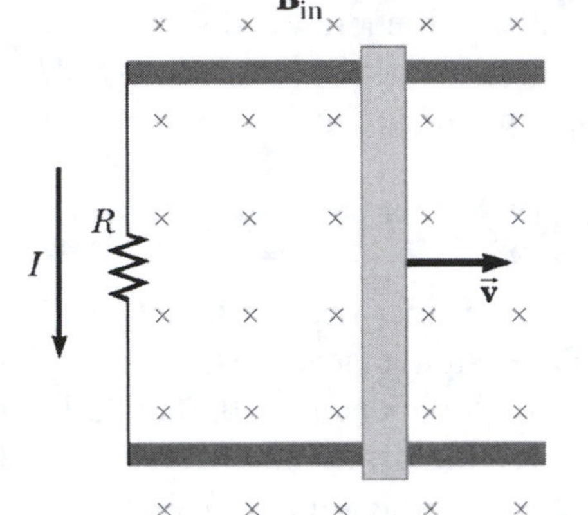

A conducting bar sliding on conducting rails and moving at constant velocity toward the right through a magnetic field directed into the page.

A **sinusoidally varying emf** is produced when a loop of wire with N turns and cross-sectional area A rotates with an angular velocity ω (measured in radians per second) in a magnetic field. For a given loop, the maximum value of the induced emf will be proportional to the angular velocity of the loop. Note the direction of the current shown in the figure.

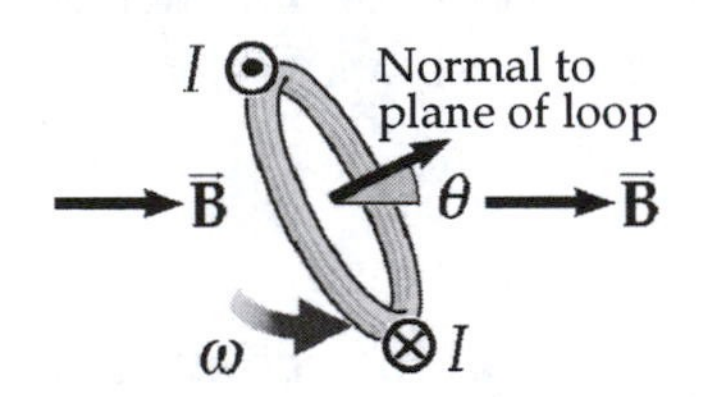

$$\mathcal{E} = NAB\omega \sin \omega t \qquad (20.7)$$

$$\mathcal{E}_{max} = NAB\omega \qquad (20.8)$$

Inductance (L) is characteristic of a conducting device (e.g., coil, solenoid, toroid, coaxial cable). The inductance of a given device, for example a coil, depends on its physical makeup: diameter, number of turns per unit length, type of material (core) on which the wire is wound, and other geometric parameters. The SI unit of inductance is the henry (H). A rate of change of current of 1 A/s in an inductor of 1 H will result in an induced voltage of 1 V.

$1\ \text{H} = 1\ \text{V}\cdot\text{s/A}$

An **inductor** is a circuit element.

Inductance is a measure of the degree to which an inductor opposes a change of current in a circuit.

The **inductance of a circuit element** can be expressed as the ratio of magnetic flux to current as shown in Equation 20.10 or as the ratio of induced emf to rate of change of current.

$$L = \frac{N\Phi_B}{I} \qquad (20.10)$$

$$L = -\frac{\mathcal{E}_L}{(\Delta I / \Delta t)}$$

The **inductance of a solenoid** can be calculated in terms of geometric quantities as shown in Equations 20.11a and 20.11b.

$$L = \frac{\mu_0 N^2 A}{\ell} \qquad (20.11a)$$

$$L = \mu_0 n^2 V \qquad (20.11b)$$

$n = N/\ell$

$V = A\ell = \textit{volume of solenoid}$

A **self-induced emf** ("back emf") will be generated in a coil when the current in the coil is changing. The emf will be proportional to the rate of change of the current and has a polarity that opposes the *change* occurring in the current. When the current increases as shown in the figure, the polarity of the self-induced emf will be as indicated by the dashed lines above the solenoid.

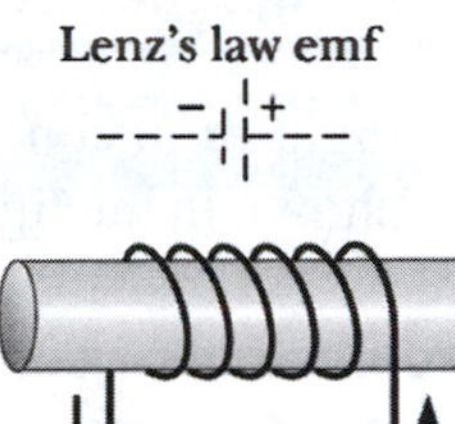

$$\mathcal{E}_L \equiv -L\frac{\Delta I}{\Delta t} \qquad (20.13)$$

A **series *RL* circuit** is shown in the figure at right. The circuit elements are a battery, resistor, inductor, and switch. *The inductor opposes any change in the current in the circuit.*

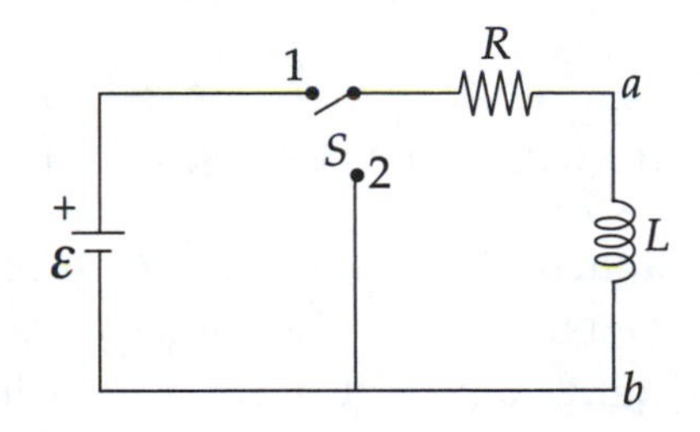

Current increases in an *RL* circuit (from $I = 0$ to $I_{max} = \mathcal{E}/R$) in a characteristic fashion when the switch in the circuit shown above is moved to position 1. This is shown in the graph at right and described by Equation 20.15 as stated below the graph.

The **time constant** (τ) is the time required for the current to reach 63.2% of its maximum value. *The maximum current is achieved in a time that is long compared to the time constant, τ.*

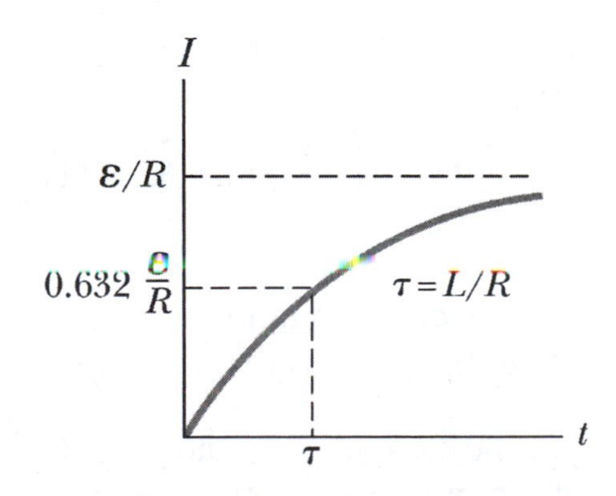

$$I = \frac{\mathcal{E}}{R}\left(1 - e^{-t/\tau}\right) \qquad (20.15)$$

The **energy stored in the magnetic field** of an inductor is proportional to the square of the current.

$$PE_L = \tfrac{1}{2}LI^2 \qquad (20.16)$$

The **energy stored in the electric field** of a charged capacitor is proportional to the square of the potential difference between the plates.

$$PE_C = \tfrac{1}{2}C(\Delta V)^2$$

SUGGESTIONS, SKILLS, AND STRATEGIES

Remember, the induced current in a circuit will have a direction which tends to maintain the flux through the circuit. Carefully follow each step in the example below.

Consider two single-turn, concentric coils *lying in the plane of the paper* as shown in the figure. The outside coil is part of a circuit containing a resistor (R), a battery ($\mathcal{E}$), and switch (S). The inner coil is not part of the circuit. When the switch is moved from "**open**" to "**closed**" the direction of the induced current in the inner coil can be predicted by Lenz's law.

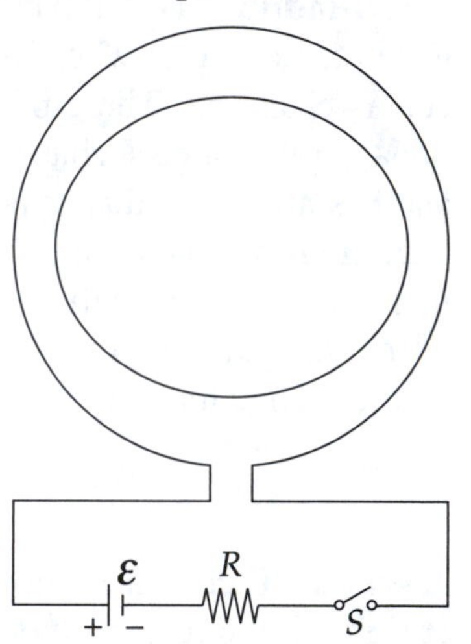

Consider the following steps:

(1) When the switch is in the "**open**" position as shown, there will be no current in the circuit.

(2) When the switch is moved to the "**closed**" position, there will be a clockwise current in the outside coil.

(3) Magnetic field lines due to current in the outside coil will be directed into the page through the area enclosed by the coil. Use right-hand rule number 2 to confirm this. **Magnetic flux into the page will penetrate the entire area enclosed by the outside coil including the area of the inner coil.**

(4) By Faraday's law, the increasing flux produces an induced emf (and current) in the inner coil.

(5) Lenz's law requires that the induced current have a direction which will tend to maintain the initial flux condition (*which in this case was zero*). By applying right-hand rule number 2 to the inner coil, you should be able to determine that the direction of the induced current in the inner coil must be counterclockwise, contributing to a flux out of the page. Try this!

As a second example you should follow steps similar to those above to predict the direction of the induced current in the inner coil when the switch is moved from "**closed**" to "**open**."

REVIEW CHECKLIST

- Calculate the emf (or current) induced in a circuit when the magnetic flux through the circuit is changing in time. The variation in flux might be due to a change in: (i) the area of the circuit, (ii) the magnitude of the magnetic field, (iii) the direction of the magnetic field, or (iv) the orientation/location of the circuit in the magnetic field. (Sections 20.1–20.2)
- Apply Lenz's law to determine the direction of an induced emf or current. You should also understand that Lenz's law is a consequence of the law of conservation of energy. (Section 20.2)
- Calculate the emf induced between the ends of a conducting bar as it moves through a region where there is a constant magnetic field (motional emf). (Section 20.3)
- Define self-inductance, L, of a circuit in terms of appropriate circuit parameters. Calculate the total magnetic energy stored in a magnetic field if you are given the values of the inductance of the device with which the field is associated and the current in the circuit. (Sections 20.5 and 20.7)
- Qualitatively describe the manner in which the instantaneous value of the current in an RL circuit changes while the current is either increasing or decreasing with time. (Section 20.6)

SOLUTIONS TO SELECTED END-OF-CHAPTER PROBLEMS

5. A long, straight wire lies in the plane of a circular coil with a radius of 0.010 m. The wire carries a current of 2.0 A and is placed along a diameter of the coil. (a) What is the net flux through the coil? (b) If the wire passes through the center of the coil and is perpendicular to the plane of the coil, what is the net flux through the coil?

Solution

The magnetic field lines produced by a current in a long, straight wire are circular, centered on the wire, and in a plane perpendicular to the wire as shown in Figure (a). Note that if the current is toward the top of the page, the field lines go into the page on the right side of the wire and come out of the page on the left side of the wire.

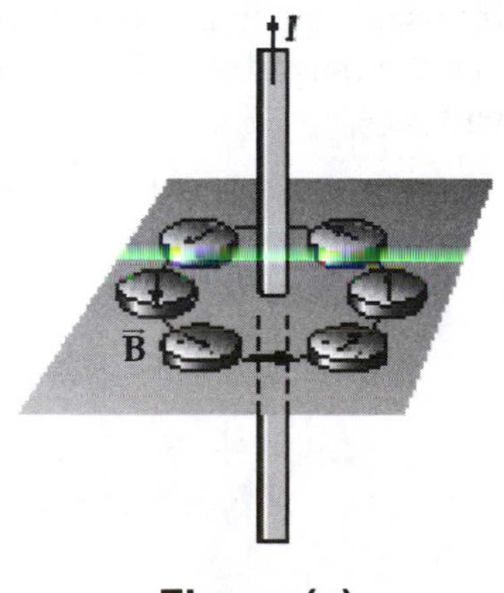

Figure (a)

(a) If the long, straight wire is along the diameter of a flat circular coil as shown in Figure (b), every magnetic field line directed into the page through the area of the coil on the right side of the long, straight wire reemerges from the page through the area of the coil on the left side of the wire. Thus, there are just as many lines passing through the area of the coil in one direction as there are passing through this area in the opposite direction, and the net flux through the area enclosed by the coil is **zero**. ◊

(b) If the wire passes through the center of the coil and is perpendicular to the plane of the coil, then all of the field lines produced by the current in the long, straight wire lie in planes parallel to the plane of the coil. Thus, none of these lines penetrate the area enclosed by the coil, and the flux through the coil is again **zero**. ◊

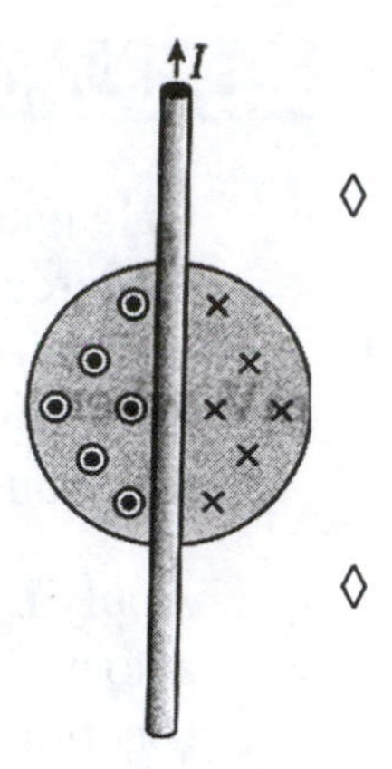

Figure (b)

---

11. A wire loop of radius 0.30 m lies so that an external magnetic field of magnitude 0.30 T is perpendicular to the loop. The field reverses its direction, and its magnitude changes to 0.20 T in 1.5 s. Find the magnitude of the average induced emf in the loop during this time.

Solution

The magnetic field is at all times perpendicular to the plane of the circular loop. Thus, the angle between the direction of the field and the line normal to the plane of the loop will be either 0° or 180°.

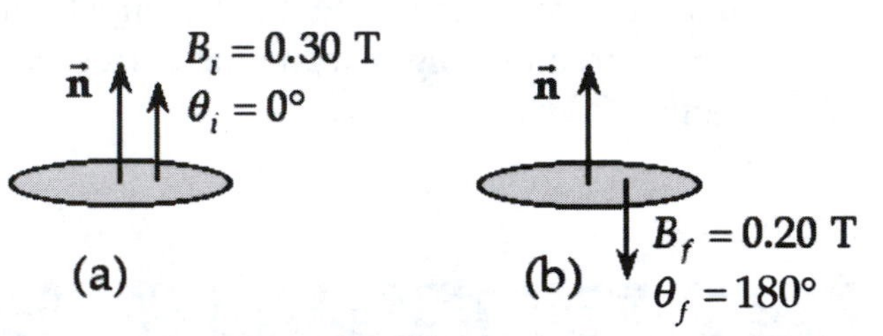

We choose the positive sense of the normal line to be in the original direction of the field as shown in Figure (a) above, so $\theta_i = 0°$. After the field reverses direction, it is anti-parallel to the chosen reference direction, and $\theta_f = 180°$ as in Figure (b). The change which occurs in the flux through the circular loop as the field reverses direction and changes magnitude is

$$\Delta\Phi_B = \Phi_{B,f} - \Phi_{B,i} = B_f A\cos\theta_f - B_i A\cos\theta_i = \left(B_f \cos\theta_f - B_i \cos\theta_i\right)A$$
$$= \left[(0.20 \text{ T})\cos 180° - (0.30 \text{ T})\cos 0°\right]\pi(0.30 \text{ m})^2 = -0.14 \text{ T}\cdot\text{m}^2 = -0.14 \text{ Wb}$$

If this change in flux occurs in a time interval $\Delta t = 1.5$ s, Faraday's law of induction tells us that the magnitude of the average induced emf in this single-turn coil during this interval will be

$$\left|\mathcal{E}_{\text{av}}\right| = \frac{N\left|\Delta\Phi_B\right|}{\Delta t} = \frac{1(0.14 \text{ Wb})}{1.5 \text{ s}} = 9.3\times10^{-2} \text{ V} = 93 \text{ mV}$$ ◊

---

15. A bar magnet is positioned near a coil of wire as shown in Figure P20.15. What is the direction of the current in the resistor when the magnet is moved (a) to the left? (b) to the right?

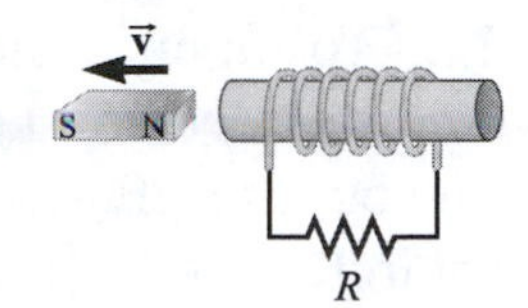

Figure P20.15

Solution

Movement of the magnet produces a change in the flux through the turns of wire making up the coil. This changing flux generates an induced emf in the coil and since a closed conducting path is present, will produce a current in the wire of the coil and through the resistance R.

(a) When the bar magnet, with the north-seeking end turned toward the coil, moves to the left, the flux passing through the loops of the coil is directed toward the right and is decreasing in magnitude. To oppose this change of flux, the induced current in the coil must generate a magnetic field directed toward the right along the axis of the coil. Imagine grasping the wire forming the leftmost turn on the coil with your right hand, such that your fingers curl around the wire and point from left to right (the desired direction of the field produced by the induced current) in the interior of the coil. Then, by right-hand rule number 2, your thumb will point along the wire in the direction of the induced current. You should find that the current will be coming out of the page over the top of the coil, going downward in the vertical segment of wire to the left of R, and flowing from **left to right** through the resistance R. ◊

(b) When the motion of the magnet is toward the right, the magnetic field due to the bar magnet is directed toward the right along the axis of the coil and increases in magnitude as the magnet comes closer. To oppose this increasing rightward flux, the induced current in the coil must generate a magnetic field directed toward the left along the axis of the coil. Imagine grasping the wire forming the leftmost turn on the coil with your right hand, so the fingers curl around the wire and point from right to left inside the coil. Then, by right-hand rule number 2, you should find that the current flows from **right to left** through the resistance R, upward in the vertical wire to the left of R, and into the page at the top of the coil. ◊

21. To monitor the breathing of a hospital patient, a thin belt is girded around the patient's chest as in Figure P20.21. The belt is a 200-turn coil. When the patient inhales, the area encircled by the coil increases by 39.0 cm^2. The magnitude of Earth's magnetic field is 50.0 mT and makes an angle of 28.0° with the plane of the coil. Assuming a patient takes 1.80 s to inhale, find the magnitude of the average induced emf in the coil during that time.

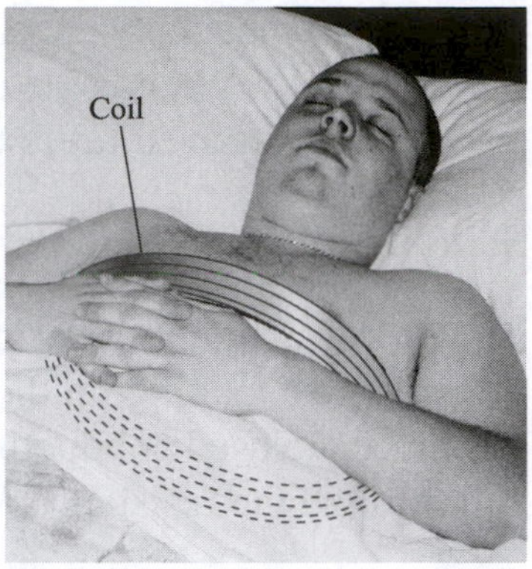

Figure P20.21

Solution

When a coil is in a magnetic field of magnitude B, the magnetic flux through each loop on the coil is given by $\Phi_B = B_\perp A = BA\cos\theta$, where A is the area enclosed the loop and θ is the angle between the direction of the magnetic field and the line perpendicular to the plane of the coil. Any event that causes a change in this flux will produce an induced emf in the coil. In this case, the flux changes because the area enclosed by a loop on the coil increases when the patient inhales. If the time the patient takes to inhale is Δt, the average rate of change of flux through each loop on the coil is $\Delta\Phi_B/\Delta t = B(\Delta A)\cos\theta/\Delta t$.

From Faraday's law of induction, for a coil having N turns, the magnitude of the average induced emf in the coil while the patient is inhaling is

$$|\mathcal{E}| = N\frac{\Delta\Phi_B}{\Delta t} = \frac{NB(\Delta A)\cos\theta}{\Delta t}$$

The external field producing flux through the coil is the Earth's magnetic field, and this field makes an angle $\phi = 28.0°$ with the plane of the coil. Thus, the angle between the magnetic field and the line perpendicular to the plane of the coil is $\theta = 90.0° - \phi = 62.0°$. We also know that the number of turns on the coil is $N = 200$, that the area enclosed by the loops of the coil increases by 39.0 cm^2 as the patient inhales, the patient takes $\Delta t = 1.80$ s to inhale, and that the magnitude of the Earth's field is $B = 50.0\ \mu\text{T}$ at the location of the patient. The average magnitude of the induced emf should then be

$$|\mathcal{E}| = \frac{(200)(50.0\times10^{-6}\ \text{T})\left[(39.0\ \text{cm}^2)(1\ \text{m}^2/10^4\ \text{cm}^2)\right]\cos 62.0°}{1.80\ \text{s}}$$

or $$|\mathcal{E}| = 1.02\times10^{-5}\ \text{V} = 10.2\ \mu\text{V}$$ ◊

24. A 2.00-m length of wire is held in an east–west direction and moves horizontally to the north with a speed of 15.0 m/s. The vertical component of Earth's magnetic field in this region is 40.0 μT directed downward. Calculate the induced emf between the ends of the wire and determine which end is positive.

Solution

When a straight conductor of length ℓ moves at speed v through a magnetic field of magnitude B, an induced emf of magnitude $|\mathcal{E}| = B\ell v$ (see Equation 20.4 in the textbook) is generated in the conductor. However, before we attempt to apply this expression to the solution of the given problem, we need to consider the cause of this emf and refine the equation somewhat.

As the conductor, and all the charges within that conductor, move through the magnetic field, a magnetic force $F_m = qvB\sin\theta = qvB_\perp$ is exerted on a particle of charge q within the conductor. It is this force that produces a separation of change within the conductor and gives rise to the induced or motional emf. Here, θ is the angle between the directions of the velocity $\vec{\mathbf{v}}$ of the conductor (and hence the charges within the conductor) and the magnetic field $\vec{\mathbf{B}}$. The factor $B\sin\theta = B_\perp$ is the component of the magnetic field perpendicular to the velocity of the conductor, and it is only this component of the field that is effective in producing the separation of charge that is the origin of the emf. Thus, we see that the expression for the motional emf in the conductor can be more clearly written as $|\mathcal{E}| = B_\perp \ell v$.

Since the velocity of our conductor is in a horizontal direction, only the vertical component of the Earth's field is effective in generating an emf in the conductor. The magnitude of this emf produced in this case is

$$|\mathcal{E}| = B_\perp \ell v = \left(40.0\times10^{-6}\ \text{T}\right)(2.00\ \text{m})(15.0\ \text{m/s}) = 1.20\times10^{-3}\ \text{V} = 1.20\ \text{mV} \qquad \Diamond$$

To determine the polarity of this emf, consider the direction of the magnetic force F_m. As the conductor (and all charges within the conductor) move northward, through a downward vertical component of Earth's field, right-hand rule number 1 tells us that positive charges within the conductor will experience a magnetic force toward the west, and negative charges experience eastward forces. Thus, free electrons within the conductor will migrate toward the eastern end of the wire, leaving the western end with a net positive charge. This means that the western end of the wire will be at a higher potential than the eastern end. ◊

29. Figure P20.29 shows a bar of mass $m = 0.200$ kg that can slide without friction on a pair of rails separated by a distance $\ell = 1.20$ m and located on an inclined plane that makes an angle $\theta = 25.0°$ with respect to the ground. The resistance of the resistor is $R = 1.00\ \Omega$, and a uniform magnetic field of magnitude $B = 0.500$ T is directed downward, perpendicular to the ground, over the entire region through which the bar moves. With what constant speed v does the bar slide along the rails?

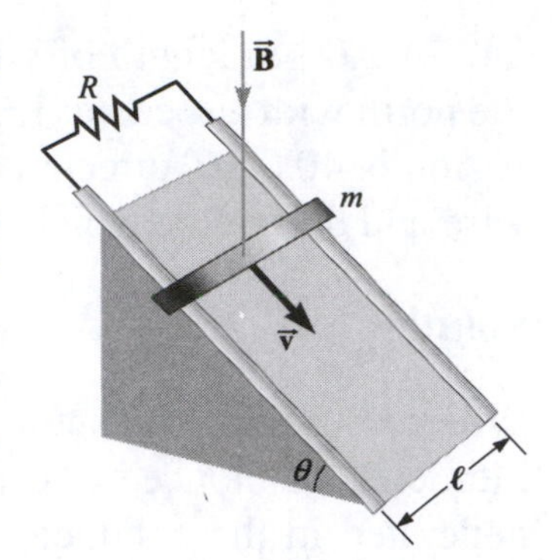

Figure P20.29

Solution

When the bar is distance s down the rails from the resistor R, the bar, the rails, and the resistor form a closed conducting loop enclosing an area $A = \ell s$. The magnetic flux passing through this enclosed area is directed downward and has magnitude $\Phi_B = BA\cos\theta = (B\cos\theta)\ell s = B_\perp \ell s$, where θ is the angle between the directions of $\vec{\mathbf{B}}$ and the line perpendicular to the area A as shown in sketch (a) below. $B_\perp = B\cos\theta$ is the component of $\vec{\mathbf{B}}$ perpendicular to this area and to the velocity $\vec{\mathbf{v}}$ of the bar.

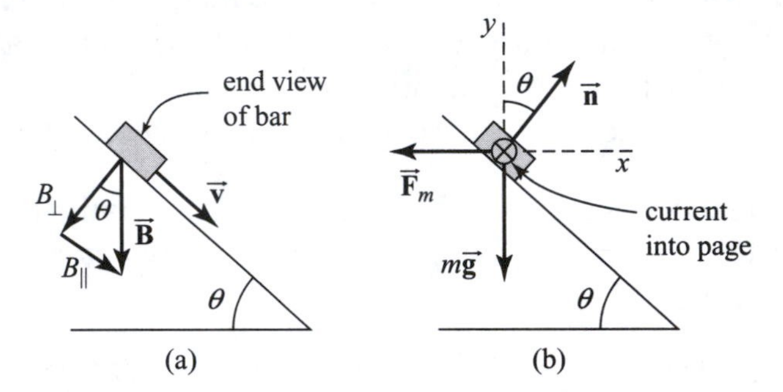

As the bar slides down the rails at speed v, the flux through area A changes at a rate

$$\frac{\Delta\Phi_B}{\Delta t} = B_\perp \ell\left(\frac{\Delta s}{\Delta t}\right) = B_\perp \ell v$$

Faraday's law then gives the emf induced in the bar as $|\mathcal{E}| = \Delta\Phi_B/\Delta t = B_\perp \ell v = B\ell v\cos\theta$, so the induced current in the closed conducting loop is $I = |\mathcal{E}|/R = (B\ell v\cos\theta)/R$. Right-hand rule number 2 tells us this current must flow into the page as shown in sketch (b) to generate a magnetic field directed upward through area A and oppose the increasing downward flux due to the external field $\vec{\mathbf{B}}$.

The bar is now a conductor carrying a current I perpendicular to a magnetic field and will experience a magnetic force of magnitude F_m directed to the left (right-hand rule number 1) as shown in sketch (b). The magnitude of this force is

$$F_m = BI\ell = B\left(\frac{B\ell v\cos\theta}{R}\right)\ell = \frac{B^2\ell^2 v\cos\theta}{R}$$

If the bar has reached terminal speed and now moves with constant velocity, then its acceleration is zero and $\Sigma F_x = \Sigma F_y = 0$. Looking at the forces shown in sketch (b), we obtain

$$\Sigma F_y = 0 \quad \Rightarrow \quad n\cos\theta = mg \qquad \text{or} \qquad n = mg/\cos\theta$$

and

$$\Sigma F_x = 0 \quad \Rightarrow \quad F_m = n\sin\theta \qquad \text{or} \qquad \frac{B^2\ell^2 v\cos\theta}{R} = \left(\frac{mg}{\cos\theta}\right)\sin\theta = mg\tan\theta$$

The constant speed of the sliding bar is therefore given by

$$v = \frac{mg\tan\theta}{B^2\ell^2\cos\theta/R} = \frac{mgR\tan\theta}{B^2\ell^2\cos\theta}$$

Using the given data: $m = 0.200$ kg, $\ell = 1.20$ m, $R = 1.00\ \Omega$, $B = 0.500$ T, and $\theta = 25.0°$, we find

$$v = \frac{(0.200\text{ kg})(9.80\text{ m/s}^2)(1.00\ \Omega)\tan(25.0°)}{(0.500\text{ T})^2(1.20\text{ m})^2\cos(25.0°)} = 2.80\text{ m/s}$$ ◊

---

33. Considerable scientific work is currently under way to determine whether weak oscillating magnetic fields such as those found near outdoor electric power lines can affect human health. One study indicated that a magnetic field of magnitude 1.0×10^{-3} T, oscillating at 60 Hz, might stimulate red blood cells to become cancerous. If the diameter of a red blood cell is 8.0 μm, determine the maximum emf that can be generated around the perimeter of the cell.

Solution

To solve this problem, one needs to recognize the similarity between the situation described here and that found in an electrical generator. In a generator, one normally has a conducting path consisting of a coil made up of N turns of wire rotating at angular frequency ω about a fixed axis. This stationary axis is perpendicular to a magnetic field which does not vary with time. However, the rotation of the coil in this constant field does cause the magnetic flux through the coil to vary sinusoidally in time. This changing magnetic flux induces a time varying emf in the coil which is given by $\mathcal{E} = \mathcal{E}_{max}\sin\omega t$, where $\mathcal{E}_{max} = NBA\omega$, with A being the area enclosed by a turn on the coil and B is the magnetic field strength.

In the situation involving the red blood cell, we have a stationary conducting path around the perimeter of the cell. This single-turn ($N = 1$) "coil" is in a magnetic field that varies sinusoidally in time with angular frequency $\omega = 2\pi f$, where $f = 60$ Hz. This arrangement causes the magnetic flux through the area enclosed by the conducting path to vary sinusoidally in time, just as we had with the rotating coil in the constant magnetic field of the generator. As with the generator, this changing magnetic flux induces a time varying emf given by $\mathcal{E} = \mathcal{E}_{max}\sin\omega t$ in the conducting path. In this case, $\mathcal{E}_{max} = (1)BA\omega$, where B is the amplitude of the oscillating magnetic field, and $A = \pi d^2/4$ is the cross-sectional area of the blood cell having diameter d.

Thus, if $B = 1.0 \times 10^{-3}$ T, $f = 60$ Hz, and $d = 8.0\ \mu$m, the maximum emf that can be generated around the perimeter of the cell is

$$\mathcal{E}_{max} = (1)(1.0 \times 10^{-3}\ \text{T})\left[\frac{\pi(8.0 \times 10^{-6}\ \text{m})^2}{4}\right][2\pi(60\ \text{Hz})] = 1.9 \times 10^{-11}\ \text{V} \qquad \Diamond$$

---

42. An emf of 24.0 mV is induced in a 500-turn coil when the current is changing at a rate of 10.0 A/s. What is the magnetic flux through each turn of the coil at an instant when the current is 4.00 A?

Solution

The magnitude of the induced emf in a coil when the current changes at a rate $\Delta I/\Delta t$ is given by

$$|\mathcal{E}| = L\left(\frac{\Delta I}{\Delta t}\right) \qquad \textbf{[1]}$$

where L is a physical property of the coil, called the self-inductance. The self-inductance of this coil is

$$L = \frac{|\mathcal{E}|}{(\Delta I/\Delta t)} = \frac{24.0 \times 10^{-3}\ \text{V}}{10.0\ \text{A/s}} = 2.40 \times 10^{-3}\ \Omega \cdot \text{s} = 2.40 \times 10^{-3}\ \text{H} = 2.40\ \text{mH}$$

The magnitude of the induced emf in a coil may also be expressed as

$$|\mathcal{E}| = N\left(\frac{|\Delta \Phi_B|}{\Delta t}\right) \qquad \textbf{[2]}$$

where N is the number of turns on the coil and Φ_B is the magnetic flux through each turn. Thus, equating equations [1] and [2], we obtain

$$N\left(\frac{|\Delta \Phi_B|}{\Delta t}\right) = L\left(\frac{\Delta I}{\Delta t}\right) \quad \text{or} \quad N|\Delta \Phi_B| = L(\Delta I)$$

Since N and L are constant for a given coil, it follows that at each instant in time, we must have $N\Phi_B = LI$. Thus, when the current in the coil is I, the magnetic flux through each turn on the coil must be $\Phi_B = LI/N$. For the given coil, the flux through each turn when the current is 4.00-A will be

$$\Phi_B = \frac{L \cdot I}{N} = \frac{(2.40 \times 10^{-3}\ \text{H})(4.00\ \text{A})}{500} = 1.92 \times 10^{-5}\ \text{T} \cdot \text{m}^2 \qquad \Diamond$$

---

45. A battery is connected in series with a 0.30-Ω resistor and an inductor as in Figure P20.43. The switch is closed at $t = 0$. The time constant of the circuit is 0.25 s and the maximum current in the circuit is 8.0 A. Find (a) the emf of the battery, (b) the inductance of the circuit, (c) the current in the circuit after one time constant has elapsed, (d) the voltage across the resistor after one time constant has elapsed, and (e) the voltage across the inductor after one time constant has elapsed.

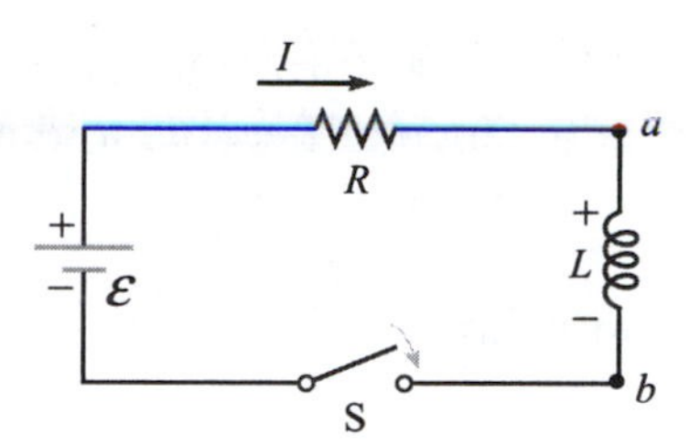

Figure P20.43

Solution

When switch S in the circuit of Figure P20.43 is closed at $t = 0$, a current given as a function of time by $I = I_{max}\left(1 - e^{-t/\tau}\right)$ begins to flow clockwise around the circuit. Here, $I_{max} = \mathcal{E}/R$ is the maximum current, reached only when $t \gg \tau$, and $\tau = L/R$ is known as the time constant of the circuit.

(a) From $I_{max} = \mathcal{E}/R$, the emf of the battery is seen to be

$$\mathcal{E} = I_{max}R = (8.0\ \text{A})(0.30\ \Omega) = 2.4\ \text{V} \qquad \lozenge$$

(b) From $\tau = L/R$, the self-inductance in this circuit is found to be

$$L = \tau R = (0.25\ \text{s})(0.30\ \Omega) = 7.5\times10^{-2}\ \Omega\cdot\text{s} = 7.5\times10^{-2}\ \text{H} = 75\ \text{mH} \qquad \lozenge$$

(c) At time $t = \tau$, the current in the circuit is

$$I = I_{max}\left(1 - e^{-\tau/\tau}\right) = I_{max}\left(1 - e^{-1}\right) = (8.0\ \text{A})(1 - 0.368) = 5.1\ \text{A} \qquad \lozenge$$

(d) The magnitude of the potential difference across the resistor at $t = \tau$ is given by Ohm's law as $|\Delta V_R| = IR$. When going through the resistor in the direction of the conventional current, the potential decreases or "drops." The voltage or potential drop across the resistor at $t = \tau$ is

$$\Delta V_R = -(5.1\ \text{A})(0.30\ \Omega) = -1.5\ \text{V} \qquad \lozenge$$

(e) Starting at point b and going clockwise around the circuit, Kirchhoff's loop rule gives $\mathcal{E} + \Delta V_R + \Delta V_L = 0$, where ΔV_R and ΔV_L are the changes in potential incurred across the resistor and across the inductor, respectively. Since we passed through the resistor in the direction of the current, $\Delta V_R = -IR = -1.5$ V. Thus, at time $t = \tau$

$$\Delta V_L = -\mathcal{E} - \Delta V_R = -2.4\ \text{V} - (-1.5\ \text{V}) = -0.90\ \text{V} \qquad \lozenge$$

50. A 300-turn solenoid has a radius of 5.00 cm and a length of 20.0 cm. Find (a) the inductance of the solenoid and (b) the energy stored in the solenoid when the current in its windings is 0.500 A.

Solution

(a) The self-inductance L of an air-filled solenoid is a physical constant determined by the permeability of free space, μ_0, the number of turns on the solenoid, and the geometric factors ℓ and A. Here, ℓ is the length of the solenoid and A is the cross-sectional area. The expression for the self-inductance in terms of these quantities is $L = \mu_0 N^2 A/\ell$. Thus, the self-inductance of the given solenoid is

$$L = \frac{\mu_0 N^2 A}{\ell} = \frac{(4\pi \times 10^{-7}\ \text{T}\cdot\text{m/A})(300)^2\left[\pi(5.00\times 10^{-2}\ \text{m})^2\right]}{20.0\times 10^{-2}\ \text{m}}$$

or $\quad L = 4.44\times 10^{-3}\ \text{H} = 4.44\ \text{mH}$ ◊

(b) The energy stored in the magnetic field of an inductor is

$$PE_L = \frac{1}{2}LI^2$$

where L is the self-inductance and I is the current flowing through the inductor. Therefore, when a current of $I = 0.500$ A exists in the windings of the solenoid described above, the stored energy is

$$PE_L = \frac{1}{2}(4.44\times 10^{-3}\ \text{H})(0.500\ \text{A})^2 = 5.55\times 10^{-4}\ \text{J} = 0.555\times 10^{-3}\ \text{J} = 0.555\ \text{mJ}$$ ◊

---

55. A rectangular coil with resistance R has N turns, each of length ℓ and width w as shown in Figure P20.55. The coil moves into a uniform magnetic field $\vec{\mathbf{B}}_{\text{in}}$ with constant velocity $\vec{\mathbf{v}}$. What are the magnitude and direction of the total magnetic force on the coil (a) as it enters the magnetic field, (b) as it moves within the field, and (c) as it leaves the field?

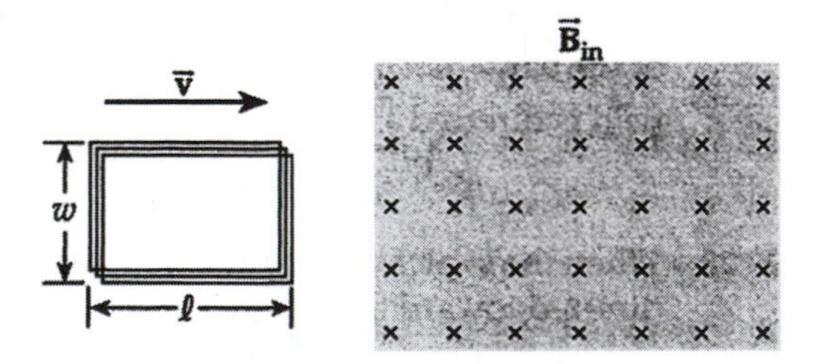

Figure P20.55

Solution

(a) During a time interval of duration Δt, after the right edge of the coil has entered the field and before the left edge reaches the field, the portion of the area enclosed by the coil which is inside the field (and has magnetic field lines passing through it) increases by an amount $\Delta A = w(v\Delta t)$. Thus, the flux through the area enclosed by each turn on the coil is directed into the page, and is increasing at a rate of $\Delta\Phi_B/\Delta t = B(\Delta A/\Delta t) = Bwv$. The magnitude of the emf induced in the coil is

$$|\mathcal{E}| = N\frac{|\Delta\Phi_B|}{\Delta t} = NBwv$$

The magnitude of the induced current is $I = |\mathcal{E}|/R = NBwv/R$, and it must flow counterclockwise around the coil producing a magnetic field directed outward through the area enclosed by the coil to oppose the increasing inward flux caused by the motion of the coil.

Note that the upper and lower edges of the coil (current carrying conductors, portions of which are in a magnetic field) experience magnetic forces that are equal in magnitude, oppositely directed, and cancel each other. However, each of the wires forming the right edge of the coil carry a current directed toward the top of the page through a magnetic field directed into the page. By right-hand rule number 1, these wires experience magnetic forces directed toward the left. The wires forming the left edge of the coil are not yet in a magnetic field, and experience no magnetic forces. Thus, the net magnetic force on the coil is directed **toward the left** with a magnitude of

$$F_{\text{total}} = NF_{\text{single wire}} = NBIw = NB(NBwv/R)w = N^2B^2w^2v/R \qquad \Diamond$$

(b) While the coil is entirely within the magnetic field, the flux through the area enclosed by the coil is constant. Hence, there is no induced emf or current in the coil, and the field exerts **zero force** on the coil. ◊

(c) As the coil is emerging from the right edge of the field, an analysis similar to that in part (a) shows that the enclosed area with flux passing through it is decreasing at a rate $\Delta A/\Delta t = -wv$. Thus, the coil will experience an induced emf of magnitude $|\mathcal{E}| = N(|\Delta\Phi_B|/\Delta t) = NB(|\Delta A|/\Delta t) = NBwv$ and carry an induced current of magnitude $I = |\mathcal{E}|/R = NBwv/R$. In this case, the induced current will flow clockwise around the loops of the coil to produce a magnetic field directed into the page through the enclosed area and oppose the decrease occurring in the inward flux due to the external field.

Again, the magnetic forces on the upper and lower edges of the coil will be equal in magnitude, opposite in direction, and will cancel. The right edge of the coil is now out of the field and experiences no magnetic force. Each of the N wires forming the left edge of the coil carry current I directed toward the top of the page through the magnetic field directed into the page. Right-hand rule number 1 then shows that each of these wires experience a magnetic force directed toward the left. The net magnetic force acting on the coil during this period is therefore directed **toward the left** with a total magnitude of

$$F_{\text{total}} = NF_{\text{single wire}} = NBIw = NB(NBwv/R)w = N^2B^2w^2v/R \qquad \Diamond$$

59. A conducting rod of length ℓ, moves on two horizontal frictionless rails, as in Figure P20.30. A constant force of magnitude 1.00 N moves the bar at a uniform speed of 2.00 m/s through a magnetic field $\vec{\mathbf{B}}$ that is directed into the page. (a) What is the current in an 8.00-Ω resistor R? (b) What is the rate of energy dissipation in the resistor? (c) What is the mechanical power delivered by the constant force?

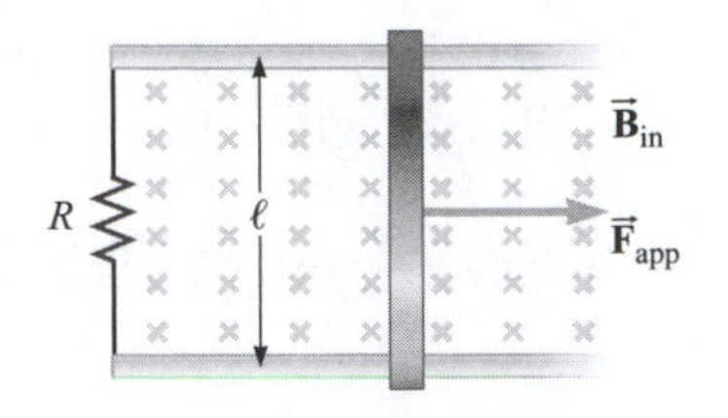

Figure P20.30

Solution

(a) As the rod of length ℓ moves to the right at speed v, the area A enclosed by the conducting path made up of the rod, rails, and the resistance R increases in size. Thus, the magnetic flux, $\Phi_B = BA$, through this conducting path is directed into the page (in the direction of the external magnetic field), and is increasing in magnitude. This increasing inward flux gives rise to an induced emf $|\mathcal{E}| = \Delta\Phi_B/\Delta t = B\ell v$, causing an induced current $I = |\mathcal{E}|/R = B\ell v/R$ to flow around this path. The external field strength may then be written as $B = IR/\ell v$.

The induced current must flow upward toward the top of the page through the rod. Then it produces a magnetic field directed out of the page through area A, opposing the increasing inward flux due to the external field.

Since the rod is carrying current I toward the top of the page through an external field directed into the page, it will experience a magnetic force of magnitude $F_m = BI\ell = \left(IR/\ell v\right)I\ell = I^2 R/v$. According to right-hand rule number 1, this magnetic force is directed toward the left, opposing the applied force, F_{app}. If the rod moves at constant speed, then $\Sigma F_x = 0$, meaning that $F_m = F_{\text{app}}$, or $I^2 R/v = F_{\text{app}}$. The induced current in the conducting path, including resistance R, is then

$$I = \sqrt{\frac{F_{\text{app}}v}{R}} = \sqrt{\frac{(1.00\ \text{N})(2.00\ \text{m/s})}{8.00\ \Omega}} = 0.500\ \text{A} \qquad \Diamond$$

(b) The power dissipated in the resistor is

$$P_{\text{dissipated}} = I^2 R = (0.500\ \text{A})^2(8.00\ \Omega) = 2.00\ \text{W} \qquad \Diamond$$

(c) The rate at which the applied force supplies energy to the system is

$$P_{\text{input}} = F_{\text{app}}v = (1.00\ \text{N})(2.00\ \text{m/s}) = 2.00\ \text{J/s} = 2.00\ \text{W} \qquad \Diamond$$

21

Alternating Current Circuits and Electromagnetic Waves

NOTES FROM SELECTED CHAPTER SECTIONS

21.1 Resistors in an AC Circuit

In an AC circuit both voltage and current vary sinsusoidally with time. If an alternating voltage is applied across a resistor, the current and voltage vary in the same manner, and both reach their maximum values at the same time. ***The current in the resistor will be in phase with the voltage.*** The voltage across a resistor is independent of the frequency and acts to limit the maximum value of the current.

The value of the rms current in an AC circuit is equal to that of a direct current which would deliver the same internal energy to a resistor as does the rms current.

21.2 Capacitors in an AC Circuit

When an alternating voltage is applied across a capacitor, the voltage reaches its maximum value one-quarter of a cycle after the current reaches its maximum value. ***The voltage across a capacitor lags the current by 90°.***

A capacitor in an AC circuit acts to limit the current. The impeding effect of a capacitor is called **capacitative reactance,** X_C**.** The value of the capacitative reactance is inversely proportional to the frequency of the applied voltage.

21.3 Inductors in an AC Circuit

When an alternating voltage is applied across an inductor (coil), the voltage reaches its maximum value one-quarter of a cycle before the current reaches its maximum value. ***The voltage across an inductor leads the current by 90°.*** The extent by which an inductor impedes the current in an AC circuit is called **inductive reactance**, X_L**.** This quantity is directly proportional to the frequency of the applied voltage.

21.4 The *RLC* Series Circuit

At any given time the current in a series *RLC* (resistor, inductor, and capacitor) circuit has the same amplitude and phase at all points in the circuit. The current through the individual elements reaches its maximum value at the same time. The voltages across the individual components (R, L, and C) do not reach their maximum values at the same time; they are out of phase with each other.

Examine the graphs below and confirm the following phase relationships:

The voltage across the resistor is in phase with the current.

The voltage across the capacitor lags the current by 90°.

The voltage across the inductor leads the current by 90°.

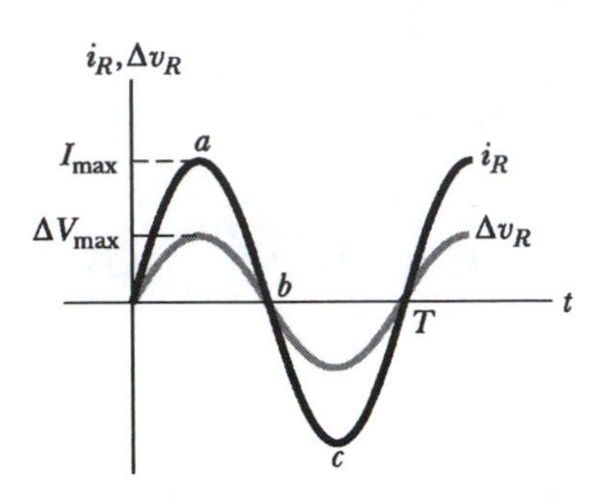

Current and voltage vs. time for resistor

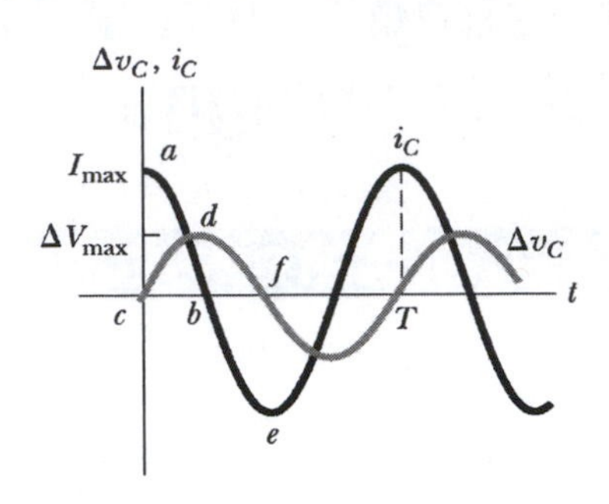

Current and voltage vs. time for capacitor

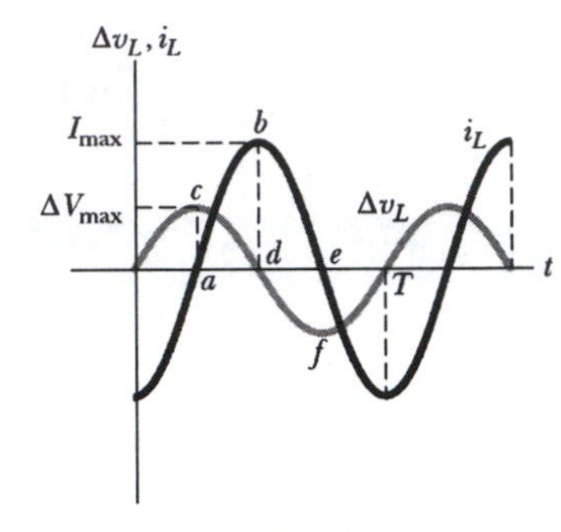

Current and voltage vs. time for inductor

The maximum applied voltage does not equal the sum of the individual maximum voltages:

$$\Delta V_{max} \neq \Delta V_{R,max} + \Delta V_{L,max} + \Delta V_{C,max}$$

The instantaneous applied voltage, however, does equal the sum of the instantaneous voltages across the individual components:

$$\Delta v = \Delta v_R + \Delta v_L + \Delta v_C$$

A **phasor diagram** illustrates the phase relationships among the voltages in an AC circuit. The maximum voltage across each circuit component is represented by a rotating vector called a phasor.

The **phasor diagram** is a very useful technique to use in the analysis of *RLC* circuits. In such a diagram, each of the quantities ΔV_R, ΔV_L, ΔV_C, and I_{max} is represented by a separate phasor (rotating vector).

A phasor diagram which describes the AC circuit of Figure (a) below is shown in Figure (b). Each phasor has a length which is proportional to the magnitude of the voltage or current which it represents and rotates counterclockwise about the common origin with a frequency which equals the frequency (f) of the alternating source.

The direction of the phasor which represents the current, shown as an open arrow in Figure (b), in the circuit is used as the *reference direction* to establish the correct phase differences among the phasors, which represent the voltage drops across the resistor, inductor, and capacitor. The *instantaneous values* Δv_R, Δv_L, Δv_C, and i are given by the *projection onto the vertical axis of the corresponding phasor.*

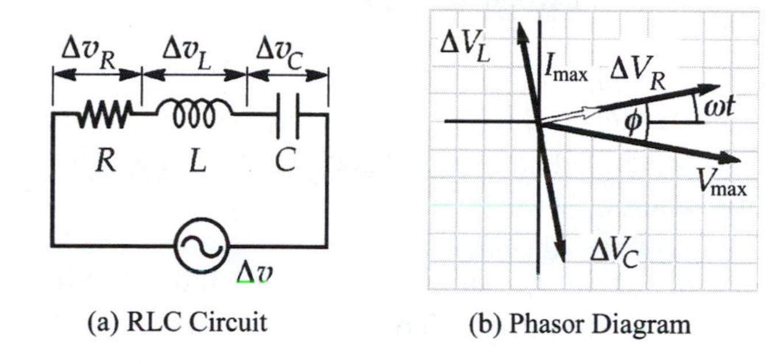

(a) RLC Circuit (b) Phasor Diagram

Consider the phasor diagram in Figure (b), assume counterclockwise rotation, and note that:

(1) ΔV_L **leads the current by 90°** (ΔV_L reaches its maximum 90° before ΔV_R).

(2) ΔV_C **lags the current by 90°** (ΔV_C reaches maximum 90° later than ΔV_R).

(3) ΔV_R *greater than* ΔV_L; the maximum voltage across the resistor is **greater than** the maximum voltage across the inductor as seen by the lengths of the respective vectors.

(4) v_R *less than* v_L; the instantaneous voltage across the resistor is **less than** the instantaneous voltage across inductor as seen by the projections of the vectors onto the vertical axis.

Remember, the instantaneous values are the projections of the maximum values onto the vertical axis.

Also notice that as time increases and the phasors rotate counterclockwise, maintaining their constant relative phase, the voltage amplitudes (ΔV_R, ΔV_L, ΔV_C) will remain constant in magnitude; but the instantaneous values (Δv_R, Δv_L, and Δv_C) will vary sinusoidally with time. For the case shown in Figure (b), the phase angle ϕ is negative (this is because $X_C > X_L$ and therefore $\Delta V_C > \Delta V_L$); hence, the current in the circuit leads the applied voltage in phase.

The maximum voltage across an *RLC* circuit (ΔV_{max}) is found by adding ΔV_R, ΔV_L, and ΔV_C as vector quantities. In the figure at right, ΔV_{max} is shown as a rotating phasor. The **phase angle** (ϕ) indicates the degree by which the maximum voltage is out of phase with the current.

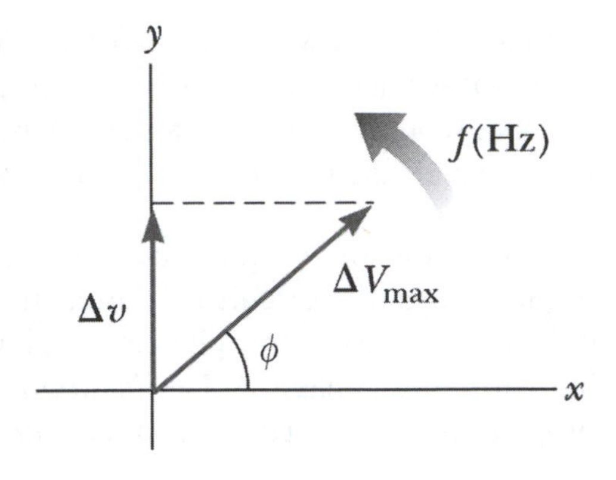

The **instantaneous voltage** (Δv) across the circuit is the component of ΔV_{max} along the y-axis. For example, when $\phi = 0$ or $180°$, $\Delta v = 0$; when $\phi = 90°$, $\Delta v = \Delta V_{max}$; and when $\phi = 270°$, $\Delta v = -\Delta V_{max}$.

The **impedance, Z,** is a parameter which represents the combined effect of R, X_C, and X_L in limiting current in an AC circuit. The value of Z depends on the values of R, L, C, and the frequency of the voltage source.

A **reactance (or impedance) triangle** showing R, X_L, X_C and Z can be drawn directly from the phasor diagram showing the voltages. This is possible because ΔV_R, ΔV_L, and ΔV_C are each proportional to the current for given values of R, X_L, and X_C.

21.5 Power in an AC Circuit

The power delivered by the generator in an AC circuit is converted to internal energy in the resistor. *There is no power loss in an ideal inductor or an ideal capacitor.* The power delivered to the circuit is determined by the power factor, $\cos \phi$.

21.6 Resonance in a Series *RLC* Circuit

As the frequency of the source (generator) increases, the resistance (R) in the circuit remains constant; inductive reactance (X_L) increases, and capacitative reactance (X_C) decreases. *At the resonance frequency, f_0 (when $X_L = X_C$), the voltage across the capacitor is equal to but 180 degrees out of phase (oppositely directed) with the voltage across the inductor. In this case, $\Delta V_L + \Delta V_C = 0$, and the impedance ($Z$) has its minimum value and is equal to R. At resonance the current in the circuit is maximum, limited only by the value of R.*

21.7 The Transformer

A transformer is a device designed to increase or decrease an AC voltage. Energy conservation requires a corresponding decrease or increase in current. In an ideal transformer, the power input to the primary equals the power output at the secondary (load). In practice, output power is less than input power; this is due in part to eddy currents induced in the transformer core. In its simplest form, a transformer consists of a primary coil of N_1 turns and a secondary coil of N_2 turns, both wound onto a common soft iron core. A step-up transformer has $N_2 > N_1$ and in a step-down transformer $N_1 > N_2$.

21.8 Maxwell's Predictions
21.9 Hertz's Confirmation of Maxwell's Predictions

Maxwell's equations are the fundamental laws governing the behavior of electric and magnetic fields. The theory Maxwell developed is based upon the following four pieces of information:

- Electric field lines originate on positive charges and terminate on negative charges. The electric field due to a point charge can be determined at a location by applying Coulomb's force law to a positive test charge placed at that location.
- Magnetic field lines always form closed loops; that is, they do not begin or end at any point.
- A varying magnetic field induces an emf and hence, an electric field. This is a statement of Faraday's law (Chapter 20).
- Magnetic fields are generated by moving charges (or currents), as summarized in Ampère's law (Chapter 19).

21.11 Properties of Electromagnetic Waves

Following is a summary of the properties of electromagnetic waves:

- **Electromagnetic waves are transverse waves** and travel through empty space with the speed of light, $c = 1/\sqrt{\epsilon_0 \mu_0}$.
- **Radiated waves exist as electric and magnetic fields which oscillate perpendicular to each other.** The plane in which the oscillations occur is perpendicular to the direction of wave propagation.
- **The ratio of $|\vec{\mathrm{E}}|$ to $|\vec{\mathrm{B}}|$ in empty space has a constant value,** the speed of light in vacuum.
- **Electromagnetic waves carry both energy and momentum.**

21.12 The Spectrum of Electromagnetic Waves

All electromagnetic waves are produced by accelerating charges. Several types of electromagnetic waves are characterized by a "typical" range of frequencies or wavelengths. The following are listed in order of increasing frequency.

- **Radio waves** ($\sim$10 km $> \lambda >$ $\sim$1 cm) are the result of electric charges accelerating through a conducting wire (antenna).
- **Microwaves** ($\sim$1 cm $> \lambda >$ $\sim$1 mm) are generated by electronic devices.
- **Infrared waves** ($\sim$1 mm $> \lambda >$ $\sim$700 nm) are produced by high-temperature objects and molecules.
- **Visible light** ($\sim$700 nm $> \lambda >$ $\sim$400 nm) is produced by the rearrangement of electrons in atoms and molecules.
- **Ultraviolet (UV) light** ($\sim$400 nm $> \lambda >$ $\sim$1 nm) is an important component of radiation from the Sun.
- **X-rays** ($\sim$10 nm $> \lambda >$ $\sim 10^{-4}$ nm) are produced when high-energy electrons strike a target of high atomic number (e.g., metal or glass).
- **Gamma rays** ($\sim 10^{-1}$ nm $> \lambda >$ $\sim 10^{-5}$ nm) are emitted by radioactive nuclei.

EQUATIONS AND CONCEPTS

A **series alternating current circuit** with a sinusoidal source of emf is shown in the figure to the right. The rectangle ▭ used here represents the circuit element(s) which, in a particular case, may be a resistor (R), a capacitor (C), an inductor (L), or some combination of the above. The frequency of the current is determined by the AC source frequency, and the output voltage is sinusoidal as shown in Equation 21.1. *Keep in mind the different notations used to designate instantaneous, maximum, and rms values of voltage and current.*

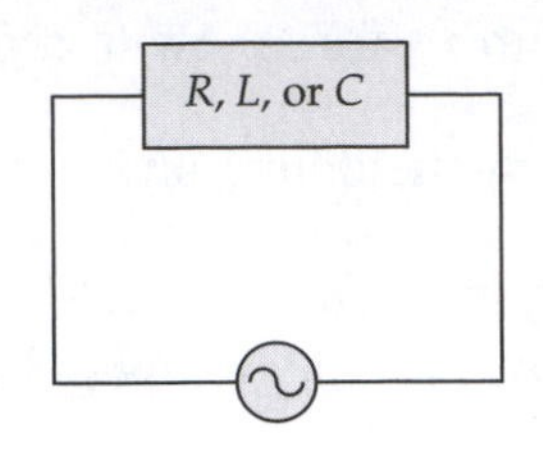

$$\Delta v = \Delta V_{max} \sin 2\pi ft \tag{21.1}$$

Δv = instantaneous voltage
ΔV_{max} = maximum voltage

Root-mean-square (rms) values of current and voltage are those values to which measuring instruments usually respond.

$$I_{rms} = \frac{I_{max}}{\sqrt{2}} = 0.707\, I_{max} \tag{21.2}$$

$$\Delta V_{rms} = \frac{\Delta V_{max}}{\sqrt{2}} = 0.707\, \Delta V_{max} \tag{21.3}$$

Current in an *RLC* circuit can be limited by:

Resistance (*R*) of the resistor

R is frequency independent

Capacitive reactance (X_C) of the capacitor

$$X_C \equiv \frac{1}{2\pi f C} \tag{21.5}$$

Inductive reactance (X_L) of the inductor

$$X_L \equiv 2\pi f L \tag{21.8}$$

The **series *RLC* circuit**, as shown at right, includes a resistor, inductor, capacitor, and a sinusoidally varying voltage source.

At any instant, the instantaneous current (i) is the same at all points in the circuit.

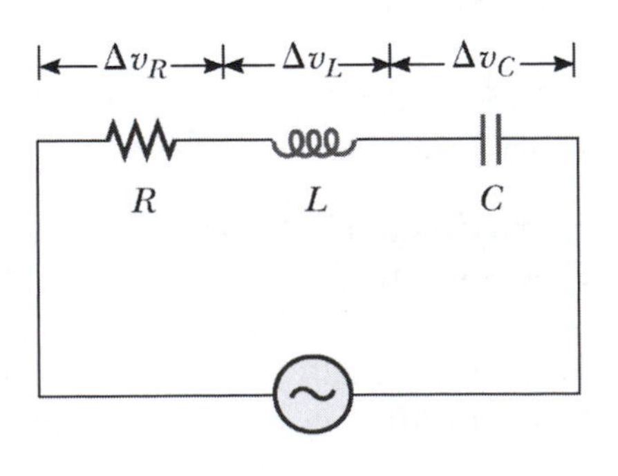

$$i = I_{max} \sin 2\pi ft$$

The **rms values of voltage and current in each circuit** element can be expressed by equations which have the form of Ohm's law. *Similar equations relate the maximum values of voltage and current.*

Resistor: $\Delta V_{R,\text{rms}} = I_{\text{rms}} R$ (21.4a)

Capacitor: $\Delta V_{C,\text{rms}} = I_{\text{rms}} X_C$ (21.6)

Inductor: $\Delta V_{L,\text{rms}} = I_{\text{rms}} X_L$ (21.9)

The **maximum (or rms) voltage** across the entire *RLC* circuit can be found in terms of:

the respective voltage values across the individual components

$$\Delta V_{\text{max}} = \sqrt{\Delta V_R^{\,2} + (\Delta V_L - \Delta V_C)^2} \qquad (21.10)$$

$$\Delta V_{\text{rms}} = \sqrt{\Delta V_{R,\text{rms}}^2 + (\Delta V_{L,\text{rms}} - \Delta V_{C,\text{rms}})^2}$$

or, the common circuit current and the values of resistance, inductive reactance, and capacitive reactance.

$$\Delta V_{\text{max}} = I_{\text{max}} \sqrt{R^2 + (X_L - X_C)^2} \qquad (21.12)$$

$$\Delta V_{\text{rms}} = I_{\text{rms}} \sqrt{R^2 + (X_L - X_C)^2}$$

Impedance (Z) is a parameter of the circuit determined by R, X_L, and X_C and is frequency dependent. *The SI unit of impedance is the ohm.* The form of Equation 21.13 can be seen from the impedance triangle.

$$Z \equiv \sqrt{R^2 + (X_L - X_C)^2} \qquad (21.13)$$

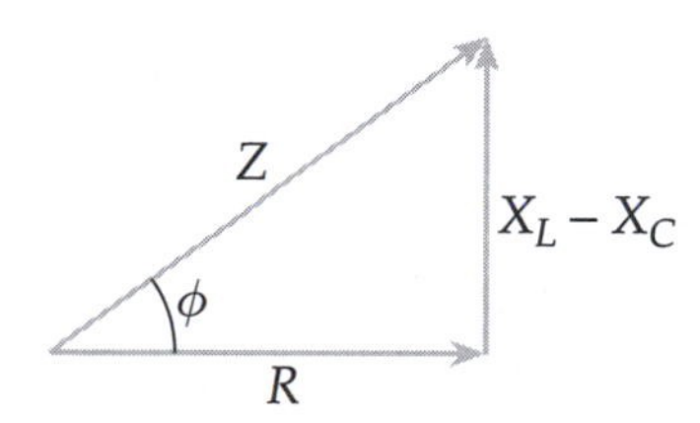

Impedance triangle

The **phase angle** (ϕ) is the degree by which the instantaneous current and instantaneous voltage are out-of-step. This parameter can be determined from the impedance triangle shown above and stated in Equation 21.15 or from a voltage triangle as expressed in Equation 21.11.

$$\tan\phi = \frac{X_L - X_C}{R} \qquad (21.15)$$

$$\tan\phi = \frac{\Delta V_L - \Delta V_C}{\Delta V_R} \qquad (21.11)$$

A **generalized form of Ohm's law** relates the maximum (or rms) voltage and maximum (or rms) current.

$$\Delta V_{\text{max}} = I_{\text{max}} Z \qquad (21.14)$$

$$\Delta V_{\text{rms}} = I_{\text{rms}} Z$$

The **average power** delivered by a generator (source of emf) to an *RLC* series circuit is directly proportional to cos ϕ, the power factor of the circuit. *There is zero power loss in ideal inductors and capacitors; the average power delivered by the source is converted to internal energy in the resistor.*

$$P_{av} = I_{rms}^2 R \quad (21.16)$$

$$P_{av} = I_{rms}\,\Delta V_{rms} \cos\phi \quad (21.17)$$

$\cos\phi =$ power factor

The **resonance frequency** (f_0) is that frequency for which $X_L = X_C$ (and $Z = R$). *At this frequency the current has its maximum value and is in phase with the applied voltage.*

$$f_0 = \frac{1}{2\pi\sqrt{LC}} \quad (21.19)$$

A **transformer** consists of a primary coil of N_1 turns and a secondary coil of N_2 turns wound on a common core. *In a step-up transformer, N_2 is greater than N_1.*

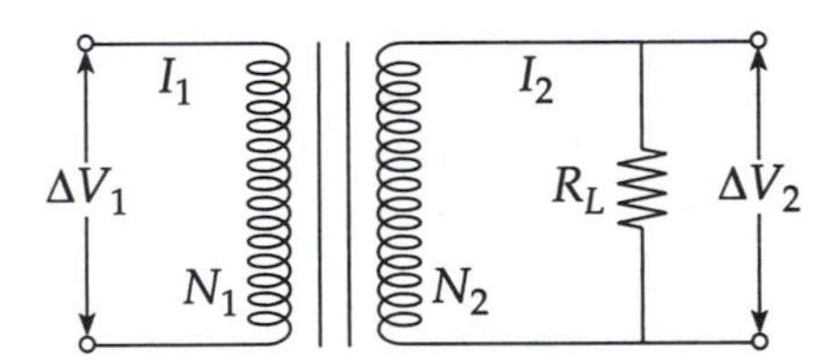

ΔV_1 = voltage across the primary
ΔV_2 = voltage across the secondary
I_1 = current in the primary
I_2 = current in the secondary

In **an ideal transformer,** the ratio of voltages is equal to the ratio of turns, and the ratio of currents is equal to the inverse of the ratio of turns. *Energy conservation requires a decrease in current to accompany an increase in voltage.*

$$\Delta V_2 = \frac{N_2}{N_1}\Delta V_1 \quad (21.22)$$

$$I_1 \Delta V_1 = I_2 \Delta V_2 \quad (21.23)$$

The **speed of an electromagnetic wave** is related to the permeability and permittivity of the medium through which it travels.

$$c = \frac{1}{\sqrt{\mu_0 \epsilon_0}} \quad (21.24)$$

The **speed of light in a vacuum** equals the ratio of the magnitudes of the associated electric and magnetic fields.

$$\frac{E}{B} = c \quad (21.26)$$

The **wave equation**: The product of frequency and wavelength of an electromagnetic wave propagating in vacuum is constant and equal to *c*.

$$c = f\lambda \quad (21.31)$$

$$c = 2.997\,92 \times 10^8 \text{ m/s} \quad (21.25)$$

Permeability constant of free space.

$$\mu_0 = 4\pi \times 10^{-7}\ \text{T} \cdot \text{m/A}$$

Permittivity of free space.

$$\epsilon_0 = 8.854\ 19 \times 10^{-12}\ \text{C}^2/\text{N} \cdot \text{m}^2$$

The **intensity (power per unit area) of electromagnetic waves** can be expressed in several alternate forms involving the maximum values of the electric and magnetic fields. The energy carried by an electromagnetic wave is shared equally by the electric and magnetic fields. *Do not confuse the symbol, I, representing intensity in Equations 21.27 and 21.28 with the same symbol used elsewhere for current.*

$$I = \frac{E_{max} B_{max}}{2\mu_0} \qquad (21.27)$$

$$I = \frac{E_{max}^2}{2\mu_0 c} = \frac{c}{2\mu_0} B_{max}^2 \qquad (21.28)$$

where

$$I = \frac{P_{av}}{A} = \text{average power per unit area}$$

The **magnitude of the momentum** delivered to a surface by an electromagnetic wave depends on the fraction of energy absorbed. *Radiation pressure is exerted on a surface as a result of momentum transfer by an electromagnetic wave.*

$$p = \frac{U}{c} \quad \text{(complete absorption)} \qquad (21.29)$$

$$p = \frac{2U}{c} \quad \text{(complete reflection)} \qquad (21.30)$$

SUGGESTIONS, SKILLS, AND STRATEGIES

PROCEDURE FOR SOLVING ALTERNATING CURRENT PROBLEMS

- First calculate as many of the unknown quantities such as X_L and X_C as possible. Be careful to use correct units; when calculating X_C, express capacitance in farads (not microfarads).

- Apply the general equation $\Delta V = IZ$ to the portion of the circuit that is of interest. That is, if you want to know the voltage drop across the combination of an inductor and a resistor, the equation reduces to $\Delta V_{max} = I_{max}\sqrt{R^2 + X_L^{\,2}}$ or $\Delta V_{rms} = I_{rms}\sqrt{R^2 + X_L^{\,2}}$.

REVIEW CHECKLIST

- Given an *RLC* series circuit in which values of resistance, inductance, capacitance, and the characteristics of the generator (source of emf) are known, calculate: (i) the rms voltage drop across each component, (ii) the rms current in the circuit, (iii) the phase angle by which the current leads or lags the voltage, (iv) the power expended in the circuit, and (f_0) the resonance frequency of the circuit. (Sections 21.1–21.6)

- Understand the important function of transformers in the process of transmitting electrical power over large distances. Make calculations of primary to secondary voltage and current ratios for an ideal transformer. (Section 21.7)
- Be aware of the important pieces of information on which Maxwell based his theory of electromagnetic waves. Summarize the properties of electromagnetic waves. Describe the relative orientation of the magnetic field, electric field, and direction of propagation for a plane electromagnetic wave. (Sections 21.8–21.9)
- Place the various types of electromagnetic waves in the correct sequence in the electromagnetic spectrum. Be aware that all forms of radiation are produced by accelerated charges and know the basis of production particular to each of the wave types. (Section 21.11–21.12)

SOLUTIONS TO SELECTED END-OF-CHAPTER PROBLEMS

1. (a) What is the resistance of a lightbulb that uses an average power of 75.0 W when connected to a 60.0-Hz power source having a maximum voltage of 170 V? (b) What is the resistance of a 100-W lightbulb?

Solution

From the definition of the rms or root-mean-square current, I_{rms}, in an AC circuit, the time-averaged power dissipated in a resistance R is given by $P_{av} = I_{rms}^2 R$. When the current varies sinusoidally in time, I_{rms} is related to the peak or maximum current by the expression $I_{rms} = I_{max}/\sqrt{2} = 0.707 I_{max}$. Since Ohm's law is valid in AC circuits, we may express the rms and peak voltages as $\Delta V_{rms} = I_{rms} R$ and $\Delta V_{max} = I_{max} R$. The average power dissipated in a resistor may then be written in a variety of ways as

$$P_{av} = I_{rms}^2 R = \frac{(\Delta V_{rms})^2}{R} = \frac{I_{max}^2 R}{2} = \frac{(\Delta V_{max})^2}{2R}$$

The resistance of a lightbulb with a power output P_{av} when connected to an AC source having a maximum voltage ΔV_{max} is therefore given as $R = (\Delta V_{max})^2 / 2P_{av}$.

(a) For a bulb with $P_{av} = 75.0$ W when $\Delta V_{max} = 170$ V, the resistance is

$$R = \frac{(170\ \text{V})^2}{2(75.0\ \text{W})} = 193\ \Omega \qquad \lozenge$$

(b) If $P_{av} = 100$ W when $\Delta V_{max} = 170$ V, the resistance of the bulb is

$$R = \frac{(170\ \text{V})^2}{2(100\ \text{W})} = 145\ \Omega \qquad \lozenge$$

12. A generator delivers an AC voltage of the form $\Delta v = (98.0\text{ V})\sin(80\pi t)$ to a capacitor. The maximum current in the circuit is 0.500 A. Find the (a) rms voltage of the generator, (b) frequency of the generator, (c) rms current, (d) reactance, and (e) value of the capacitance.

Solution

(a) The general expression for a sinusoidal AC voltage is $\Delta v = \Delta V_{max}\sin(\omega t)$, with $\omega = 2\pi f$, where f is the frequency of the voltage fluctuations. Comparing this to the given voltage, we see that $\Delta V_{max} = 98.0$ V and $\omega = 80\pi$ rad/s. The root-mean-square (rms) value of this sinusoidal voltage is then

$$\Delta V_{rms} = \frac{\Delta V_{max}}{\sqrt{2}} = \frac{98.0\text{ V}}{\sqrt{2}} = 69.3\text{ V} \qquad \Diamond$$

(b) Since $\omega = 2\pi f = 80\pi$ rad/s, the frequency of the generator is

$$f = \frac{\omega}{2\pi} = \frac{80\pi\text{ rad/s}}{2\pi\text{ rad}} = 40\text{ Hz} \qquad \Diamond$$

(c) When the applied voltage, and hence the current, vary sinusoidally in time, the maximum current and the rms current are related by the expression $I_{rms} = I_{max}/\sqrt{2}$. The rms current in the given case is therefore

$$I_{rms} = \frac{I_{max}}{\sqrt{2}} = \frac{0.500\text{ A}}{\sqrt{2}} = 0.354\text{ A} \qquad \Diamond$$

(d) When an AC voltage is applied to a capacitor, the rms voltage and the rms current are related by an expression similar to Ohm's law, namely $\Delta V_{rms} = I_{rms}X_C$, where $X_C = 1/\omega C = 1/2\pi f C$ is the reactance of the capacitor. The reactance of the capacitor in the given circuit is then

$$X_C = \frac{\Delta V_{rms}}{I_{rms}} = \frac{\Delta V_{max}/\sqrt{2}}{\Delta I_{max}/\sqrt{2}} = \frac{\Delta V_{max}}{I_{max}} = \frac{98.0\text{ V}}{0.500\text{ A}} = 196\ \Omega \qquad \Diamond$$

(e) From $X_C = 1/\omega C$, the capacitance of the capacitor in the given circuit is

$$C = \frac{1}{\omega X_C} = \frac{1}{(80\pi\text{ rad/s})(196\ \Omega)} = 2.03\times10^{-5}\text{ F} = 20.3\ \mu\text{F} \qquad \Diamond$$

14. An AC generator has an output rms voltage of 78.0 V at a frequency of 80.0 Hz. If the generator is connected across a 25.0-mH inductor, find the (a) inductive reactance, (b) rms current, and (c) maximum current in the circuit.

Solution

(a) The rms AC voltage across an induction coil and the rms current through that inductor are related by an expression similar to Ohm's law, namely $\Delta V_{rms} = I_{rms} X_L$, where $X_L = \omega L = 2\pi f L$ is the inductive reactance of the coil. Here, f is the frequency of the AC voltage and current. The inductive reactance of a 25.0-mH inductor to an 80.0-Hz AC voltage is

$$X_L = 2\pi f L = 2\pi(80.0\ \text{Hz})(25.0\times10^{-3}\ \text{H})$$
$$= 2\pi(80.0\ \text{s}^{-1})(25.0\times10^{-3}\ \Omega\cdot\text{s}) = 12.6\ \Omega \quad \Diamond$$

(b) The rms current in this inductor when an 80.0-Hz AC voltage with $\Delta V_{rms} = 78.0$ V is applied will be

$$I_{rms} = \frac{\Delta V_{rms}}{X_L} = \frac{78.0\ \text{V}}{12.6\ \Omega} = 6.19\ \text{A} \quad \Diamond$$

(c) Ordinary AC voltages and current vary sinusoidally in time. In that case, the relations between the maximum and the rms values are $\Delta V_{max} = \sqrt{2}(\Delta V_{rms})$ and $I_{max} = \sqrt{2}(I_{rms})$. The maximum current in this circuit is then

$$I_{max} = \sqrt{2}(I_{rms}) = \sqrt{2}(6.19\ \text{A}) = 8.75\ \text{A} \quad \Diamond$$

---

21. A resistor ($R = 9.00\times10^2\ \Omega$), a capacitor ($C = 0.250\ \mu$F), and an inductor ($L = 2.50$ H) are connected in series across a 2.40×10^2-Hz AC source for which $\Delta V_{max} = 1.40\times10^2$ V. Calculate (a) the impedance of the circuit, (b) the maximum current delivered by the source, and (c) the phase angle between the current and voltage. (d) Is the current leading or lagging the voltage?

Solution

(a) The impedance of a series *RLC* circuit is given by $Z = \sqrt{R^2 + (X_L - X_C)^2}$, where R is the total resistance in the circuit. The inductive reactance is $X_L = 2\pi f L$, where L is the total self-inductance and f is the frequency of the power source. The capacitive reactance is $X_C = 1/2\pi f C$ for a circuit containing total capacitance C. For the given circuit,

$$R = 9.00\times10^2\ \Omega = 900\ \Omega$$

$$X_L = 2\pi(240\ \text{Hz})(2.50\ \text{H}) = 3.77\times10^3\ \Omega$$

and $X_C = \dfrac{1}{2\pi(240 \text{ Hz})(0.250\times10^{-6} \text{ F})} = 2.65\times10^3\ \Omega$

so the impedance is

$$Z = \sqrt{(900\ \Omega)^2 + (3.77\times10^3\ \Omega - 2.65\times10^3\ \Omega)^2} = 1.44\times10^3\ \Omega = 1.44\ \text{k}\Omega \quad \Diamond$$

(b) The maximum current delivered to this circuit by the source is

$$I_{\max} = \frac{\Delta V_{\text{source,max}}}{Z} = \frac{1.40\times10^2\ \text{V}}{1.44\times10^3\ \Omega} = 9.72\times10^{-2}\ \text{A} = 0.097\ 2\ \text{A} \quad \Diamond$$

(c) The difference in phase between the applied voltage and the current in this circuit is

$$\phi = \tan^{-1}\left(\frac{X_L - X_C}{R}\right) = \tan^{-1}\left(\frac{3.77\times10^3\ \Omega - 2.65\times10^3\ \Omega}{900\ \Omega}\right) = +51.2° \quad \Diamond$$

(d) Since the inductive reactance exceeds the capacitive reactance, the phase angle ϕ is positive. This means that the applied voltage leads the current, or the *current lags behind the voltage* in this circuit. ◊

---

30. An AC source operating at 60 Hz with a maximum voltage of 170 V is connected in series with a resistor ($R = 1.2$ kΩ) and a capacitor ($C = 2.5\ \mu$F). (a) What is the maximum value of the current in the circuit? (b) What are the maximum values of the potential difference across the resistor and the capacitor? (c) When the current is zero, what are the magnitudes of the potential difference across the resistor, the capacitor, and the AC source? How much charge is on the capacitor at this instant? (d) When the current is at a maximum, what are the magnitudes of the potential differences across the resistor, the capacitor, and the AC source? How much charge is on the capacitor at this instant?

Solution

(a) The capacitive reactance for this circuit is

$$X_C = \frac{1}{2\pi f C} = \frac{1}{2\pi(60\ \text{Hz})(2.5\times10^{-6}\ \text{F})} = 1.1\times10^3\ \Omega = 1.1\ \text{k}\Omega$$

Therefore, the impedance of this *RC* circuit (with zero inductive reactance) is

$$Z = \sqrt{R^2 + X_C^2} = \sqrt{(1.2\times10^3\ \Omega)^2 + (1.1\times10^3\ \Omega)^2} = 1.6\times10^3\ \Omega = 1.6\ \text{k}\Omega$$

and the maximum value of the current in the circuit is

$$I_{\max} = \frac{\Delta V_{\text{source,max}}}{Z} = \frac{170\ \text{V}}{1.6\times10^3\ \Omega} = 0.11\ \text{A} \quad \Diamond$$

(b) The maximum values of the voltages across the resistor and the capacitor are

$$\Delta V_{R,\max} = RI_{\max} = (1.2\times 10^{3}\ \Omega)(0.11\ \text{A}) = 1.3\times 10^{2}\ \text{V}$$ ◊

and $\Delta V_{C,\max} = X_C I_{\max} = (1.1\times 10^{3}\ \Omega)(0.11\ \text{A}) = 1.2\times 10^{2}\ \text{V}$ ◊

(c) When the instantaneous current is $i = 0$, the instantaneous voltage across the resistor is $\Delta v_R = iR = 0$. ◊

The voltage across a capacitor is 90° or a quarter cycle out of phase with the current. Thus, when the instantaneous current is zero, the magnitude of the instantaneous voltage across the capacitor is a maximum, or $|\Delta v_C| = \Delta V_{C,\max} = 1.2\times 10^{2}\ \text{V}$. ◊

The stored charge is then $q = C|\Delta v_C| = (2.5\ \mu\text{F})(1.2\times 10^{2}\ \text{V}) = 3.0\times 10^{2}\ \mu\text{C}$. ◊

Applying Kirchhoff's loop rule at any instant to the single loop of this series circuit gives $\Delta v_{\text{source}} + \Delta v_R + \Delta v_C = 0$. Thus, at the instant when $i = 0$, $\Delta v_{\text{source}} + 0 + \Delta V_{C,\max} = 0$ and $|\Delta v_{\text{source}}| = \Delta V_{C,\max} = 1.2\times 10^{2}\ \text{V}$. ◊

(d) When the instantaneous current is a maximum, the instantaneous voltage across the resistor ($\Delta v_R = iR$) will be a maximum, while the instantaneous voltage across the capacitor (a quarter cycle out of phase with the current) will be zero.

Thus, at this instant, we have $|\Delta v_R| = I_{\max} R = \Delta V_{R,\max} = 1.3\times 10^{2}\ \text{V}$ ◊

$|\Delta v_C| = 0$ and $q = C|\Delta v_C| = 0$ ◊

Applying Kirchhoff's loop rule to the circuit at this instant gives

$$\Delta v_{\text{source}} + \Delta V_{R,\max} + 0 = 0 \qquad \text{or} \qquad |\Delta v_{\text{source}}| = \Delta V_{R,\max} = 1.3\times 10^{2}\ \text{V}$$ ◊

---

34. A series *RLC* circuit has a resistance of 22.0 Ω and an impedance of 80.0 Ω. If the rms voltage applied to the circuit is 160 V, what average power is delivered to the circuit?

Solution

The rms current in this circuit will be

$$I_{\text{rms}} = \frac{\Delta V_{\text{rms}}}{Z} = \frac{160\ \text{V}}{80.0\ \Omega} = 2.0\ \text{A}$$

In an *RLC* circuit, the current is a quarter cycle out of phase with both the voltage across the inductor and the voltage across the capacitor. This means that each of these circuit elements absorb energy on one half of the cycle and return that energy back to the source on the other half of the cycle. Thus, the time-average power consumption by either a pure inductor or a pure capacitor is $P_L = P_C = 0$. However, the current is always in phase with the voltage across the resistor, and this element consumes energy on all parts of the cycle, giving it a time-average power consumption of $P_R = I_{\text{rms}}^2 R$. Therefore, the average power delivered to this circuit by the source is

$$P_{\text{av}} = P_R + P_L + P_C = I_{\text{rms}}^2 R + 0 + 0 = (2.0\ \text{A})^2(22.0\ \Omega) = 88\ \text{W} \qquad \Diamond$$

An alternate way of computing the average power delivered to an AC circuit is through the expression $P_{\text{av}} = I_{\text{rms}} \Delta V_{\text{rms}} \cos\phi$, where I_{rms} is the rms current in the circuit, ΔV_{rms} is the rms value of the applied voltage, and $\cos\phi$ is called the power factor. Here, ϕ is the phase difference between the current and the applied voltage.

The impedance of a series *RLC* circuit is $Z = \sqrt{R^2 + (X_L - X_C)^2}$. Thus, the total reactance may be written as $|X_L - X_C| = \sqrt{Z^2 - R^2}$. The magnitude of the phase difference between the applied voltage and the current in this circuit is then

$$|\phi| = \tan^{-1}\left(\frac{|X_L - X_C|}{R}\right) = \tan^{-1}\left(\frac{\sqrt{Z^2 - R^2}}{R}\right) = \tan^{-1}\left(\frac{\sqrt{(80.0)^2 - (22.0)^2}}{22.0}\right) = 74.0°$$

The alternate expression for the average power delivered to the circuit gives

$$P_{\text{av}} = I_{\text{rms}} \Delta V_{\text{rms}} \cos\phi = (2.0\ \text{A})(160\ \text{V})\cos(74.0°) = 88\ \text{W} \qquad \Diamond$$

the same result as obtained by the first approach above.

---

39. The AM band extends from approximately 500 kHz to 1 600 kHz. If a 2.0-μH inductor is used in a tuning circuit for a radio, what are the extremes that a capacitor must reach in order to cover the complete band of frequencies?

Solution

A radio is tuned to a particular station when the resonance frequency of the tuning circuit in the radio matches the broadcast frequency of the desired station. Commonly, the tuning circuit is a series *RLC* circuit in which the antenna serves as the power source, and the capacitance of the capacitor is varied to adjust the resonance frequency. Resonance occurs in the circuit when $X_L = X_C$, or $2\pi f L = 1/2\pi f C$. Thus, the resonance frequency is given by

$$f_0 = \frac{1}{2\pi\sqrt{LC}}$$

To have $f_0 = f_{\text{min}} = 500 \text{ kHz} = 5.00 \times 10^5 \text{ Hz}$, the required capacitance in the tuning circuit is

$$C = C_{\text{max}} = \frac{1}{4\pi^2 L\left(f_{0,\text{min}}\right)^2} = \frac{1}{4\pi^2\left(2.0\times10^{-6}\text{ H}\right)\left(5.00\times10^5\text{ Hz}\right)^2}$$

or $\quad C = C_{\text{max}} = 5.1\times10^{-8}\text{ F} = 51\text{ nF}$ ◊

To tune in a station at the maximum frequency in the AM radio band $\left(f_{\text{max}} = 1\,600 \text{ kHz} = 1.6\times10^6 \text{ Hz}\right)$, the capacitance in the tuning circuit for this radio must be

$$C = C_{\text{min}} = \frac{1}{4\pi^2 L\left(f_{0,\text{max}}\right)^2} = \frac{1}{4\pi^2\left(2.0\times10^{-6}\text{ H}\right)\left(1.60\times10^6\text{ Hz}\right)^2}$$

or $\quad C = C_{\text{min}} = 4.9\times10^{-9}\text{ F} = 4.9\text{ nF}$ ◊

---

45. An AC power generator produces 50 A (rms) at 3 600 V. The voltage is stepped up to 100 000 V by an ideal transformer, and the energy is transmitted through a long-distance power line that has a resistance of 100 Ω. What percentage of the power delivered by the generator is dissipated as heat in the power line?

Solution

The voltage applied to the primary coil of the transformer by the generator is $\Delta V_{1,\text{rms}} = 3600$ V, and the rms current in the primary coil is $I_{1,\text{rms}} = 50$ A. The rms voltage across the secondary coil of the transformer is $\Delta V_{2,\text{rms}} = 100\,000$ V.

In an ideal transformer, *the power input to the primary equals the power output at the secondary*, or

$$I_{1,\text{rms}}\Delta V_{1,\text{rms}} = I_{2,\text{rms}}\Delta V_{2,\text{rms}}$$

Thus, the current in the circuit connected to the secondary of the transformer must be

$$I_{2,\text{rms}} = \frac{I_{1,\text{rms}}\Delta V_{1,\text{rms}}}{\Delta V_{2,\text{rms}}} = \frac{(50\text{ A})(3\,600\text{ V})}{100\,000\text{ V}} = 1.8\text{ A}$$

and the power loss in the resistance of the long-distance distribution line is

$$P_{\text{loss, line}} = I_{2,\text{rms}}^2 R_{\text{line}} = (1.8\text{ A})^2(100\ \Omega) = 3.2\times10^2\text{ W}$$

The power delivered to the primary of the transformer by the generator is

$$P_{\text{input}} = I_{1,\text{rms}}\Delta V_{1,\text{rms}} = (50\text{ A})(3\,600\text{ V}) = 1.8\times10^5\text{ W}$$

The percentage of the input power that is lost in the distribution line is therefore

$$\% \text{ Loss} = \frac{P_{\text{loss,line}}}{P_{\text{input}}} \times 100\% = \left(\frac{3.2 \times 10^2 \text{ W}}{1.8 \times 10^5 \text{ W}} \right) \times 100\% = 0.18\% \quad \Diamond$$

---

53. Oxygenated hemoglobin absorbs weakly in the red (hence its red color) and strongly in the near infrared, whereas deoxygenated hemoglobin has the opposite absorption. This fact is used in a "pulse oximeter" to measure oxygen saturation in arterial blood. The device clips onto the end of a person's finger and has two light-emitting diodes—a red (660 nm) and an infrared (940 nm)—and a photocell that detects the amount of light transmitted through the finger at each wavelength. (a) Determine the frequency of each of these light sources. (b) If 67% of the energy of the red source is absorbed in the blood, by what factor does the amplitude of the electromagnetic wave change? *Hint:* The intensity of the wave is equal to the average power per unit area as given by Equation 21.28.

Solution

(a) The speed of electromagnetic waves in vacuum is $c = 3.00 \times 10^8$ m/s, so the relation between the wavelength, frequency, and speed of propagation of such waves is $\lambda f = c$, or $f = c/\lambda$.

Red light, with wavelength $\lambda_{\text{red}} = 660 \text{ nm} = 660 \times 10^{-9} \text{ m} = 6.60 \times 10^{-7} \text{ m}$, is an electromagnetic wave having a frequency of

$$f_{\text{red}} = \frac{c}{\lambda_{\text{red}}} = \frac{3.00 \times 10^8 \text{ m}}{6.60 \times 10^{-7} \text{ m}} = 4.55 \times 10^{14} \text{ Hz} \quad \Diamond$$

The infrared radiation is an electromagnetic wave with wavelength $\lambda_{\text{IR}} = 940$ nm $= 9.40 \times 10^{-7}$ m and frequency

$$f_{\text{IR}} = \frac{c}{\lambda_{\text{IR}}} = \frac{3.00 \times 10^8 \text{ m}}{9.40 \times 10^{-7} \text{ m}} = 3.19 \times 10^{14} \text{ Hz} \quad \Diamond$$

(b) If 67% of the incident intensity, I_i, of the electromagnetic wave is absorbed in passing through the blood, the transmitted intensity is $I_t = I_i - 0.67I_i = 0.33I_i$.

The intensity of an electromagnetic wave is proportional to the square of the amplitude of either the electric or magnetic component of the wave, $I = (1/2\mu_o c)E_{\text{max}}^2 = (c/2\mu_o)B_{\text{max}}^2$, or $E_{\text{max}} = \sqrt{(2\mu_0 c)I}$ and $B_{\text{max}} = \sqrt{(2\mu_0/c)I}$. Thus, if $I_{\text{transmitted}}/I_{\text{incident}} = 0.33$, we have

$$\frac{E_{\text{max},t}}{E_{\text{max},i}} = \frac{B_{\text{max},t}}{B_{\text{max},i}} = \sqrt{\frac{I_{\text{transmitted}}}{I_{\text{incident}}}} = \sqrt{0.33} = 0.57 \quad \Diamond$$

That is, the transmitted amplitude of either the electric or magnetic component of the wave is 57% of the incident amplitude of that component. ◊

---

57. A microwave oven is powered by an electron tube called a magnetron that generates electromagnetic waves of frequency 2.45 GHz. The microwaves enter the oven and are reflected by the walls. The standing-wave pattern produced in the oven can cook food unevenly, with hot spots in the food at antinodes and cool spots at nodes, so a turntable is often used to rotate the food and distribute the energy. If a microwave oven is used with a cooking dish in a fixed position, the antinodes can appear as burn marks on foods such as carrot strips or cheese. The separation distance between the burns is measured to be 6.00 cm. Calculate the speed of the microwaves from these data.

Solution

In a standing-wave pattern, the distance between successive nodes or between successive antinodes is

$$d_{\text{N-N}} = d_{\text{A-A}} = \frac{\lambda}{2}$$

where λ is the wavelength of the radiation producing the standing-wave pattern. Since the burn marks on the food coincide with the locations of antinodes in the standing-wave pattern formed by microwaves inside the oven, and the space between these marks is found to be $d_{\text{A-A}} = 6.00$ cm, we determine the wavelength of the microwaves to be

$$\lambda = 2d_{\text{A-A}} = 2(6.00\text{ cm}) = 12.0\text{ cm} = 0.120\text{ m}$$

The relation between the frequency, wavelength, and speed of propagation of a wave is $\lambda f = v$. Using the wavelength computed from the data on the burn mark locations and the known frequency of the microwave radiation ($f = 2.45$ GHz), the calculated speed for the microwaves is

$$v = \lambda f = (0.120\text{ m})\left(2.45\times10^{9}\text{ Hz}\right) = 2.94\times10^{8}\text{ m/s}$$ ◊

In air, the speed of propagation for microwaves, and all other forms of electromagnetic radiation, is very nearly the same as their speed in a vacuum, $c = 3.00\times10^{8}$ m/s. The percentage error in our calculated value for the speed of the microwaves is then

$$\%\text{ error} = \left(\frac{c-v}{c}\right)\times100\% = \left(\frac{3.00\times10^{8} - 2.94\times10^{8}}{3.00\times10^{8}}\right)\times100\% = 2.00\%$$

71. As a way of determining the inductance of a coil used in a research project, a student first connects the coil to a 12.0-V battery and measures a current of 0.630 A. The student then connects the coil to a 24.0-V (rms), 60.0-Hz generator and measures an rms current of 0.570 A. What is the inductance?

Solution

When a DC voltage source (such as a battery) is connected to the coil, the inductive reactance of the coil, $X_L = 2\pi f L$, is zero because the constant current in the coil does not generate an induced emf. Under these conditions, only the resistance of its windings is effective in limiting the current through the coil. Thus, the resistance of this coil must be

$$R = \frac{\Delta V_{\text{DC}}}{I_{\text{DC}}} = \frac{12.0 \text{ V}}{0.630 \text{ A}} = 19.0 \ \Omega$$

When an AC voltage is applied to the coil, both its inductive reactance and its resistance play a role in limiting the current. The impedance of the coil to the applied AC voltage is

$$Z = \frac{\Delta V_{\text{rms}}}{I_{\text{rms}}} = \frac{24.0 \text{ V}}{0.570 \text{ A}} = 42.1 \ \Omega$$

Since, with zero capacitive reactance present, the impedance is given by $Z = \sqrt{R^2 + X_L^2}$, the inductive reactance of this coil at a frequency of $f = 60.0$ Hz is

$$X_L = \sqrt{Z^2 - R^2} = \sqrt{(42.1 \ \Omega)^2 - (19.0 \ \Omega)^2} = 37.6 \ \Omega$$

The self-inductance of the coil is then given by

$$L = \frac{X_L}{2\pi f} = \frac{37.6 \ \Omega}{2\pi (60.0 \text{ Hz})} = 9.97 \times 10^{-2} \text{ H} = 99.7 \text{ mH}$$ ◊

76. The U.S. Food and Drug Administration limits the radiation leakage of microwave ovens to no more than $5.0\ \text{mW/cm}^2$ at a distance of 2.0 in. A typical cell phone, which also transmits microwaves, has a peak output power of about 2.0 W. (a) Approximating the cell phone as a point source, calculate the radiation intensity of a cell phone at a distance of 2.0 in. How does the answer compare with the maximum allowable microwave oven leakage? (b) The distance from your ear to your brain is about 2 in. What would the radiation intensity in your brain be if you used a Bluetooth headset, keeping the phone in your pocket, 1.0 m away from your brain? Most headsets are so-called Class 2 devices with a maximum output power of 2.5 mW.

Solution

(a) The intensity of the radiation at distance r from a point source which is radiating total power P is $I = P/A = P/4\pi r^2$. Therefore, if we approximate the cell phone as a point source, the intensity of the radiation at distance $r = 2.0$ in when it is radiating 2.0 W of power is

$$I = \frac{P}{4\pi r^2} = \frac{2.0\ \text{W}}{4\pi\left[(2.0\ \text{in})(2.54\ \text{cm/1 in})\right]^2}$$

$$= 6.2\times 10^{-3}\ \text{W/cm}^2 = 6.2\ \text{mW/cm}^2 \qquad \lozenge$$

Comparing this result to the FDA maximum allowable microwave oven leakage at a distance of 2.0 inches, we see it exceeds that standard by

$$\%\ \text{excess} = \left(\frac{I - I_{\text{max}}}{I_{\text{max}}}\right)\times 100\%$$

$$= \left(\frac{6.2\ \text{mW/cm}^2 - 5.0\ \text{mW/cm}^2}{5.0\ \text{mW/cm}^2}\right)\times 100\% = 24\% \qquad \lozenge$$

(b) If the cell phone is kept in the pocket at a distance of 1.0 m from the brain, and a Bluetooth headset (emitting 2.5 mW of power) is used at a distance of 2.0 inches from the brain, the total intensity of microwave radiation at the brain from the two devices would be

$$I_{\text{total}} = I_{\text{phone}} + I_{\text{headset}} = \frac{P_{\text{phone}}}{4\pi r_{\text{phone}}^2} + \frac{P_{\text{headset}}}{4\pi r_{\text{headset}}^2}$$

or $$I_{\text{total}} = \frac{2.0\times 10^3\ \text{mW}}{4\pi\left(1.0\times 10^2\ \text{cm}\right)^2} + \frac{2.5\ \text{mW}}{4\pi\left[2.0\ \text{in}(2.54\ \text{cm/1 in})\right]^2} = 2.4\times 10^{-2}\ \text{mW/cm}^2 \qquad \lozenge$$

This radiation level of $0.024\ \text{mW/cm}^2$ is well below the FDA maximum for leakage from a microwave oven.

22

Reflection and Refraction of Light

NOTES FROM SELECTED CHAPTER SECTIONS

22.1 The Nature of Light

Thomas Young showed that light exhibits interference behavior, Maxwell predicted that light is a form of high-frequency electromagnetic waves, and Einstein explained the photoelectric effect on the assumption that light is composed of "corpuscles" or discontinuous quanta of energy called photons.

In view of these developments we conclude that:

> Light must be regarded as having a dual nature; in some cases light acts as a wave, and in others it acts as a particle.
>
> Light travels in a straight line in a homogeneous medium until it strikes a boundary between two different materials.
>
> Light beams can be represented by a technique called the ray approximation in which a ray of light is shown as a line drawn along the direction of travel of the beam.

22.2 Reflection and Refraction
22.3 The Law of Refraction

Specular reflection occurs when light is reflected from a smooth surface; **diffuse reflection** occurs at rough surfaces. In this and following chapters reflection will be considered to be specular reflection.

When light is refracted:

> The path of a light ray through a refracting surface is reversible.
>
> The index of refraction is characteristic of a particular material.
>
> The frequency does not change as light travels from one medium into another.

A line drawn perpendicular to a surface at the point where an incident ray strikes the surface is called the normal line. *Angles of incidence, reflection, and refraction are measured relative to the normal.* The incident, reflected, and refracted rays, and the normal line all lie in the same plane, and $\theta_1 = \theta_1'$.

For a given angle of incidence, the angle of refraction depends on the optical properties of the media above and below the boundary. The figure on the right illustrates a situation in which the speed of light is greater in the medium above the boundary than in the medium below the boundary. In such cases the angle of refraction, as shown, will be smaller than the angle of incidence.

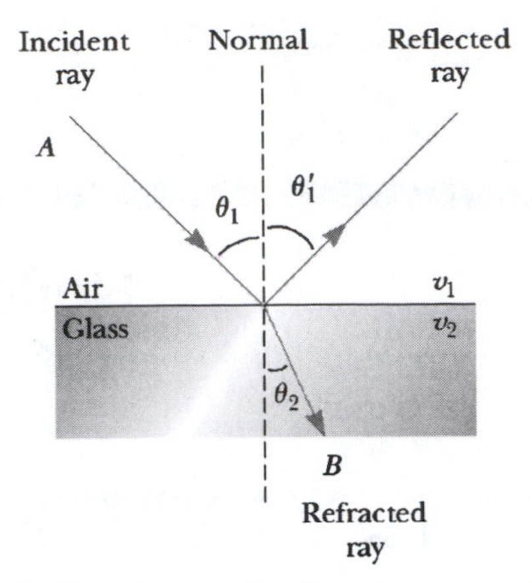

Reflection and refraction at a plane surface

θ_1 = angle of incidence

θ_1' = angle of reflection

θ_2 = angle of refraction

22.4 Dispersion and Prisms

An important property of the index of refraction (n) is that its value in a refractive material depends on the wavelength. This phenomenon is called dispersion. Since n is a function of the wavelength, when a light beam is incident on the surface of a refracting material, different wavelengths are bent at different angles. *The index of refraction of a given material decreases with increasing wavelength.* This means that blue light bends more than red light, when passing into a refracting material. In a prism, dispersion occurs at one surface as light enters the prism and again at a second surface as it leaves.

22.6 Huygens's Principle

A wave front is a surface which passes through those points in a wave that have both the same phase and amplitude. Huygens's technique for determining the successive positions of a wave front is based on the following geometric construction: All points on a wave front can be considered as point sources for the production of spherical secondary waves (wavelets). The wavelets propagate forward with speeds characteristic of waves in the particular medium. At any later time, the new position of the wave front is found by constructing a surface tangent to the set of wavelets.

22.7 Total Internal Reflection

When light is incident on a boundary between two media, some of the light is always reflected; the remaining light is refracted into the second medium. Total internal reflection is possible only when the light is initially traveling in the medium of greater index of refraction ($n_1 > n_2$). As the angle of incidence increases, the intensity of the reflected beam increases and that of the refracted beam decreases. If $n_1 > n_2$ there exists an angle of incidence called the critical angle, θ_c, beyond which no refraction occurs. When $\theta_1 \geq \theta_c$, the incident light is totally reflected back into the medium of greater index of refraction, and the intensity of the refracted beam is zero.

EQUATIONS AND CONCEPTS

The law of reflection states that the angle of reflection (the angle measured between the reflected ray and the normal) equals the angle of incidence (the angle between the incident ray and the normal). *The incident ray, reflected ray, and normal line are in the same plane.*

$$\theta_1' = \theta_1 \tag{22.2}$$

Normal
Incident ray
Reflected ray
θ_1 θ_1'

The **index of refraction** of a transparent medium equals the ratio of the speed of light in vacuum to the speed of light in the medium. The index of refraction of a given medium can be expressed as the ratio of the wavelength of light in vacuum to the wavelength in that medium. *The frequency of a wave is characteristic of the source; as light travels from one medium into another of different index of refraction, the frequency remains constant, but the wavelength changes.* In the figure on the right λ_1 is greater than λ_2; therefore v_1 will be greater than v_2 because the relation $f = (v/\lambda)$ must be the same in both media.

$$n \equiv \frac{\text{speed of light in vacuum}}{\text{speed of light in medium}} = \frac{c}{v} \tag{22.4}$$

$$n = \frac{\lambda_0}{\lambda_n} \tag{22.7}$$

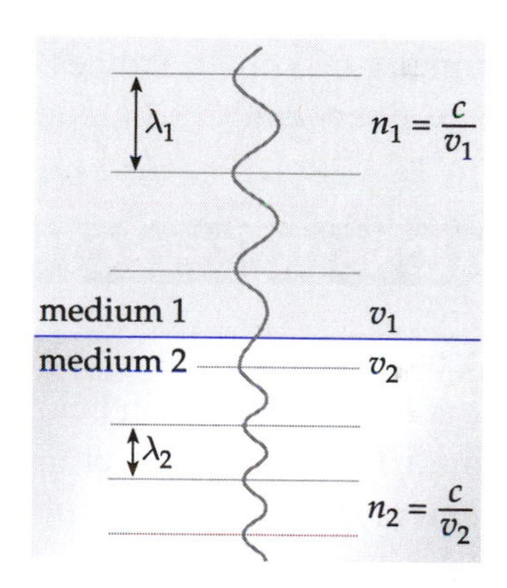

Snell's law of refraction can be expressed in terms of the speeds of light in the media on either side of the refracting surface or in terms of the indices of refraction of the two media. *As illustrated in the figure, the angles θ_1 and θ_2 are measured from the normal line to the respective ray*

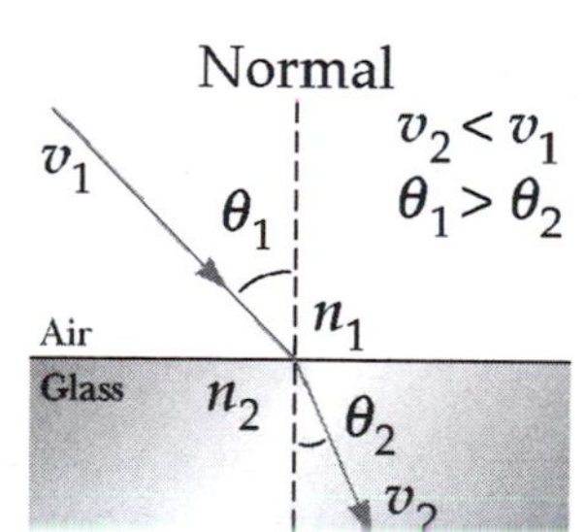

$$\frac{\sin\theta_2}{\sin\theta_1} = \frac{v_2}{v_1} = \text{constant} \tag{22.3}$$

$$n_1 \sin\theta_1 = n_2 \sin\theta_2 \tag{22.8}$$

The **critical angle** is the minimum angle of incidence for which total internal reflection can occur. *Total internal reflection is possible only when a light ray is directed from a medium of high index of refraction into a medium of lower index of refraction.*

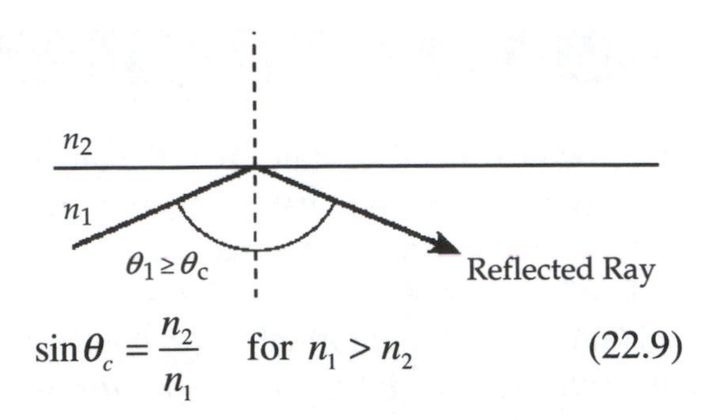

$$\sin\theta_c = \frac{n_2}{n_1} \quad \text{for } n_1 > n_2 \qquad (22.9)$$

REVIEW CHECKLIST

- When refraction occurs at a plane surface, use Snell's law to calculate the index of refraction of either medium, the angle of refraction, or the angle of incidence when the other variables in the equation are known. (Section 22.3)
- Describe the process of dispersion of a beam of white light as it passes through a prism. (Sections 22.4 and 22.5)
- Describe the conditions under which total internal reflection of a light ray is possible, and calculate the critical angle for internal reflection at a boundary between two optical media of known indices of refraction. Describe the application of internal reflection to fiber optics techniques. (Section 22.7)

SOLUTIONS TO SELECTED END-OF-CHAPTER PROBLEMS

4. (a) Calculate the wavelength of light in vacuum that has a frequency of 5.45×10^{14} Hz. (b) What is its wavelength in benzene? (c) Calculate the energy of one photon of such light in vacuum. Express the answer in electron volts. (d) Does the energy of the photon change when it enters the benzene? Explain.

Solution

(a) In all types of waves, the wavelength, frequency, and the speed of propagation are related by the expression $\lambda f = v$. For light waves, and all other types of electromagnetic waves, the speed of propagation in vacuum is given by $v = c = 3.00\times10^8$ m/s. Thus, the wavelength in vacuum, λ_0, of light that has a frequency of $f = 5.45\times10^{14}$ Hz is

$$\lambda_0 = \frac{c}{f} = \frac{3.00\times10^8 \text{ m/s}}{5.45\times10^{14} \text{ Hz}} = 5.50\times10^{-7} \text{ m} = 550\times10^{-9} \text{ m} = 550 \text{ nm} \qquad \lozenge$$

(b) When a wave passes from one medium into another, the speed of propagation and the wavelength generally undergo a change. However, the frequency of the wave remains fixed at the original value. The speed of propagation of light in a medium other than vacuum is given by $v = c/n$, where n is the index of refraction of the material making up that medium. The wavelength of light having frequency f in a material with refractive index n is therefore

$$\lambda = \frac{v}{f} = \frac{c/n}{f} = \frac{c/f}{n} = \frac{\lambda_0}{n}$$

From Table 22.1, the index of refraction of benzene is $n = 1.501$, so the wavelength of the given light wave in benzene is

$$\lambda = \frac{550 \text{ nm}}{1.501} = 366 \text{ nm}$$ ◊

(c) The energy of a photon of light having frequency f is given by $E = hf$, where $h = 6.63 \times 10^{-34}$ J·s is known as Planck's constant. Thus, the energy of a photon of the given light is

$$E = (6.63 \times 10^{-34} \text{ J}\cdot\text{s})(5.45 \times 10^{14} \text{ Hz}) = (3.61 \times 10^{-19} \text{ J})\left(\frac{1 \text{ eV}}{1.60 \times 10^{-19} \text{ J}}\right)$$

$$= 2.26 \text{ eV}$$ ◊

(d) Since the energy of a photon is directly proportional to the frequency of the light, and the frequency is unchanged as the light enters the benzene, the energy of the photon is also unchanged. ◊

---

13. A ray of light is incident on the surface of a block of clear ice at an angle of 40.0° with the normal. Part of the light is reflected, and part is refracted. Find the angle between the reflected and refracted light.

Solution

When light reflects from the boundary between different media, the angle of reflection always equals the angle of incidence. That is, the angle the reflected ray makes with the normal to the boundary is the same and the angle the incident ray made with this normal line. In this case, the angle of incidence as the light approaches the surface of the ice is given as $\theta_1 = 40.0°$. Thus, the angle of reflection at this surface is $\phi = 40.0°$.

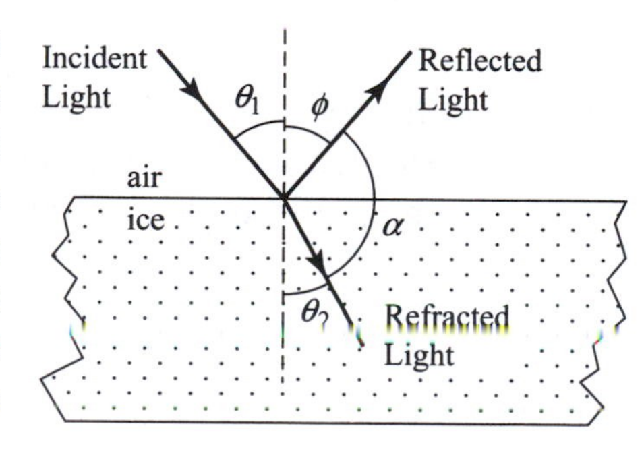

Snell's law gives the relation between the angle of incidence and the angle of refraction as light crosses the boundary between materials having indices of refraction n_1 and n_2 as $n_1 \sin\theta_1 = n_2 \sin\theta_2$. Table 22.1 in the textbook gives the refractive indices of air and ice as

$n_{air} = 1.000$ and $n_{ice} = 1.309$. Therefore, if the angle the incident ray makes with the normal line in air is $\theta_1 = 40.0°$, the angle between the refracted ray and the normal in the ice is given by $n_{ice} \sin\theta_2 = n_{air} \sin\theta_1$, or

$$\theta_2 = \sin^{-1}\left(\frac{n_{air} \sin\theta_1}{n_{ice}}\right) = \sin^{-1}\left(\frac{(1.000)\sin(40.0°)}{1.309}\right) = 29.4°$$

Observe from the sketch given above that the angle of reflection, the angle of refraction, and the angle between the reflected and the refracted rays add to 180°. Thus, the angle between the reflected and the refracted rays is given by

$$\alpha = 180.0° - \phi - \theta_2 = 180.0° - 40.0° - 29.4° = 110.6°$$ ◊

---

18. A ray of light strikes a flat, 2.00-cm-thick block of glass ($n = 1.50$) at an angle of 30.0° with respect to the normal (Fig. P22.18). (a) Find the angle of refraction at the top surface. (b) Find the angle of incidence at the bottom surface and the refracted angle. (c) Find the lateral distance d by which the light beam is shifted. (d) Calculate the speed of light in the glass and (e) the time required for the light to pass through the glass block. (f) Is the travel time through the block affected by the angle of incidence? Explain.

Solution

(a) When the angle of incidence is $\theta_1 = 30.0°$, the angle of refraction, θ_2, at the top surface of the glass block is given by Snell's law as

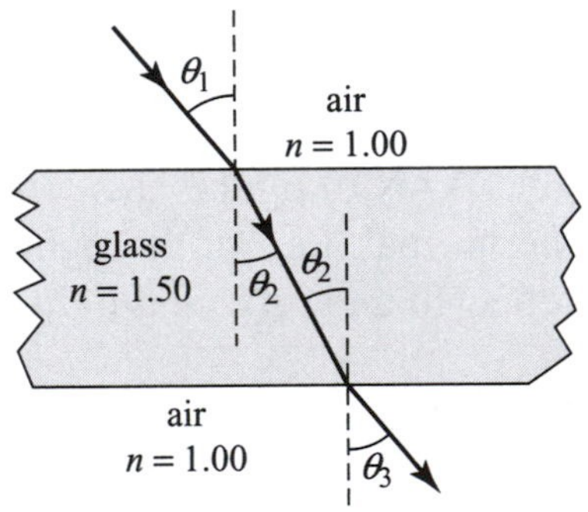

$$\theta_2 = \sin^{-1}\left(\frac{n_{air} \sin\theta_1}{n_{glass}}\right) = \sin^{-1}\left(\frac{(1.00)\sin(30.0°)}{1.50}\right)$$

or $\theta_2 = 19.5°$ ◊

(b) Since the top and bottom surfaces of the glass block are parallel, the normal lines to these surfaces are also parallel. Hence, the angle of incidence at the bottom surface is the same as the angle of refraction at the top surface. ◊

Applying Snell's law to the refraction at the bottom surface then gives the angle of refraction at this surface as

$$\theta_3 = \sin^{-1}\left(\frac{n_{glass} \sin\theta_2}{n_{air}}\right) = \sin^{-1}\left(\frac{(1.50)\sin(19.5°)}{1.00}\right) = 30.0°$$ ◊

Note that this means the ray emerging from the glass is parallel to the incident ray. This will always be true when the surfaces of the intermediate medium are parallel and the same material (air in this case) exists above and below this layer of intermediate material.

(c) The sketch at the right shows the lateral displacement d of the light ray as it passes through the glass block. This displacement may be expressed as $d = \overline{AC}\sin\alpha$, where the distance $\overline{AC}$ is the hypotenuse of triangle ABC, and $\alpha = \theta_1 - \theta_2 = 30.0° - 19.5° = 10.5°$.

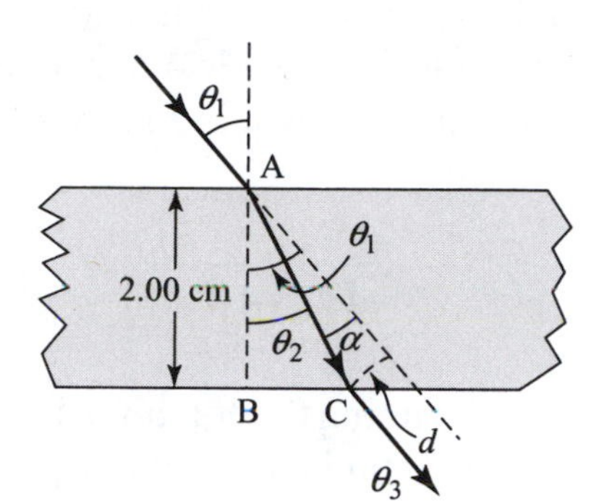

From triangle ABC, we find

$$\overline{AC} = \frac{2.00 \text{ cm}}{\cos\theta_2} = \frac{2.00 \text{ cm}}{\cos(19.5°)} = 2.12 \text{ cm}$$

Thus, $d = \overline{AC}\sin\alpha = (2.12 \text{ cm})\sin(10.5°) = 0.386 \text{ cm}$. ◊

(d) From the definition of index of refraction, the speed of light in this glass is

$$v = \frac{c}{n_{\text{glass}}} = \frac{3.00\times10^8 \text{ m/s}}{1.50} = 2.00\times10^8 \text{ m/s}$$ ◊

(e) The time required for the light to pass through the glass block is given by

$$t = \frac{\overline{AC}}{v} = \frac{2.12 \text{ cm}}{2.00\times10^8 \text{ m/s}}\left(\frac{1 \text{ m}}{10^2 \text{ cm}}\right) = 1.06\times10^{-10} \text{ s}$$ ◊

(f) The distance $\overline{AC}$ the light travels in the glass block depends on the value of the angle of refraction θ_2, which is a function of the angle of incidence θ_1. Hence, changing the angle of incidence will affect the distance $\overline{AC}$ and also the required travel time $t = \overline{AC}/v$. ◊

23. A person looking into an empty container is able to see the far edge of the container's bottom, as shown in Figure P22.23a. The height of the container is h, and its width is d. When the container is completely filled with a fluid of index of refraction n and viewed from the same angle, the person can see the center of a coin at the middle of the container's bottom, as shown in Figure P22.23b. (a) Show that the ratio h/d is given by

$$\frac{h}{d}=\sqrt{\frac{n^2-1}{4-n^2}}$$

(b) Assuming the container has a width of 8.00 cm and is filled with water, use the expression above to find the height of the container.

Solution

(a) Before the container is filled, the ray's path is as shown in Figure (a) at the right. From this figure, observe that

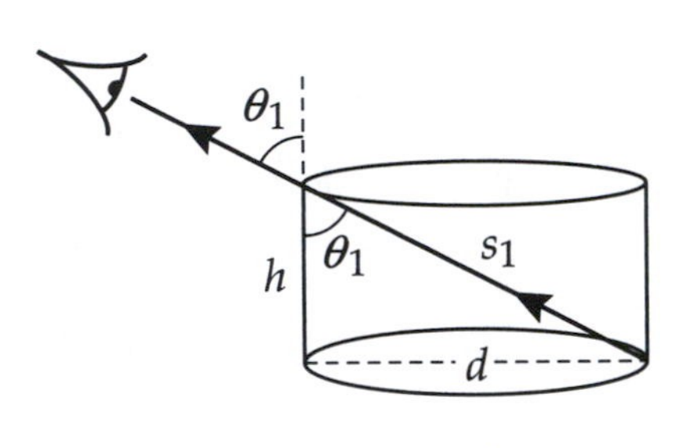

(a)

$$\sin\theta_1=\frac{d}{s_1}=\frac{d}{\sqrt{h^2+d^2}}=\frac{1}{\sqrt{(h/d)^2+1}}$$

After the container is filled, the ray's path is shown in Figure (b). From this figure, we see

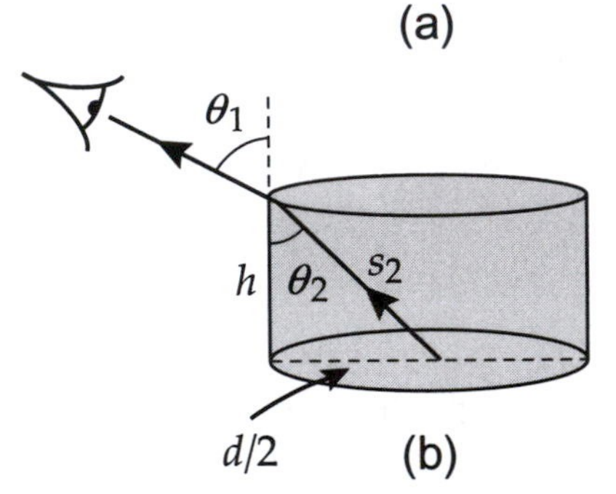

(b)

$$\sin\theta_2=\frac{d/2}{s_2}=\frac{d/2}{\sqrt{h^2+(d/2)^2}}=\frac{1}{\sqrt{4(h/d)^2+1}}$$

From Snell's law, $n_{\text{air}}\sin\theta_1=n\sin\theta_2$, or

$$\frac{1.00}{\sqrt{(h/d)^2+1}}=\frac{n}{\sqrt{4(h/d)^2+1}}$$

and $4(h/d)^2+1=n^2(h/d)^2+n^2$. Simplifying, this gives $(4-n^2)(h/d)^2=n^2-1$, or

$$\frac{h}{d}=\sqrt{\frac{n^2-1}{4-n^2}}$$ ◊

(b) If $d=8.00$ cm and $n=n_{\text{water}}=1.333$, then

$$h=(8.00\text{ cm})\sqrt{\frac{(1.333)^2-1}{4-(1.333)^2}}=4.73\text{ cm}$$ ◊

29. The index of refraction for red light in water is 1.331 and that for blue light is 1.340. If a ray of white light enters the water at an angle of incidence of 83.00°, what are the underwater angles of refraction for the (a) blue and (b) red components of the light?

Solution

As light travels from one medium having index of refraction n_1 into a second medium with index of refraction n_2, Snell's law gives the relation among the indices of refraction, the angle of incidence, and the angle of refraction. This law states that

$$n_1 \sin\theta_1 = n_2 \sin\theta_2$$

and the angle of refraction is found to be

$$\theta_2 = \sin^{-1}\left(\frac{n_1 \sin\theta_1}{n_2}\right)$$

In this case, the angle of incidence is $\theta_1 = 83.00°$, and the index of refraction of the first medium is $n_1 = 1.000$, for both the red and blue light.

(a) With an index of refraction of $n_{2,\text{blue}} = 1.340$ for the blue light in water, the underwater angle of refraction for the blue component of this light is

$$\theta_{2,\text{blue}} = \sin^{-1}\left(\frac{n_1 \sin\theta_1}{n_{2,\text{blue}}}\right) = \sin^{-1}\left(\frac{(1.000)\sin(83.00°)}{1.340}\right) = 47.79° \quad \Diamond$$

(b) The water has an index of refraction of $n_{2,\text{red}} = 1.331$ for red light, giving an underwater angle of refraction of

$$\theta_{2,\text{red}} = \sin^{-1}\left(\frac{n_1 \sin\theta_1}{n_{2,\text{red}}}\right) = \sin^{-1}\left(\frac{(1.000)\sin(83.00°)}{1.331}\right) = 48.22° \quad \Diamond$$

32. The index of refraction for violet light in silica flint glass is 1.66 and that for red light is 1.62. What is the angular dispersion of visible light passing through an equilateral prism of apex angle 60.0° if the angle of incidence is 50.0°? (See Fig. P22.32.)

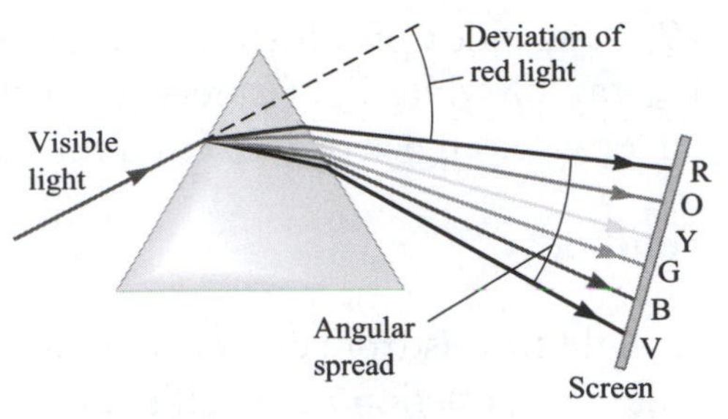

Figure P22.32

Solution

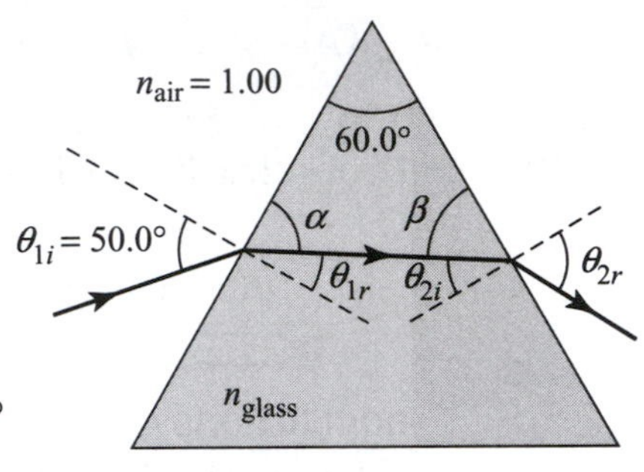

Consider the sketch at the right showing the path of a light ray through the prism, with angles of incidence and refraction of θ_{1i} and θ_{1r} at the first surface. At the second surface, these angles are θ_{2i} and θ_{2r}. Considering the geometry, we find the relation between θ_{2i} and θ_{1r} as follows:

$$\theta_{2i} = 90.0° - \beta = 90.0° - (180.0° - 60.0° - \alpha) = \alpha - 30.0°$$

But $\alpha = 90.0° - \theta_{1r}$, so $\theta_{2i} = 60.0° - \theta_{1r}$.

With $n_{air} = 1.00$ and $\theta_{1i} = 50.0°$, applying Snell's law to the refraction at the first surface gives the angle of refraction at that surface as

$$\theta_{1r} = \sin^{-1}\left(\frac{n_{air}\sin\theta_{1i}}{n_{glass}}\right) = \sin^{-1}\left[\frac{\sin(50.0°)}{n_{glass}}\right] \quad \textbf{[1]}$$

The angle of refraction at the second surface is found from Snell's law as

$$\theta_{2r} = \sin^{-1}\left(\frac{n_{glass}\sin\theta_{2i}}{n_{air}}\right) = \sin^{-1}\left[n_{glass}\sin(60.0° - \theta_{1r})\right] \quad \textbf{[2]}$$

Using Equations [1] and [2] with $n_{glass} = 1.66$ for violet light gives

$$(\theta_{1r})_{violet} = \sin^{-1}\left[\frac{\sin(50.0°)}{1.66}\right] = 27.5° \text{ and } (\theta_{2r})_{violet} = \sin^{-1}[1.66\sin(60.0° - 27.5°)] = 63.1°$$

With $n_{glass} = 1.62$ for red light, Equations [1] and [2] yield

$$(\theta_{1r})_{red} = \sin^{-1}\left[\frac{\sin(50.0°)}{1.62}\right] = 28.2° \text{ and } (\theta_{2r})_{red} = \sin^{-1}[1.62\sin(60.0° - 28.2°)] = 58.6°$$

The dispersion produced by the prism is then

$$Dispersion = (\theta_{2r})_{violet} - (\theta_{2r})_{red} = 63.1° - 58.6° = 4.5°$$

◊

40. A beam of laser light with wavelength 612 nm is directed through a slab of glass having index of refraction 1.78. (a) For what minimum incident angle would a ray of light undergo total internal reflection? (b) If a layer of water is placed over the glass, what is the minimum angle of incidence on the glass-water interface that will result in total internal reflection at the water-air interface? (c) Does the thickness of the water layer or glass affect the result? (d) Does the index of refraction of the intervening layer affect the result?

Solution

(a) The minimum angle of incidence for which total internal reflection occurs is the critical angle. At the critical angle, the angle of refraction is 90.0° as shown in the figure at the right. From Snell's law, $n_g \sin\theta_i = n_a \sin 90.0°$, the critical angle for the glass-air interface is found to be

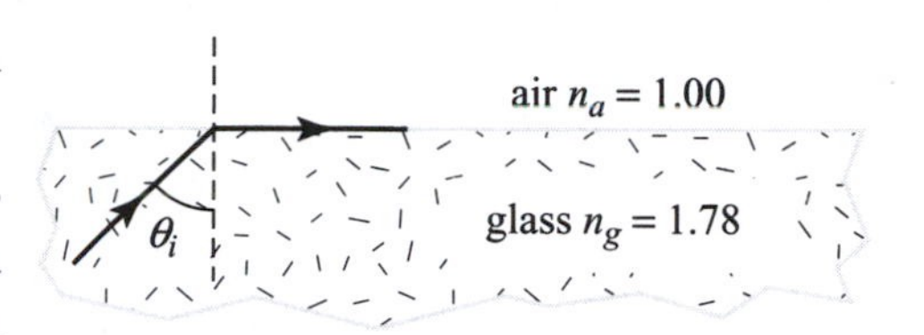

$$\theta_i = \theta_c = \sin^{-1}\left(\frac{n_a \sin 90.0°}{n_g}\right) = \sin^{-1}\left(\frac{1.00}{1.78}\right) = \boxed{34.2°}$$ ◊

(b) When the slab of glass has a layer of water on top, we want the angle of incidence at the water-air interface to equal the critical angle for that combination of media. At this angle, Snell's law gives

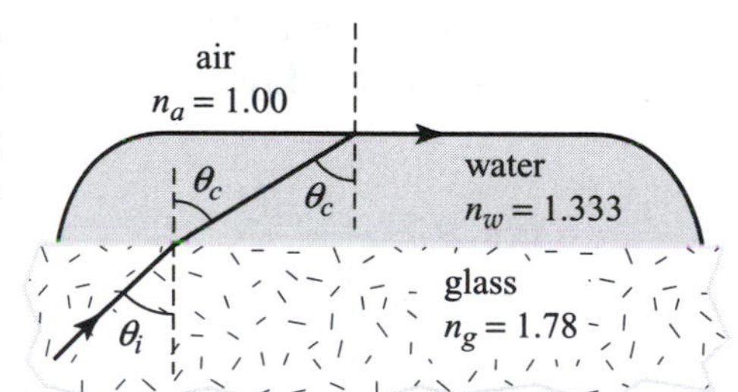

$$n_w \sin\theta_c = n_a \sin 90° = 1.00$$

and $\sin\theta_c = 1.00/n_w$

Now, considering the refraction at the glass-water interface, Snell's law gives $n_g \sin\theta_i = n_w \sin\theta_c$. Combining this with the result for $\sin\theta_c$ from above, we find the required angle of incidence in the glass to be

$$\theta_i = \sin^{-1}\left(\frac{n_w \sin\theta_c}{n_g}\right) = \sin^{-1}\left(\frac{n_w(1.00/n_w)}{n_g}\right) = \sin^{-1}\left(\frac{1.00}{n_g}\right) = \sin^{-1}\left(\frac{1.00}{1.78}\right) = \boxed{34.2°}$$ ◊

(c) Notice that the thickness of the water layer was not needed in the calculations of the angles of incidence in Part (b) above. Thus, that property of the water layer does not affect the result. While the thickness of the water layer has a role in determining the point where the light ray strikes the air-water interface, it does not affect the required angle of incidence. ◊

(d) In the calculation of part (b), observe that the index of refraction of the water canceled. Thus, the refractive index of the intervening layer does not affect the required angle of incidence at the surface of the glass. This will always be true when the upper and lower surfaces of the intervening layer are parallel to each other. ◊

43. The light beam in Figure P22.43 strikes surface 2 at the critical angle. Determine the angle of incidence, θ_1.

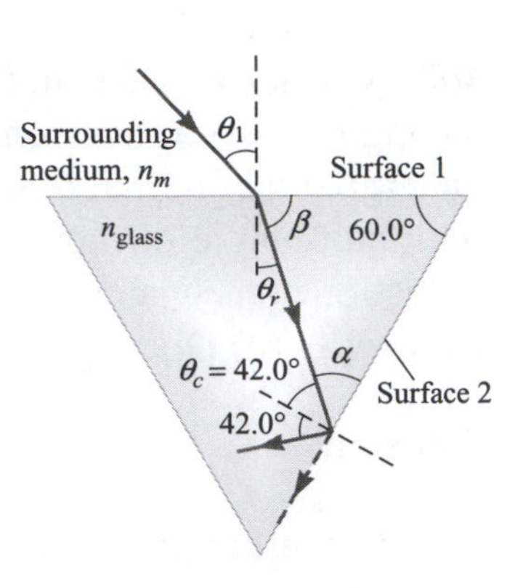

Figure P22.43 (modified)

Solution

As light attempts to go from a medium having refractive index n_1 into a medium with refractive index n_2, where $n_1 > n_2$, total internal reflection occurs if the angle of incidence is equal to or greater than a critical angle, θ_c, given by

$$\sin\theta_c = \frac{n_2}{n_1}$$

Where the light strikes surface 2 in Figure P22.43, the angle of incidence equals the critical angle and the indices of refraction are $n_1 = n_{\text{glass}}$, $n_2 = n_m$. Thus, we have that

$$\sin 42.0° = \frac{n_m}{n_{\text{glass}}}$$

At the upper surface of the prism, where the light is coming from the surrounding medium into the glass, Snell's law gives $n_m \sin\theta_1 = n_{\text{glass}} \sin\theta_r$, or

$$\sin\theta_1 = \frac{\sin\theta_r}{n_m / n_{\text{glass}}}$$

Combining this with the previous result gives

$$\sin\theta_1 = \frac{\sin\theta_r}{\sin 42.0°}$$

Observe in Figure P22.43 that angle α is the complement of 42.0°, so $\alpha = 48.0°$. Also, $\alpha + \beta + 60.0° = 180°$ since they are the interior angles in a triangle. This gives

$$\beta = 180° - \alpha - 60.0° = 180° - 48.0° - 60.0° = 72.0°$$

Finally, observe that angle β is the complement of the angle of refraction at surface 1, giving $\theta_r = 90.0° - \beta = 18.0°$. Snell's law then gives the angle of incidence at surface 1 of the prism as

$$\theta_1 = \sin^{-1}\left(\frac{\sin\theta_r}{\sin 42.0°}\right) = \sin^{-1}\left(\frac{\sin 18.0°}{\sin 42.0°}\right) = 27.5° \qquad \lozenge$$

49. As shown in Figure P22.49, a light ray is incident normal to one face of a 30°–60°–90° block of flint glass (a prism) that is immersed in water. (a) Determine the exit angle θ_3 of the ray. (b) A substance is dissolved in the water to increase the index of refraction n_2. At what value of n_2 does total internal reflection cease at point P?

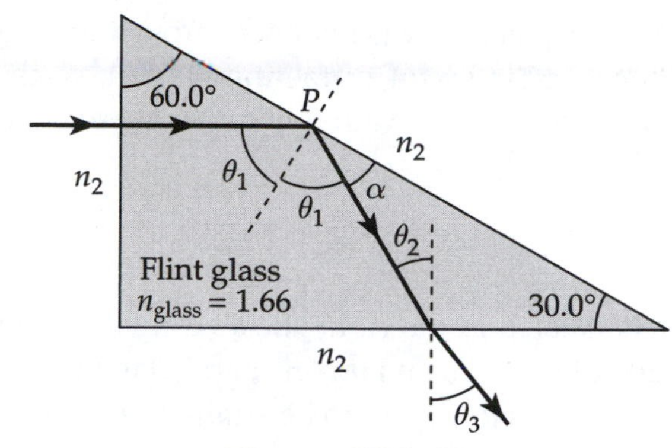

Figure P22.49 (modified)

Solution

(a) The horizontal ray strikes the first surface of the prism at normal incidence and passes with zero deviation into the glass. Notice that, inside the glass, this horizontal ray is parallel to the base of the prism, and hence makes a 30.0° angle with the diagonal face of the prism. This means that the angle of incidence at point P is $\theta_1 = 90.0° - 30.0° = 60.0°$. The angle of reflection at point P is also $\theta_1 = 60.0°$, so $\alpha = 90.0° - \theta_1 = 30.0°$. Since the sum of the interior angles of a triangle is 180.0°, we have $\alpha + (\theta_2 + 90.0°) + 30.0° = 180.0°$, or the angle of incidence at the lower surface of the prism is $\theta_2 = 180° - 120° - \alpha = 30.0°$.

With water as the surrounding medium ($n_2 = 1.333$), and $n_{glass} = 1.66$ for flint glass, Snell's law gives

$$\theta_3 = \sin^{-1}\left(\frac{n_{glass}\sin\theta_2}{n_2}\right) = \sin^{-1}\left(\frac{1.66\sin 30.0°}{1.333}\right) = 38.5° \qquad \Diamond$$

(b) Total internal reflection ceases at point P when the angle of incidence at this point equals the critical angle. That is, when

$$\sin\theta_1 = \sin\theta_c = \frac{n_2}{n_{glass}}$$

or $\quad n_2 = n_{glass}\sin\theta_1 = (1.66)\sin 60.0° = 1.44 \qquad \Diamond$

53. A piece of wire is bent through an angle θ. The bent wire is partially submerged in benzene (index of refraction = 1.50), so that, to a person looking along the dry part, the wire appears to be straight and makes an angle of 30.0° with the horizontal. Determine the value of θ.

Solution

The sketch at the right shows a light ray that leaves the lower end of the wire and travels parallel to the wire in the benzene. If the wire is to appear to be straight to a person looking along the dry part of the wire, the lower end of the wire must appear to be aligned with the dry portion of the wire, or lie along the dashed line in the sketch. Thus, when the shown ray from the lower end reaches the air-benzene interface, it must refract and travel parallel to the dry portion of the wire to enter the observer's eye.

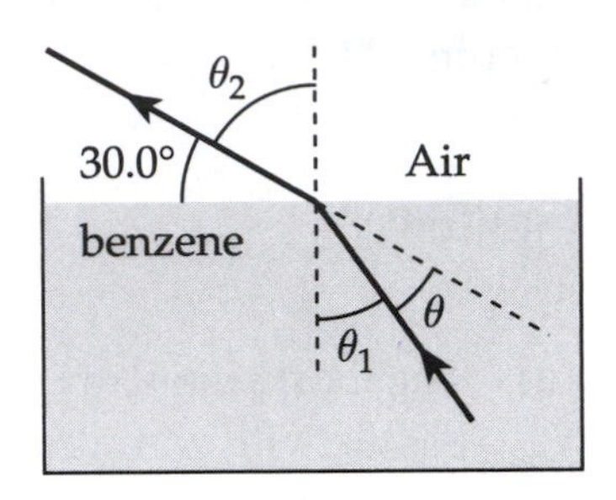

If the dry part of the wire, and hence the refracted ray in the sketch, are directed at 30.0° above the benzene surface, the angle of refraction is $\theta_2 = 90.0° - 30.0° = 60.0°$. Snell's law then gives the angle of incidence at the air-benzene interface as

$$\theta_1 = \sin^{-1}\left[\frac{n_{\text{air}} \sin\theta_2}{n_{\text{benzene}}}\right] = \sin^{-1}\left[\frac{(1.00)\sin 60.0°}{1.50}\right] = 35.3°$$

Since $\theta_2 = \theta_1 + \theta$ (vertical angles), the angle through which the wire is bent (or deviates from the dashed straight line path) is seen to be

$$\theta = \theta_2 - \theta_1 = 60.0° - 35.3° = 24.7°$$ ◊

---

57. A light ray enters a rectangular block of plastic at an angle $\theta_1 = 45.0°$ and emerges at an angle $\theta_2 = 76.0°$, as shown in Figure P22.57. (a) Determine the index of refraction of the plastic. (b) If the light ray enters the plastic at a point $L = 50.0$ cm from the bottom edge, how long does it take the light ray to travel through the plastic?

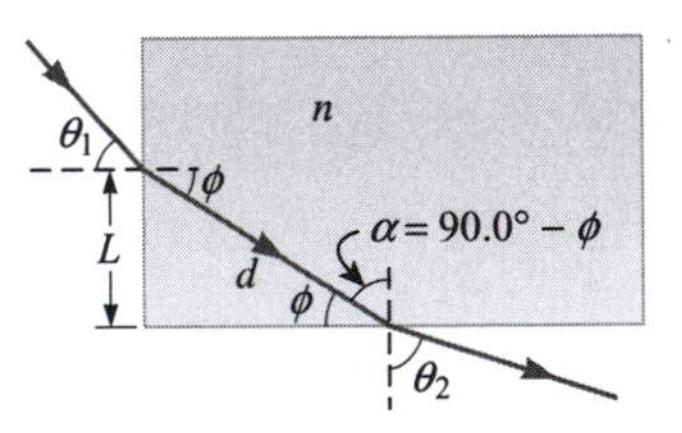

Figure P22.57 (modified)

Solution

(a) Consider the figure given above and apply Snell's law to the refraction at the left end of the plastic block. With the block surrounded by air ($n_{\text{air}} = 1.00$), this yields

$$n \sin\phi = n_{\text{air}} \sin\theta_1 = (1.00)\sin 45.0° \quad [1]$$

Next, we apply Snell's law to the refraction at the bottom of the block, recognizing that $\sin\alpha = \sin(90.0° - \phi) = \cos\phi$, and obtain

$$n\cos\phi = n_{\text{air}}\sin\theta_2 = (1.00)\sin 76.0° \qquad \textbf{[2]}$$

Squaring both Equations [1] and [2] and adding the results, gives

$$n^2\left(\sin^2\phi + \cos^2\phi\right) = \sin^2(45.0°) + \sin^2(76.0°)$$

Using the trigonometric identity $\sin^2\theta + \cos^2\theta = 1$, we find the index of refraction for the plastic to be

$$n = \sqrt{\sin^2(45.0°) + \sin^2(76.0°)} = 1.20 \qquad \Diamond$$

(b) Observe from the above figure that the distance d the light travels inside the plastic is given by $d = L/\sin\phi$. Equation [1] gives $\sin\phi = \sin 45.0°/n$, so $d = nL/\sin 45.0°$, and the time required for the light to travel this distance is $t = d/v = nL/v\sin(45.0°)$, where $v = c/n$ is the speed of light in the plastic. If $L = 50.0$ cm $= 0.500$ m, the time required for the light to travel through the plastic is

$$t = \frac{nL}{(c/n)\sin(45.0°)} = \frac{n^2L}{c\cdot\sin(45.0°)} = \frac{(1.20)^2(0.500\text{ m})}{(3.00\times10^8\text{ m/s})\sin(45.0°)}$$

or $\quad t = 3.39\times10^{-9}\text{ s} = 3.39\text{ ns} \qquad \Diamond$

---

61. A thick piece of Lucite ($n = 1.50$) has the shape of a quarter circle of radius $R = 12.0$ cm as shown in the side view of Figure P22.61. A light ray traveling in air parallel to the base of the Lucite is incident at a distance $h = 6.00$ cm above the base and emerges out of the Lucite at an angle θ with the horizontal. Determine the value of θ.

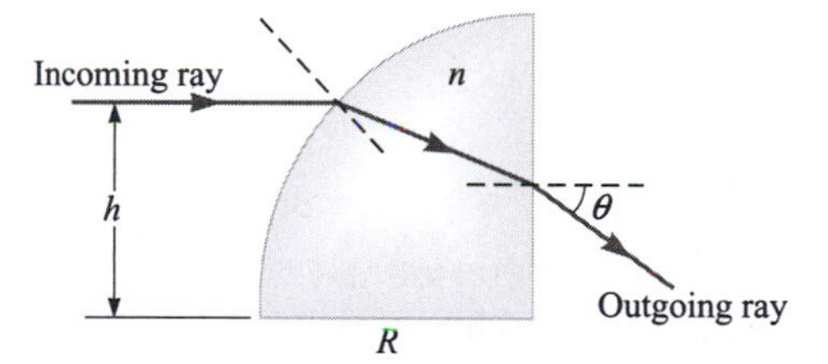

Figure P22.61

Solution

Note that the normal line to the surface at the point where the light enters the Lucite is a radius line of the circular arc. From triangle PAC in the modified version of Figure P22.61 given at the right, observe that the angle of incidence at point P is given by

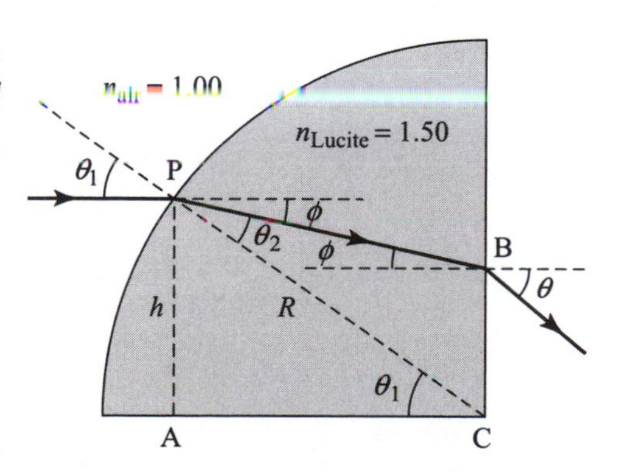

$$\theta_1 = \sin^{-1}\left(\frac{h}{R}\right) = \sin^{-1}\left(\frac{6.00\text{ cm}}{12.0\text{ cm}}\right) = 30.0°$$

Snell's law then gives the angle of refraction at point P as

$$\theta_2 = \sin^{-1}\left(\frac{n_{\text{air}}\sin\theta_1}{n_{\text{Lucite}}}\right) = \sin^{-1}\left(\frac{1.00\sin(30.0°)}{1.50}\right) = 19.5°$$

Looking at point P, we notice that the total angle $(\theta_2 + \phi)$ and angle θ_1 are vertical angles formed by two intersecting straight lines. Thus, $(\theta_2 + \phi) = \theta_1$, and the angle of incidence at point B is given by

$$\phi = \theta_1 - \theta_2 = 30.0° - 19.5° = 10.5°$$

The angle of refraction at point B, where the light emerges from the Lucite, is then given by Snell's law as

$$\theta = \sin^{-1}\left(\frac{n_{\text{Lucite}}\sin\phi}{n_{\text{air}}}\right) = \sin^{-1}\left[\frac{(1.50)\sin(10.5°)}{1.00}\right] = 15.9° \qquad \lozenge$$

23

Mirrors and Lenses

NOTES FROM SELECTED CHAPTER SECTIONS

In equations and diagrams, the **object distance**, p, is the distance from the object to the mirror (or lens). The **image distance**, q, is the distance from the mirror (or lens) to the location of the image.

23.1 Flat Mirrors

Images formed by flat mirrors have the following properties:

- The image is as far behind the mirror as the object is in front.
- The image is unmagnified, virtual, and upright.
- The image has *apparent* left-right reversal.

23.2 Images Formed by Concave Mirrors

A **spherical mirror** is a reflecting surface which has the shape of a segment of a sphere. If the inner surface is reflecting, the mirror is **concave**; if the outer surface of the sphere is reflecting, then the mirror is **convex**.

The **principal axis** of a spherical mirror is along a line drawn from the center (or apex) of the mirror to the **center of curvature** of the mirror. The **focal point** of a concave mirror is on the front side of the mirror at the point along the principal axis at which incoming parallel light rays converge or focus after reflection. For a convex mirror, the focal point is on the back side of the mirror at the location from which incoming parallel rays appear to diverge after reflection. The **focal length** (the distance from the mirror to the focal point) is positive for a concave mirror and negative for a convex mirror.

A real image is formed at a point when reflected light rays actually pass through the point.

A virtual image is formed at a point when reflected light rays appear to diverge from the point.

23.3 Convex Mirrors and Sign Conventions

A convex mirror is a diverging mirror; light rays from an object (incident on the front side of the mirror) are reflected so as to appear to be coming from the image position (on the back side of the mirror). *Images of real objects formed by convex mirrors are always virtual, upright, and smaller than the object. This is true for all object distances.*

A concave mirror is a converging mirror. The nature of an image (real or virtual, upright or inverted, smaller or larger than the object) formed by a concave mirror depends on the object distance, p, relative to the focal length, f. Possible object locations and image outcomes are given in the table below:

If the object distance is...	Then the image will be
$p > 2f$	real, inverted, smaller than object
$p = 2f$	real, inverted, same size as object
$f < p < 2f$	real, inverted, larger than object
$p = f$	reflected rays are parallel, no image
$p < f$	virtual, upright, larger than object

A **ray diagram** is a graphical technique (scale drawing) that can be used to determine the location, nature, and relative size of an image formed by a spherical mirror. The positions of the object and focal point are shown along the principal axis of the mirror. **See the example ray diagrams for mirrors in Suggestions, Skills, and Strategies.**

The point of intersection of any two of the following three rays in a ray diagram for mirrors locates the image:

1. Ray #1 is drawn from the top of the object parallel to the principal axis and is reflected back through the focal point of a concave mirror. In the case of a convex mirror, the reflected ray appears to diverge from the focal point.

2. Ray #2 is drawn from the top of the object through the focal point of a concave mirror (toward the focal point of a convex mirror) and is reflected parallel to the principal axis.

3. Ray #3 is drawn from the top of the object through the center of curvature and is reflected back on itself.

The point at which the rays intersect in the case of a concave mirror (or the point from which they appear to diverge in the case of a convex mirror) locates the top of the image.

23.4 Images Formed by Refraction

A **real object** is one which is located on the front side of a refracting surface. A **virtual object** exists when the incident light rays are converging toward a point located on the back side of a refracting surface. The front side is defined as the side of the surface in which light rays originate. The location and character of an image formed by a refracting surface depends on the shape of the surface (convex, concave, or flat) and the relative values of the indexes of refraction of the media on the two sides of the refracting surface.

The following cases for real objects are illustrated in Suggestions, Skills, and Strategies:

(a) An object located beyond the focal point on the front side of a convex surface where $n_1 < n_2$ (n_1 in the region of the incident ray and n_2 in the region of the refracted ray) will result in a real image on the back side of the refracting surface.

(b) An object located on the front side of a concave surface where $n_1 < n_2$ will result in a virtual image on the front side of the refracting surface.

(c) An object located on the front side of a convex surface where $n_1 > n_2$ will result in a virtual image on the front side of the refracting surface.

(d) An object located on the front side of a flat surface will result in a virtual image on the front side of the surface, with $|q| < p$ if $n_2 < n_1$ and with $|q| > p$ if $n_2 > n_1$.

23.6 Thin Lenses

A converging lens is thicker in the middle than at the rim and has a positive focal length (when surrounded by a medium with an index of refraction smaller than that of the lens). A diverging lens is thicker at the rim and has a negative focal length.

A thin lens has two focal points as illustrated below. Note the reversal of locations of F_1 and F_2 for the two lens types.

The following three rays form the ray diagram for a thin lens:

1. Ray #1 is drawn from the object parallel to the principal axis. After being refracted by the lens, this ray passes through focal point F_1 in the case of a converging lens and appears to diverge from F_1 in the case of a diverging lens.

2. Ray #2 is drawn through the center of the lens. This ray continues in a straight line.

3. Ray #3 is drawn through focal point F_2 (toward F_2 in the case of a diverging lens) and emerges from the lens parallel to the principal axis.

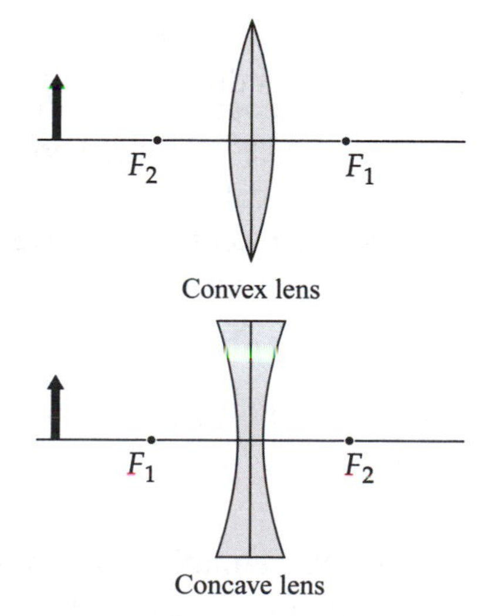

In the figures above, consider the arrow to be an object and construct a ray diagram for each of the two cases shown.

See the examples of ray diagrams for thin lenses in Suggestions, Skills, and Strategies.

When a combination of two lenses is used to form an image, the image of the first lens is treated as the object for the second lens. The overall magnification of the combination is the product of the magnifications of the separate lenses.

23.7 Lens and Mirror Aberrations

Aberrations are responsible for the formation of imperfect images by lenses and mirrors. **Spherical aberration** is due to the fact that parallel incident light rays that are different distances from the optical axis are focused at different points following refraction or reflection. **Chromatic aberration** arises from the fact that light of different wavelengths focuses at different points when refracted by a lens. This occurs because the index of refraction is a function of wavelength.

EQUATIONS AND CONCEPTS

Lateral magnification of a flat mirror is defined as the ratio of image height to the object height. *This ratio always has a value of +1 for a flat mirror since the upright image is always the same size as the object.*

$$M \equiv \frac{\text{image height}}{\text{object height}} = \frac{h'}{h} \qquad (23.1)$$

Lateral magnification of a spherical mirror can be expressed either as the ratio of the image size to the object size or as the negative of the ratio of the image distance to the object distance. *When the image is inverted both M and h' will have negative values.*

$$M = \frac{h'}{h} = -\frac{q}{p} \qquad (23.2)$$

The **focal length of a spherical mirror** is the distance from the vertex of the mirror to the focal point, F, located midway between the center of curvature and the vertex of the mirror. *For a concave mirror, the focal point is in front of the mirror, and f is positive. For a convex mirror, the focal point is back of the mirror, and f is negative.*

$$f = \frac{R}{2} \qquad (23.5)$$

R = radius of curvature

The **mirror equation** is used to determine the location of an image formed by reflection of paraxial rays from a spherical surface.

$$\frac{1}{p} + \frac{1}{q} = \frac{1}{f} \quad \begin{pmatrix} p = \text{object distance} \\ q = \text{image distance} \end{pmatrix} \qquad (23.6)$$

An image formed by a spherical refracting surface of radius R, separating two media with indices of refraction n_1 and n_2, can be real or virtual depending on the value of R. In Equation 23.7, n_1 is the index of refraction on the side of the incident rays, and n_2 is the index of refraction on the side of the refracted rays.

$$\frac{n_1}{p} + \frac{n_2}{q} = \frac{n_2 - n_1}{R} \quad (23.7)$$

$$M = \frac{h'}{h} = -\frac{n_1 q}{n_2 p} \quad (23.8)$$

A **flat refracting surface** ($R = \infty$) will form an image on the same side of the surface as the object regardless of the relative values of n_1 and n_2. *You should review the example ray diagrams and sign conventions for refracting surfaces in Suggestions, Skills, and Strategies.*

$$q = -\frac{n_2}{n_1} p \quad (23.9)$$

Several **combinations of lens parameters and image characteristics for a thin lens** are shown in the following equations.

Features of the geometry of a thin lens are shown in the figures at right. A lens can be considered "thin" when the lens thickness is much less that the radii of the surfaces (R_1 and R_2). A given radius is positive if the center of curvature is in back of the lens and negative if the center of curvature is in front of the lens. In the figure, R_1 is positive and R_2 is negative. A convex lens (bottom left) converges incident parallel light rays to the focal point (F_2). Parallel light rays entering a concave lens (bottom right) appear to diverge from the focal point (F_1).

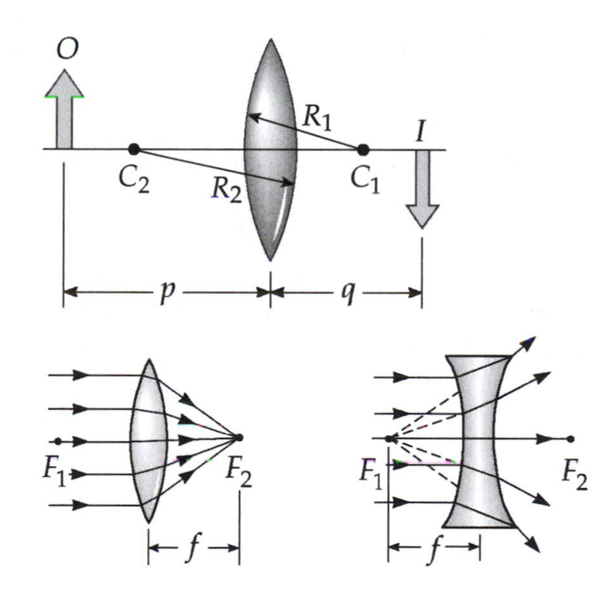

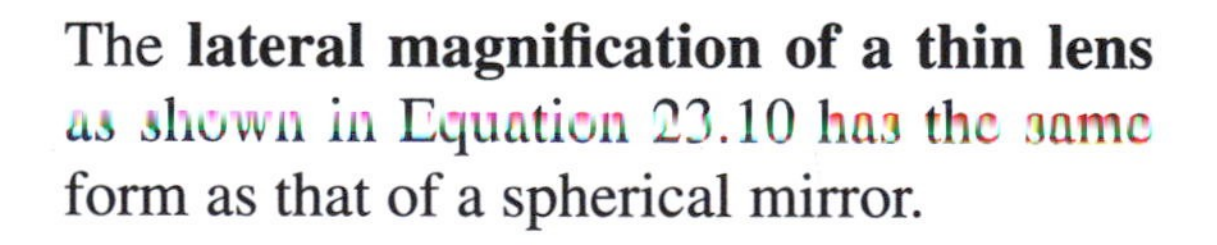

The **lateral magnification of a thin lens** as shown in Equation 23.10 has the same form as that of a spherical mirror.

$$M = \frac{h'}{h} = -\frac{q}{p} \quad (23.10)$$

The **thin-lens equation** can be used to find the image location when the focal length and object distance are known.

$$\frac{1}{p} + \frac{1}{q} = \frac{1}{f} \qquad (23.11)$$

The overall magnification of an image formed by two optical elements (e.g., lens, mirror, or lens-mirror combination) equals the product of the magnifications due to the two individual elements.

$$M = M_1 M_2$$

The **lens-maker's equation** can be used to calculate the focal length of a thin lens in terms of the physical characteristics of the lens. *If the lens is surrounded by a medium other than air, the index of refraction given in Equation 23.12 must be the ratio of the index of refraction of the lens to that of the surrounding medium.*

$$\frac{1}{f} = (n-1)\left(\frac{1}{R_1} - \frac{1}{R_2}\right) \qquad (23.12)$$

For a convex lens R_1 is positive and R_2 is negative.

SUGGESTIONS, SKILLS, AND STRATEGIES

A major portion of this chapter is devoted to the development and presentation of equations which can be used to determine the location and nature of images formed by various optical components acting either singly or in combination. It is essential that these equations be used with the correct algebraic sign associated with each quantity involved. You must understand clearly the sign conventions for mirrors, refracting surfaces, and lenses. Sign conventions to be used with spherical mirrors, thin lenses, and refracting surfaces are summarized on the following pages.

SIGN CONVENTIONS FOR SPHERICAL MIRRORS

Equations: $\frac{1}{p} + \frac{1}{q} = \frac{1}{f} = \frac{2}{R}$ $\quad M = \frac{h'}{h} = -\frac{q}{p}$

The front side of the mirror is the region on which light rays are incident and reflected.

p is + if the object is in front of the mirror (real object). p is – if the object is in back of the mirror (virtual object).
q is + if the image is in front of the mirror (real image). q is – if the image is in back of the mirror (virtual image).
Both f and R are + if the center of curvature is in front (concave mirror). Both f and R are – if the center of curvature is in back (convex mirror).
If M is positive, the image is upright. If M is negative, the image is inverted.

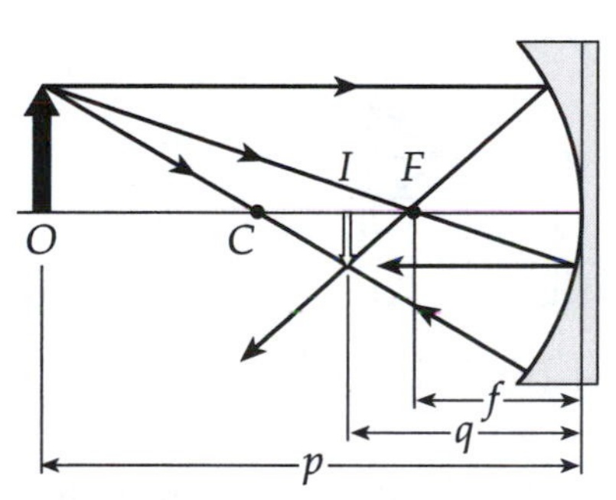

(a) Concave Mirror
$p > 2f$: $q+, f+, R+$
Image real, inverted, diminished

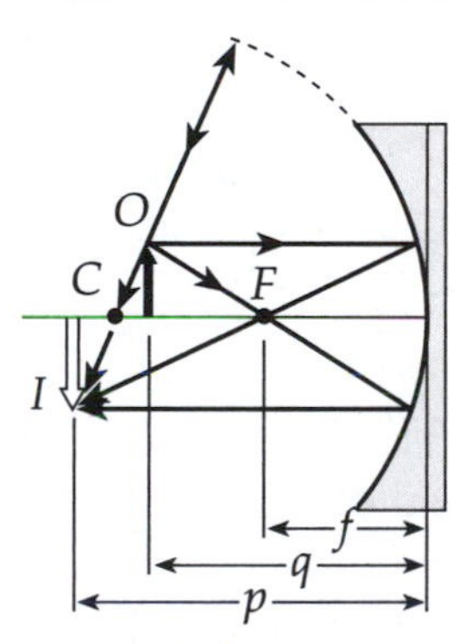

(b) Concave Mirror
$2f > p > f$: $q+, f+, R+$
Image real, inverted, enlarged

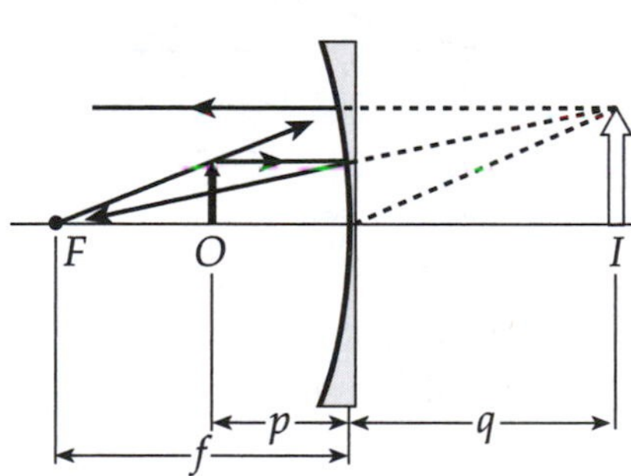

(c) Concave Mirror
$p < f$: $q-, f+, R+$
Image virtual, upright, enlarged

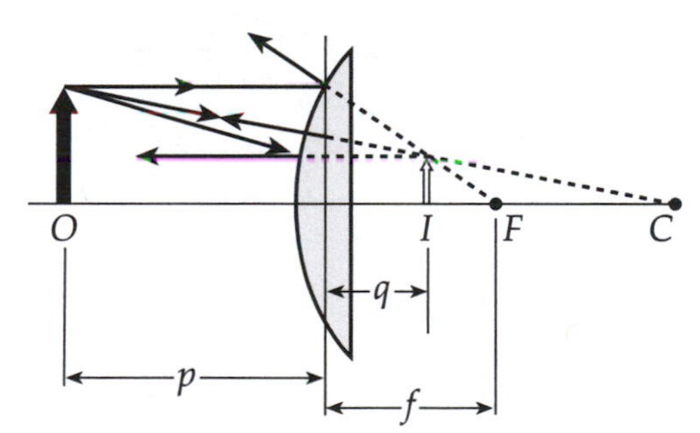

(d) Convex Mirror
$p+, q-, f-, R-$
Image virtual, upright, diminished
for any object distance

SIGN CONVENTIONS FOR REFRACTING SURFACES

Equations: $\dfrac{n_1}{p}+\dfrac{n_2}{q}=\dfrac{n_2-n_1}{R}$ $\qquad M=\dfrac{h'}{h}=-\dfrac{n_1 q}{n_2 p}$

In the following table, the **front** side of the surface is the side **from which the light is incident**.

p is + if the object is in front of the surface (real object). p is – if the object is in back of the surface (virtual object).
q is + if the image is in back of the surface (real image). q is – if the image is in front of the surface (virtual image).
R is + if the center of curvature is in back of the surface. R is – if the center of curvature is in front of the surface.
n_1 refers to the index of refraction of the first medium (before refraction). n_2 is the index of refraction of the second medium (after refraction).

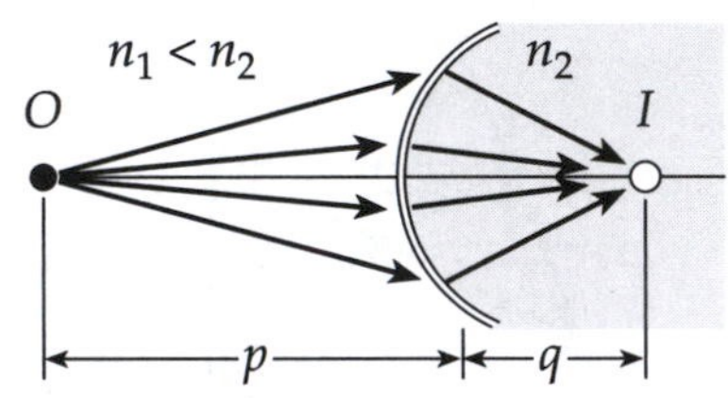

(a) p + (real object)
q + (real image)
R + (convex to incident light)

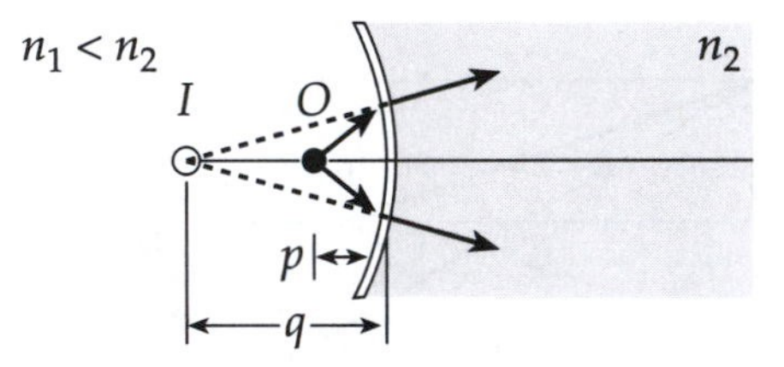

(b) p + (real object)
q – (virtual image)
R – (concave to incident light)

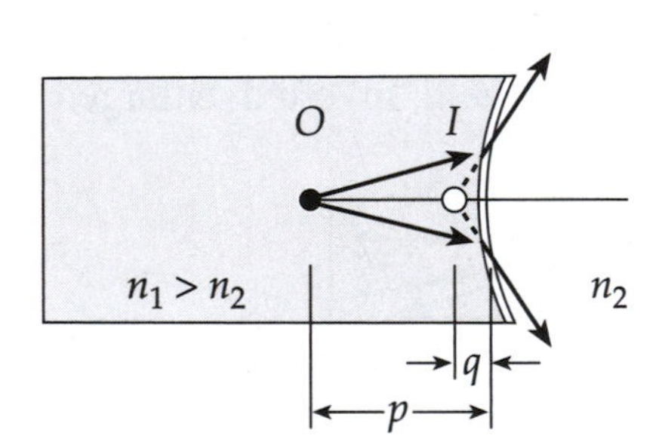

(c) p + (real object)
q – (virtual image)
R + (concave to incident light)

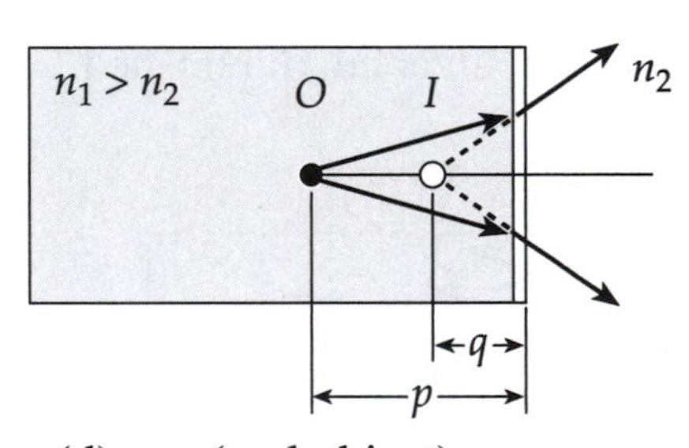

(d) p + (real object)
q – (virtual image)

SIGN CONVENTIONS FOR THIN LENSES

Equations: $\dfrac{1}{p}+\dfrac{1}{q}=\dfrac{1}{f}=(n-1)\left(\dfrac{1}{R_1}-\dfrac{1}{R_2}\right)$ $\quad M=\dfrac{h'}{h}=-\dfrac{q}{p}$

In the following table, the **front** of the lens is the **side from which the light is incident.**

p is + if the object is in front of the lens. p is – if the object is in back of the lens.
q is + if the image is in back of the lens. q is – if the image is in front of the lens.
f is + if the lens is thickest at the center. f is – if the lens is thickest at the edges.
R_1 and R_2 are + if the center of curvature is in back of the lens. R_1 and R_2 are – if the center of curvature is in front of the lens.

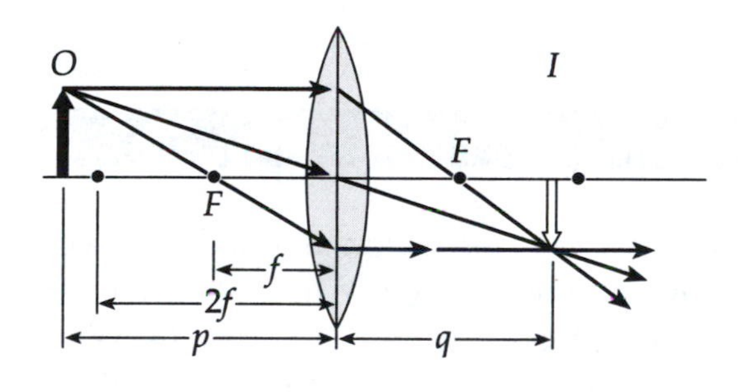

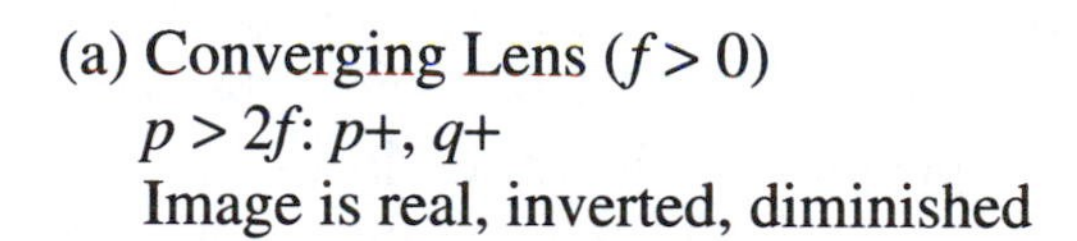

(a) Converging Lens ($f > 0$)
$p > 2f$: $p+$, $q+$
Image is real, inverted, diminished

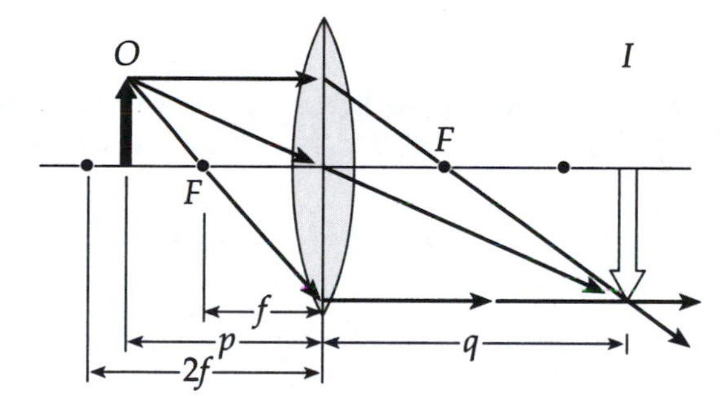

(b) Converging Lens ($f > 0$)
$2f > p > f$: $p+$, $q+$
Image is real, inverted, enlarged

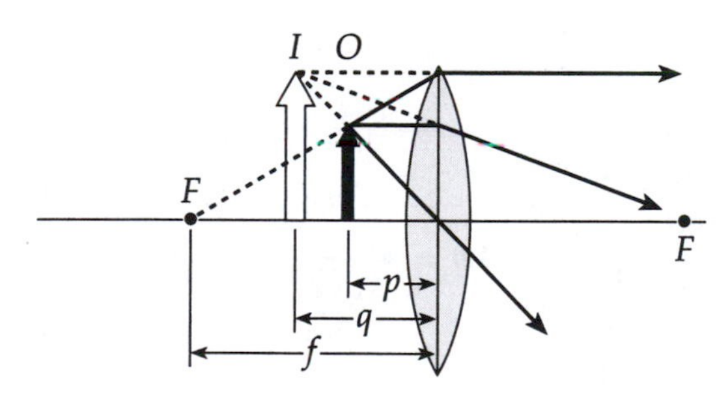

(c) Converging Lens ($f > 0$)
$p < f$: $p+$, $q-$
Image is virtual, upright, enlarged

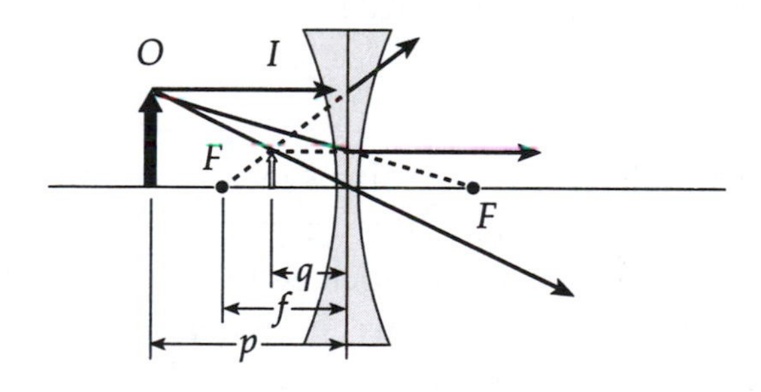

(d) Diverging Lens ($f < 0$)
$p+$, $q-$
Image is virtual, upright, diminished

REVIEW CHECKLIST

- You should know the sign conventions used with mirrors, lenses, and refracting surfaces. (Sections 23.3, 23.4, and 23.6)
- Understand the manner in which the algebraic signs associated with calculated quantities correspond to the nature of the image: real or virtual, upright or inverted. (Sections 23.3, 23.4, and 23.6)
- Calculate the location of the image of a specified object as formed by a flat mirror, spherical mirror, plane refracting surface, spherical refracting surface, or thin lens. Determine the magnification and character of the image in each case. (Sections 23.3, 23.4, and 23.6).
- Construct ray diagrams to determine the location and nature of the image of a given object when the geometric characteristics of the optical device (lens, refracting surface, or mirror) are known. (Sections 23.3, 23.4, and 23.6)

SOLUTIONS TO SELECTED END-OF-CHAPTER PROBLEMS

2. Two plane mirrors stand facing each other, 3.00 m apart, and a woman stands between them. The woman faces one of the mirrors from a distance of 1.00 m and holds her left arm out to the side of her body with the palm of her left hand facing the closer mirror. (a) What is the apparent position of the closest image of her left hand, measured perpendicularly from the surface of the mirror in front of her? (b) Does it show the palm of her hand or the back of her hand? (c) What is the position of the next image? (d) Does it show the palm of her hand or the back of her hand? (e) What is the position of the third closest image? (f) Does it show the palm of her hand or the back of her hand? (g) Which of the images are real and which are virtual?

Solution

We shall refer to the mirror located 1.0 m in front of the woman as M_F and the mirror located 2.0 m behind her as M_B. For any object located in front of a plane (or flat) mirror, the mirror forms an upright, virtual image that is the same size as the object, and is located as far behind the mirror as the object is in front of the mirror.

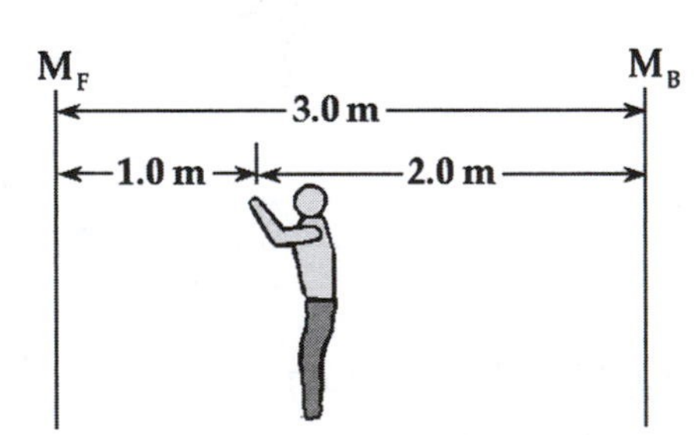

(a) The woman's hand is located 1.0 m in front of the mirror M_F. This mirror forms an upright, virtual image $I_{palm,1}$ located 1.0 m behind M_F. ◊

(b) The image $I_{palm,1}$ is an image of the woman's palm, which faces M_F. ◊

(c) The back of the woman's hand is located 2.0 m in front of mirror M_B, which forms an image $I_{back,1}$ 2.0 m behind that mirror. Image $I_{back,1}$ is an upright, virtual image, which serves as an object for mirror M_F. This object is located $d = 2.0\text{ m} + 3.0\text{ m} = 5.0\text{ m}$ in front of M_F, and M_F forms an upright, virtual image, $I_{back,2}$, 5.0 m behind M_F. This is the second closest image seen by the woman. ◊

(d) The image $I_{back,2}$, is an image of the back of the woman's hand which faces mirror M_B.

(e) The image $I_{palm,1}$ [see part (a)] serves as an object for mirror M_B. Since this object is located distance $d' = 1.0\text{ m} + 3.0\text{ m} = 4.0\text{ m}$ in front of M_B, the mirror forms an upright, virtual image $I_{palm,2}$ located 4.0 m behind M_B. Image $I_{palm,2}$ is located distance $d'' = 4.0\text{ m} + 3.0\text{ m} = 7.0\text{ m}$ in front of mirror M_F and serves as an object for that mirror. Mirror M_F then forms an upright, virtual image, $I_{palm,3}$, at a distance of 7.0 m behind the mirror. This is the third closest image the woman can see looking into mirror M_F. ◊

(f) The image $I_{palm,3}$ is an image of the palm.

(g) All images seen by the woman are virtual images, located behind mirror M_F. ◊

---

8. To fit a contact lens to a patient's eye, a *keratometer* can be used to measure the curvature of the cornea—the front surface of the eye. This instrument places an illuminated object of known size at a known distance p from the cornea, which then reflects some light from the object, forming an image of it. The magnification M of the image is measured by using a small viewing telescope that allows a comparison of the image formed by the cornea with a second calibrated image projected into the field of view by a prism arrangement. Determine the radius of curvature of the cornea when $p = 30.0$ cm and $M = 0.013\,0$.

Solution

When an object is located distance p from a spherical reflecting surface, an image is formed at distance q from that surface. The lateral magnification associated with this image formation is given by

$$M = \frac{h'}{h} = -\frac{q}{p}$$

where h and h' are the sizes of the object and image, respectively. The magnification is positive for upright images and negative for inverted images.

If, using the keratometer, the magnification of the image formed by the cornea of the eye is found to be $M = +0.013\,0$ when $p = 30.0$ cm, we find that the image and object distances must be related by $M = -q/p = 0.013\,0$. Thus, image distance q must be

$$q = -0.013\,0p = -0.013\,0(30.0\text{ cm}) = -0.390\text{ cm}$$

The negative sign here tells us that the image is a virtual image, located behind the surface of the cornea. Now, knowing the values of both the object and image distances, we may use the mirror equation, $1/p+1/q=2/R=1/f$, to calculate the radius of curvature of the cornea of this eye. Solving the mirror equation for the radius gives

$$R=\frac{2pq}{p+q}=\frac{2(+30.0\text{ cm})(-0.390\text{ cm})}{30.0\text{ cm}+(-0.390\text{ cm})}=-0.790\text{ cm}$$

The negative sign associated with this result tells us that the cornea has a convex shape, with the center of curvature located behind the surface. The magnitude of the radius of this convex surface is $|R|=0.790$ cm. ◊

---

13. A concave makeup mirror is designed so that a person 25 cm in front of it sees an upright image magnified by a factor of two. What is the radius of curvature of the mirror?

Solution

When a real object is located in front of a concave mirror, the nature of the image formed depends on the location of the object relative to the center of curvature (at distance R from the mirror) and the focal point (at distance $f=R/2$ from the mirror). When the object is located outside the center of curvature, or the object distance is $p>R$, the image is real, inverted, and diminished in size. If the image is between the center of curvature and the focal point, or $R>p>f$, the image is real, inverted, and larger than the object. If the object is located inside the focal point, or $p<f$, the image is virtual, upright, and larger than the object.

In this case, the person sees an upright, enlarged image that is twice the size of the object. Therefore, the magnification is $M=+2.0$, positive because the image is upright, and 2.0 because the image is twice as large as the object. Since the person is a real object, located 25 cm *in front* of the mirror, and the image is upright, this image must be virtual (all real images that a concave mirror forms of real objects are inverted).

The lateral magnification is $M=-q/p$, so the image distance in this case must be

$$q=-Mp=-(+2.0)(+25\text{ cm})=-50\text{ cm}$$

with the negative sign confirming that the image is virtual (located behind the mirror). We can now use the mirror equation to determine the radius of curvature. This equation states that $1/p+1/q=2/R$, or

$$R=\frac{2pq}{p+q}=\frac{2(25\text{ cm})(-50\text{ cm})}{25\text{ cm}-50\text{ cm}}=+1.0\times10^{2}\text{ cm}=+1.0\text{ m}$$ ◊

---

19. A spherical mirror is to be used to form an image, five times as tall as an object, on a screen positioned 5.0 m from the mirror. (a) Describe the type of mirror required. (b) Where should the mirror be positioned relative to the object?

Solution

(a) Since the image is formed on a screen located in front of the mirror, the image is real, and the image distance is positive ($q = +5.0$ m).

We could also have determined that the image is real by recognizing that the light reflecting from the mirror must actually come to a focus on the screen in order to form an image on it. Since real images are located where the light actually comes to a focus, while virtual images are located where the light has not traveled but still appears to diverge away from, we again conclude that the image is real.

Of the different types of mirrors (flat mirrors, concave mirrors, and convex mirrors), only a concave mirror will form a real image of a real object. Recognizing this, we conclude that the mirror used to form this image is a concave mirror. ◊

(b) With both $p > 0$ and $q > 0$, the magnification given by $M = -q/p$ will be negative, indicating that the image is inverted.

If the image is five times the size of the object, then

$$|M| = \frac{q}{p} = 5 \qquad \text{and} \qquad q = 5p$$

The object distance is then $p = q/5 = (5.0\text{ m})/5 = 1.0$ m, or the mirror should be 1.0 m from the object, with the object between the mirror and the screen. ◊

(a – again) Now that we know both the image distance and the required object distance, we can give greater detail on the characteristics of the mirror needed. The mirror equation, $1/p + 1/q = 2/R = 2/f$, gives the required focal length of the concave mirror as

$$f = \frac{pq}{p+q} = \frac{(1.0\text{ m})(5.0\text{ m})}{1.0\text{ m} + 5.0\text{ m}} = 0.83\text{ m}$$ ◊

23. A paperweight is made of a solid glass hemisphere with index of refraction 1.50. The radius of the circular cross section is 4.0 cm. The hemisphere is placed on its flat surface, with the center directly over a 2.5-mm-long line drawn on a sheet of paper. What length of line is seen by someone looking vertically down on the hemisphere?

Solution

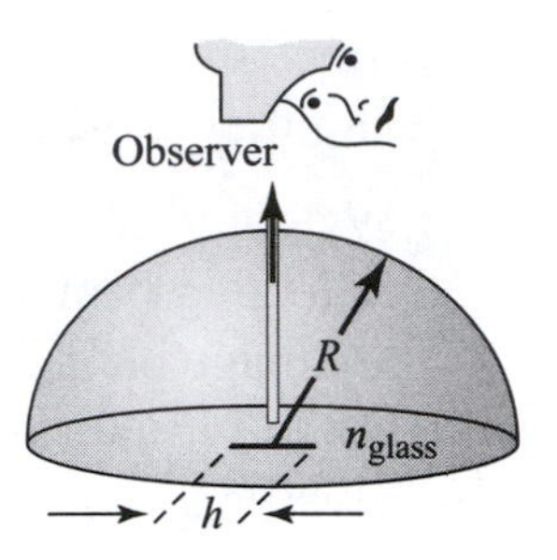

The center of curvature of the convex surface of the hemisphere is on the side of the surface from which the light is coming (i.e., in front of the surface). Thus, the radius of curvature is negative by the sign convention adopted in Section 23.4 of the text, and $R = -4.0$ cm. A 2.5-mm-long line located at the center of curvature serves as a real object for the spherical refracting surface. The object distance is then $p = +4.0$ cm.

The index of refraction on the side of the surface the light is coming from is $n_1 = n_{\text{glass}} = 1.50$, and that on the side the light is going toward is $n_2 = n_{\text{air}} = 1.00$. The location of the image is found from

$$\frac{n_1}{p} + \frac{n_2}{q} = \frac{n_2 - n_1}{R} \qquad \text{which becomes} \qquad \frac{1.00}{q} = \frac{1.00 - 1.50}{(-4.0 \text{ cm})} - \frac{1.50}{(+4.0 \text{ cm})}$$

Solving gives $q = -4.0$ cm, so the image is virtual and located at the same position as the object.

When light is refracted by a single spherical surface, the magnification is given by

$$M = \frac{\text{image size}}{\text{object size}} = \frac{h'}{h} = -\frac{n_1 q}{n_2 p}$$

The size of the image formed by the hemispherical paper weight is

$$h' = Mh = \left(-\frac{n_1 q}{n_2 p}\right)h = -\frac{(1.50)(-4.0 \text{ cm})}{(1.00)(+4.0 \text{ cm})}(2.5 \text{ mm}) = +3.8 \text{ mm}$$

Thus, the image is upright, virtual, located at the same position as the object, and is 3.8 mm long. ◊

27. A jellyfish is floating in a water-filled aquarium 1.00 m behind a flat pane of glass 6.00 cm thick and having an index of refraction of 1.50. (a) Where is the image of the jellyfish located? (b) Repeat the problem when the glass is so thin that its thickness can be neglected. (c) How does the thickness of the glass affect the answer to part (a)?

Solution

In the sketch at the right, the jellyfish is at point O, a distance p in front of the water-glass interface. Refraction by this plane surface forms a virtual image I' at distance $|q'|$ from the inner glass surface. This image, located distance $p' = |q'| + t$ in front of the outer surface of the glass, serves as a real object for refraction at the glass-air interface. This refraction produces a virtual image I located distance $|q|$ behind that plane surface.

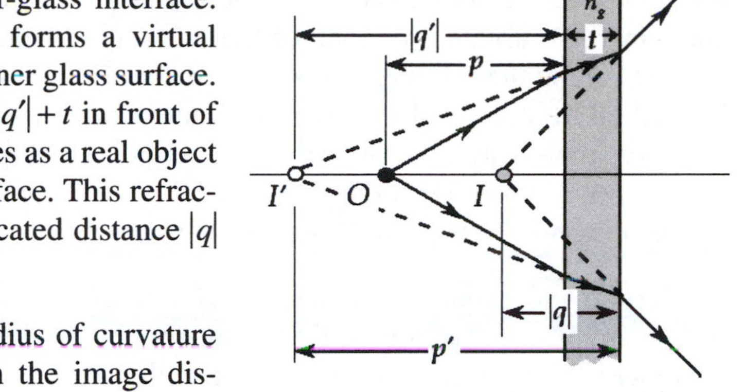

A plane surface has an infinite radius of curvature ($R \to \infty$), so the relation between the image distance and object distance for refraction at a single surface (Equation 23.7 in the textbook) reduces to $q = -(n_2/n_1)p$. Thus, the image distances in the above sketch are $q' = -(n_g/n_w)p$, or $|q'| = (n_g/n_w)p$, and

$$q = -\left(\frac{n_a}{n_g}\right)p' = -\left(\frac{n_a}{n_g}\right)(|q'|+t) = -\left(\frac{n_a}{n_g}\right)\left[\left(\frac{n_g}{n_w}\right)p+t\right] = -\left[\left(\frac{n_a}{n_w}\right)p+\left(\frac{n_a}{n_g}\right)t\right] \quad \textbf{[1]}$$

(a) When the jellyfish is located distance $p = 1.00\text{ m} = 100\text{ cm}$ in front of glass having thickness $t = 6.00\text{ cm}$, the final image distance is

$$q = -\left[\left(\frac{1.00}{1.333}\right)(100\text{ cm}) + \left(\frac{1.00}{1.50}\right)(6.00\text{ cm})\right] = -79.0\text{ cm}$$

Therefore, the image of the jellyfish viewed by the observer is 79.0 cm behind the outer surface of the glass pane. ◊

(b) If the glass is so thin that its thickness can be neglected, we let $t \to 0$, and Equation [1] from above becomes $q = -(n_a/n_w)p$. Under these conditions, the final image distance is $q = -(1.00/1.333)(100\text{ cm}) = -75.0\text{ cm}$, or the jellyfish appears to be 75.0 cm behind the outer surface of the glass. ◊

(c) From Equation [1], we see that increasing t increases the value of $|q|$, so the jellyfish appears to be farther behind the front surface of the glass. ◊

29. A contact lens is made of plastic with an index of refraction of 1.50. The lens has an outer radius of curvature of +2.00 cm and an inner radius of curvature of +2.50 cm. What is the focal length of the lens?

Solution

A contact lens is made to fit over the front of the eye and has a convex-concave shape as shown in the sketch at the right. As light travels from left to right through this lens, notice that the center of curvature of each of the refracting surfaces are located behind the surface. That is, the center of curvature C is located on the side of the surface the light is going toward as it crosses the surface, not on the side the light is coming from. Thus, by the sign convention of Table 23.2 in the textbook, the radius of curvature is positive for both surfaces of this lens.

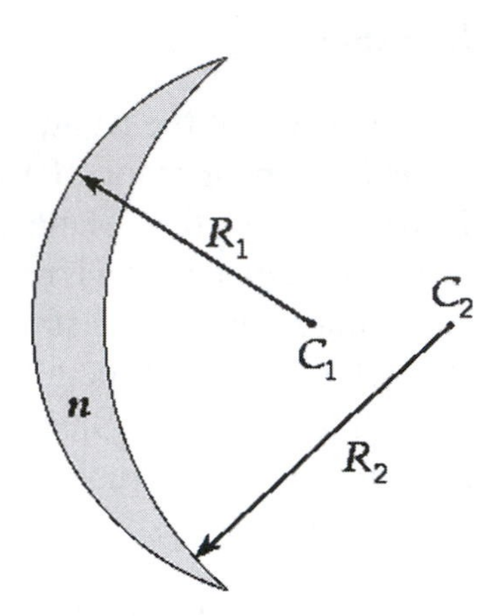

For the lens under consideration, $R_1 = +2.00$ cm, $R_2 = +2.50$ cm, and the index of refraction for the lens material is $n = 1.50$. The lens-maker's equation gives the focal length of a thin lens, surrounded by air, in terms of the radii of its surfaces and the index of refraction of the lens material as

$$\frac{1}{f} = (n-1)\left(\frac{1}{R_1} - \frac{1}{R_2}\right)$$

Thus, for the described lens,

$$\frac{1}{f} = (1.50-1)\left(\frac{1}{2.00 \text{ cm}} - \frac{1}{2.50 \text{ cm}}\right) = (0.50)\left[\frac{2.50 \text{ cm} - 2.00 \text{ cm}}{(2.00 \text{ cm})(2.50 \text{ cm})}\right] = \frac{(0.50)^2}{5.00 \text{ cm}}$$

and the focal length is found to be

$$f = \frac{5.00 \text{ cm}}{0.25} = +20 \text{ cm}$$ ◊

33. A diverging lens has a focal length of magnitude 20.0 cm. (a) Locate the images for object distances of (i) 40.0 cm, (ii) 20.0 cm, and (iii) 10.0 cm. For each case, state whether the image is (b) real or virtual and (c) upright or inverted. (d) For each case, find the magnification.

Solution

(a) The focal length of a diverging lens is negative, so the focal length of the given lens is $f = -20.0$ cm. For an object distance p, the thin-lens equation, $1/p + 1/q = 1/f$ gives the image distance as $q = pf/(p - f)$. Thus the image distance for each of the listed cases are:

(i) $p = 40.0 \text{ cm}$ gives $q = \dfrac{(40.0 \text{ cm})(-20.0 \text{ cm})}{40.0 \text{ cm} - (-20.0 \text{ cm})} = -13.3 \text{ cm}$, or the image is located 13.3 cm in front of the lens. ◊

(ii) When $p = 20.0 \text{ cm}$, $q = \dfrac{(20.0 \text{ cm})(-20.0 \text{ cm})}{20.0 \text{ cm} - (-20.0 \text{ cm})} = -10.0 \text{ cm}$, and the image is located 10.0 cm in front of the lens. ◊

(iii) If $p = 10.0 \text{ cm}$, the image distance is $q = \dfrac{(10.0 \text{ cm})(-20.0 \text{ cm})}{10.0 \text{ cm} - (-20.0 \text{ cm})} = -6.67 \text{ cm}$. In this case, the image is located 6.67 cm in front of the lens. ◊

(b) Observe that in each of these cases, the image distance is negative, so the image is located in front of the lens (or on the side of the lens from which the light approaches). Thus, this diverging lens is forming a virtual image in each case. ◊

(c) For each of the cases, the object distance is positive ($p > 0$), and the image distance is negative ($q < 0$). Hence, the ratio (q/p) will be negative, and the lateral magnification produced by the lens, $M = -(q/p)$, will be positive in each of these cases. This means that the image is *upright* is all three cases. ◊

(d) The value of the lateral magnification in each case is given by:

(i) $M = -q/p = -(-13.3 \text{ cm})/(+40.0 \text{ cm}) = +0.333$ ◊

(ii) $M = -q/p = -(-10.0 \text{ cm cm})/(+20.0 \text{ cm}) = +0.500$ ◊

(iii) $M = -q/p = -(-6.67 \text{ cm cm})/(+10.0 \text{ cm}) = +0.667$ ◊

45. Lens L_1 in Figure P23.45 has a focal length of 15.0 cm and is located a fixed distance in front of the film plane of a camera. Lens L_2 has a focal length of 13.0 cm, and its distance d from the film plane can be varied from 5.00 cm to 10.0 cm. Determine the range of distances for which objects can be focused on the film.

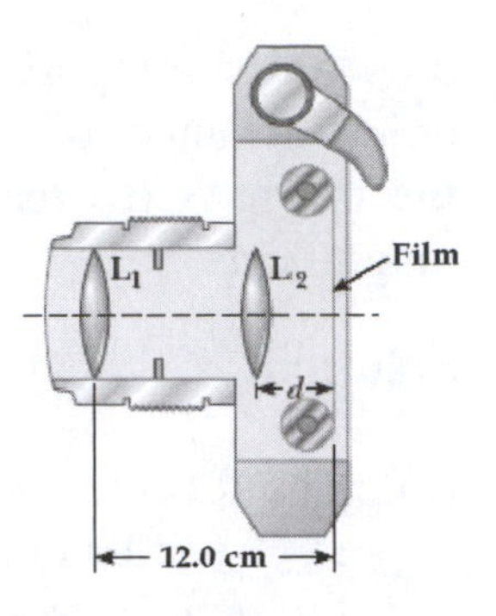

Figure P23.45

Solution

Note: Final answers to this problem are highly sensitive to round-off error. To avoid this, we retain extra digits in some intermediate answers and round only the final answers to the correct number of significant figures.

Both lenses are convergent lenses, so their focal lengths are $f_1 = +15.0\text{ cm}$ and $f_2 = +13.0\text{ cm}$. The image formed by lens L_1 serves as the object for lens L_2. Since L_2 forms a real image on the film, the image distance for this lens is $q_2 = +d$, and the thin-lens equation gives its object distance as

$$p_2 = \frac{q_2 f_2}{q_2 - f_2} = \frac{d(13.0\text{ cm})}{d - 13.0\text{ cm}}$$

Since the largest possible value of d is less than 13.0 cm, the above equation shows that $p_2 < 0$. Thus, the image that lens L_1 tries to form is located to the right of L_2 and serves as a virtual object for L_2. Since L_1 is a fixed distance of 12.0 cm from the film, the image distance for L_1 is

$$q_1 = 12.0\text{ cm} + \left(|p_2| - d\right)$$

When lens L_2 is distance $d = 5.00\text{ cm}$ in front of the film,

$$p_2 = \frac{(5.00\text{ cm})(13.0\text{ cm})}{5.00\text{ cm} - 13.0\text{ cm}} = -8.125\text{ cm, and}$$

$$q_1 = 12.0\text{ cm} + (8.125\text{ cm} - 5.00\text{ cm}) = 15.125\text{ cm}$$

The object distance for L_1 in this camera setting is

$$p_1 = \frac{q_1 f_1}{q_1 - f_1} = \frac{(15.125\text{ cm})(15.0\text{ cm})}{15.125\text{ cm} - 15.0\text{ cm}} = 1820\text{ cm} = 18.2\text{ m} \qquad \Diamond$$

When lens L_2 is distance $d = 10.0\text{ cm}$ in front of the film, we have

$$p_2 = \frac{(10.0\text{ cm})(13.0\text{ cm})}{10.0\text{ cm} - 13.0\text{ cm}} = -43.3\text{ cm, and}$$

$$q_1 = 12.0\text{ cm} + (43.3\text{ cm} - 10.0\text{ cm}) = 45.3\text{ cm}$$

The object distance for L_1 with this camera setting is

$$p_1 = \frac{q_1 f_1}{q_1 - f_1} = \frac{(45.3 \text{ cm})(15.0 \text{ cm})}{45.3 \text{ cm} - 15.0 \text{ cm}} = 22.4 \text{ cm} = 0.224 \text{ m} \qquad \lozenge$$

The camera can focus images on the film of objects located anywhere between 0.224 m and 18.2 m in front of the camera. ◊

---

51. The lens and the mirror in Figure P23.51 are separated by 1.00 m and have focal lengths of +80.0 cm and −50.0 cm, respectively. If an object is placed 1.00 m to the left of the lens, where will the final image be located? State whether the image is upright or inverted, and determine the overall magnification.

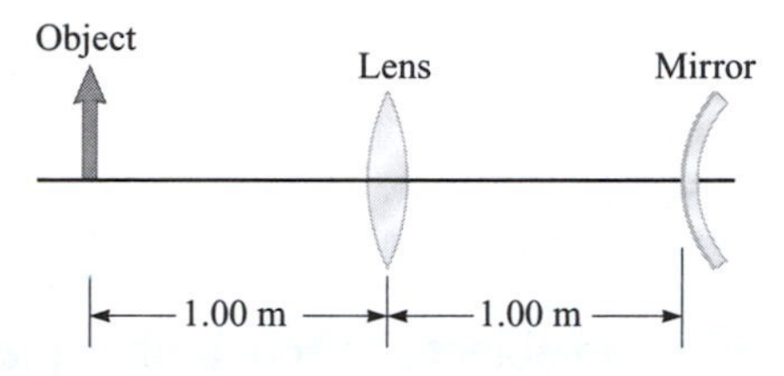

Figure P23.51

Solution

As light from the object makes its first pass through the lens, it is going left to right. With an object distance of $p_1 = +1.00 \text{ m} = 100 \text{ cm}$, and a focal length of $f_L = +80.0 \text{ cm}$ for the lens, the thin-lens equation gives the image distance as

$$q_1 = \frac{p_1 f_L}{p_1 - f_L} = \frac{(+100 \text{ cm})(+80.0 \text{ cm})}{100 \text{ cm} - 80.0 \text{ cm}} = +400 \text{ cm}$$

Thus, the lens attempts to form a real image 400 cm to the right of the lens, or 300 cm to the right of the mirror. However, the mirror redirects the light by reflecting it back toward the lens. The image the lens tried to form serves as a virtual object for the mirror (virtual since it is located behind the mirror), with object distance $p_2 = -300 \text{ cm}$. Since the mirror has a focal length of $f_M = -50.0 \text{ cm}$, the mirror equation gives the image distance for the mirror as

$$q_2 = \frac{p_2 f_M}{p_2 - f_M} = \frac{(-300 \text{ cm})(-50.0 \text{ cm})}{-300 \text{ cm} - (-50.0 \text{ cm})} = -60.0 \text{ cm}$$

Hence, the mirror forms a virtual image located 60.0 cm behind the mirror, or 160 cm to the right of the lens.

As the light reflected by the mirror now passes through the lens, going right to left, the image formed by the mirror serves as a real object for the lens, with an object distance of $p_3 = +160 \text{ cm}$. The final image formed by the lens will then have an image distance of

$$q_3 = \frac{p_3 f_L}{p_3 - f_L} = \frac{(+160 \text{ cm})(+80.0 \text{ cm})}{160 \text{ cm} - 80.0 \text{ cm}} = +160 \text{ cm}$$

Therefore, this final image is a real image located 160 cm to the left of the lens. ◊

The overall magnification produced by this optical system is $M_{\text{total}} = M_1 M_2 M_3$, where M_1, M_2, and M_3 are the magnifications produced by each of the processes the light undergoes. For the first refraction through the lens, $M_1 = -q_1/p_1$. For the reflection by the mirror, $M_2 = -q_2/p_2$, and for the final refraction by the lens, $M_3 = -q_3/p_3$. The overall magnification is then

$$M_{\text{total}} = \left(\frac{-q_1}{p_1}\right)\left(\frac{-q_2}{p_2}\right)\left(\frac{-q_3}{p_3}\right) = \left[\frac{-(+400\text{ cm})}{100\text{ cm}}\right]\left[\frac{-(-60.0\text{ cm})}{-300\text{ cm}}\right]\left[\frac{-(+160\text{ cm})}{160\text{ cm}}\right]$$

or $$M_{\text{total}} = -0.800$$ ◊

Since the overall magnification is negative, the final image is inverted. ◊

---

57. An object 2.00 cm high is placed 40.0 cm to the left of a converging lens having a focal length of 30.0 cm. A diverging lens having a focal length of –20.0 cm is placed 110 cm to the right of the converging lens. (a) Determine the final position and magnification of the final image. (b) Is the image upright or inverted? (c) Repeat parts (a) and (b) for the case in which the second lens is a converging lens having a focal length of +20.0 cm.

Solution

With the object located 40.0 cm in front of the converging first lens, $p_1 = +40.0$ cm and $f_1 = +30.0$ cm. The thin-lens equation gives the image distance for this lens as

$$q_1 = \frac{p_1 f_1}{p_1 - f_1} = \frac{(40.0\text{ cm})(30.0\text{ cm})}{40.0\text{ cm} - 30.0\text{ cm}} = +120\text{ cm}$$

and the magnification due to this lens is $M_1 = -q_1/p_1 = -120\text{ cm}/40.0\text{ cm} = -3.00$.

The real image the first lens tries to form 120 cm beyond that lens serves as the object for the second lens. Since the two lenses are only 110 cm apart, the location of this object is to the right of the second lens as shown in the diagram below. Thus, this is a virtual object and the object distance for the second lens is $p_2 = 110\text{ cm} - q_1 = -10.0$ cm.

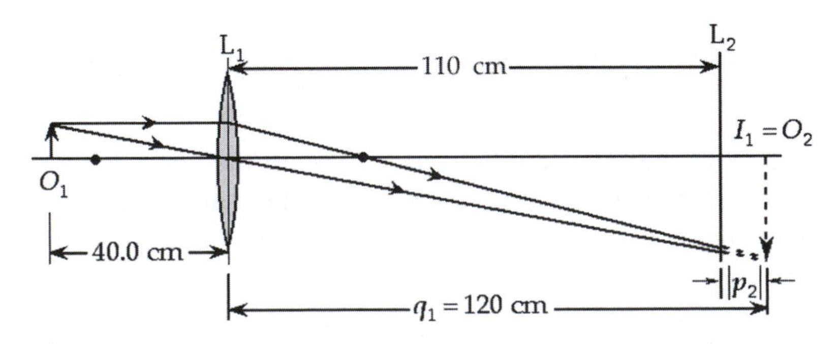

(a) If $f_2 = -20.0$ cm, the image distance and magnification for the second lens is

$$q_2 = \frac{p_2 f_2}{p_2 - f_2} = \frac{(-10.0 \text{ cm})(-20.0 \text{ cm})}{-10.0 \text{ cm} - (-20.0 \text{ cm})} = +20.0 \text{ cm}$$

and $M_2 = -q_2/p_2 = -20.0 \text{ cm}/(-10.0 \text{ cm}) = +2.00$.

Thus, the final image is located 20.0 cm to the right of lens L_2, and the overall magnification is $M_{\text{total}} = M_1 M_2 = (-3.00)(+2.00) = -6.00$. ◊

(b) The final image is real ($q_2 > 0$), inverted ($M_{\text{total}} < 0$), located 20.0 cm to the right of the second lens, and $|h'| = h\,|M| = (2.00 \text{ cm})(6.00) = 12.0$ cm high. ◊

(c) If $f_2 = +20.0$ cm, the image distance and magnification for the second lens is

$$q_2 = \frac{p_2 f_2}{p_2 - f_2} = \frac{(-10.0 \text{ cm})(20.0 \text{ cm})}{-10.0 \text{ cm} - 20.0 \text{ cm}} = +6.67 \text{ cm}$$

and $M_2 = -q_2/p_2 = -6.67 \text{ cm}/(-10.0 \text{ cm}) = +0.667$.

The overall magnification is then $M = M_1 M_2 = (-3.00)(+0.667) = -2.00$. ◊

In this case, the final image is real ($q_2 > 0$), inverted ($M < 0$), located 6.67 cm beyond the second lens, and $|h'| = h\,|M| = (2.00 \text{ cm})(2.00) = 4.00$ cm high. ◊

---

63. The lens-maker's equation applies to a lens immersed in a liquid if n in the equation is replaced by n_1/n_2 Here n_1 refers to the refractive index of the lens material and n_2 is that of the medium surrounding the lens. (a) A certain lens has focal length of 79.0 cm in air and a refractive index of 1.55. Find its focal length in water. (b) A certain mirror has focal length of 79.0 cm in air. Find its focal length in water.

Solution

(a) The general form of the lens-maker's equation, applicable to the case where a medium of refractive index n_2 surrounds a lens with refractive index n_1, is

$$\frac{1}{f} = \left(\frac{n_1}{n_2} - 1\right)\left(\frac{1}{R_1} - \frac{1}{R_2}\right)$$

When $n_1 = 1.55$ and $n_2 = n_{\text{air}} = 1.00$, we have

$$\frac{1}{f_{\text{air}}} = \left(\frac{1.55}{1.00} - 1\right)\left(\frac{1}{R_1} - \frac{1}{R_2}\right) = 0.550\left(\frac{1}{R_1} - \frac{1}{R_2}\right) \qquad [1]$$

where $f_{air} = 79.0$ cm. When the lens is in water, with $n_2 = n_{water} = 1.33$, the focal length is given by

$$\frac{1}{f_{water}} = \left(\frac{1.55}{1.33} - 1\right)\left(\frac{1}{R_1} - \frac{1}{R_2}\right) = 0.165\left(\frac{1}{R_1} - \frac{1}{R_2}\right) \qquad \textbf{[2]}$$

Dividing Equation [1] by Equation [2] gives

$$\frac{f_{water}}{f_{air}} = \frac{0.550}{0.165} = 3.33$$

Thus, the focal length in water is $f_{water} = 3.33 f_{air} = 3.33(79.0 \text{ cm}) = 263$ cm. ◊

(b) The formation of an image by a mirror is governed by the law of reflection, which is independent of both the material making up the mirror and the surrounding medium. Thus, the focal length of a mirror is determined only by its geometric properties and is independent of the composition of the mirror or the surroundings. This means that, for the mirror, $f_{water} = f_{air} = 79.0$ cm. ◊

24

Wave Optics

NOTES FROM SELECTED CHAPTER SECTIONS

24.1 Conditions for Interference

In order to observe **sustained interference** in light waves, the following conditions must be met:

- The **sources must be coherent;** they must maintain a constant phase with respect to each other.
- The **sources must be monochromatic;** they must have identical wavelengths.
- The **superposition principle must apply.**

24.2 Young's Double-Slit Experiment

The figure at right is a schematic diagram (not to scale) of the experimental setup of Young's double-slit experiment. The source S_0 is a monochromatic source that is divided into two coherent sources by passing through slits S_1 and S_2. The resulting interference pattern is observed on a viewing screen as bright and dark fringes corresponding to constructive and destructive interference of light from slits S_1 and S_2.

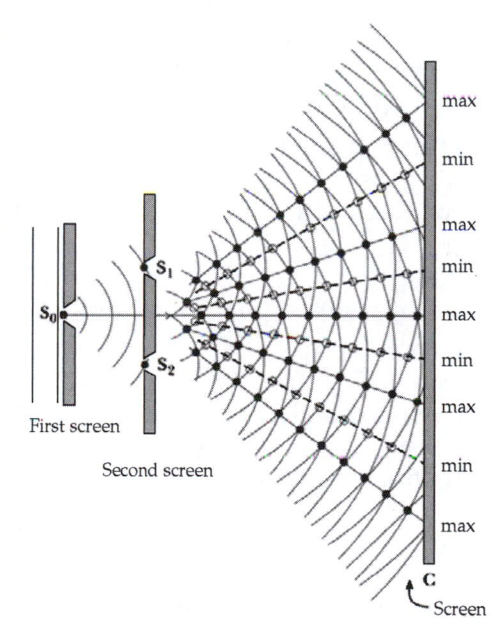

Constructive interference (a bright fringe) occurs when the path difference equals either zero or an integral multiple of the wavelength.

Destructive interference (a dark fringe) results when the path difference equals an odd multiple of a half wavelength.

24.3 Change of Phase Due to Reflection

Consider a light wave traveling in a medium with an index of refraction n_1 and experiencing partial reflection at the surface of a medium with an index of refraction n_2 as shown in the figure below.

- At Surface A, where $n_1 < n_2$, the reflected ray experiences a phase change of 180°.
- At Surface B, where $n_1 > n_2$, the reflected ray experiences no phase change.
- There is no phase change in the transmitted ray regardless of the relative values of n_1 and n_2.

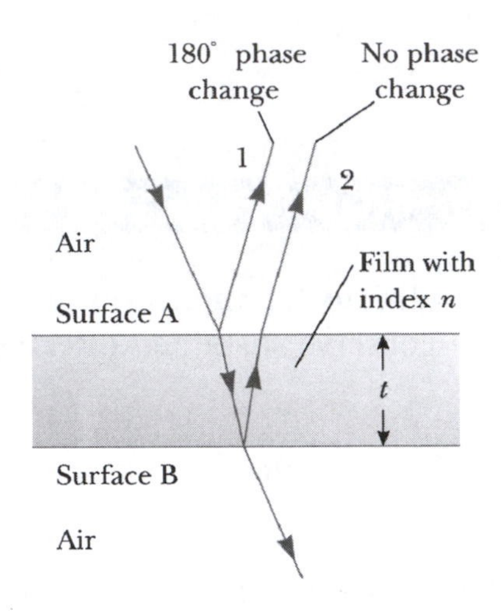

Phase change upon reflection

24.4 Interference in Thin Films

In order to predict constructive or destructive interference in thin films you must consider:

- the difference in path length traveled by the two interfering waves (reflected from the top and bottom surfaces).
- any expected changes in phase due to reflection.
- the wavelength of the light within the film.

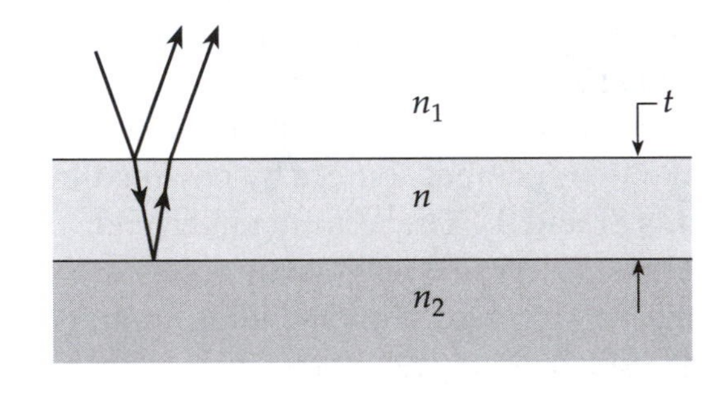

There are two general cases to consider (see figure above):

(a) Reflection resulting in a phase change at only one surface of the film:

n_1 and n_2 both less than n (phase change occurs only at the top surface), or

n_1 and n_2 both greater than n (phase change occurs only at the bottom surface)

Constructive interference will occur under these conditions when the path difference ($2t$) is an odd number of half wavelengths *as measured in the film* (i.e., $t = (m + \frac{1}{2})\ \lambda/2n$, where λ/n is the wavelength in the film).

Destructive interference will occur under these conditions when the path difference ($2t$) equals an integral number of wavelengths *as measured in the film* (i.e., $t = m\lambda/2n$).

(b) Reflection resulting in phase changes at both top and bottom surfaces of the film (or at neither surface):

$n_1 < n$ and $n < n_2$ (phase change at both surfaces), or

$n_1 > n$ and $n > n_2$ (no phase change at either surface)

In these cases, any phase changes that occur are offsetting, and interference of the reflected rays depends only on the difference in distance traveled by the two reflected rays and the index of refraction of the film.

Constructive interference will occur when the path difference ($2t$) equals an integral number of wavelengths *as measured in the film*; the film thickness must be an integral number of half wavelengths; that is, $t = m\lambda / 2n$.

Destructive interference in this case will be observed when the path difference equals an odd number of half wavelengths *as measured in the film*; that is, when $t = \left(m + \frac{1}{2}\right) \lambda/2n$.

24.6 Diffraction
24.7 Single-Slit Diffraction

Diffraction occurs when light waves deviate (or spread) from their initial direction of travel when passing through small openings, around obstacles, or by sharp edges.

The diffraction pattern produced by a single narrow slit consists of a broad, intense central band (the central maximum), flanked by a series of narrower and less intense secondary bands (called secondary maxima) alternating with a series of dark bands, or minima.

In the case of **single-slit diffraction**, each portion of the slit acts as a source of waves, and light from one portion of the slit can interfere with light from another portion. The resultant intensity on the screen depends on angle θ, which determines the direction between the perpendicular to the plane of the slit and the direction to a point on the screen.

One type of diffraction, called **Fraunhofer diffraction**, occurs when the rays reaching the observing screen are approximately parallel.

24.8 The Diffraction Grating

A diffraction grating, consisting of many equally spaced parallel slits, separated by a distance d, will produce a diffraction pattern. *There will be a series of principal maxima (bright lines) for each wavelength component in the incident light.*

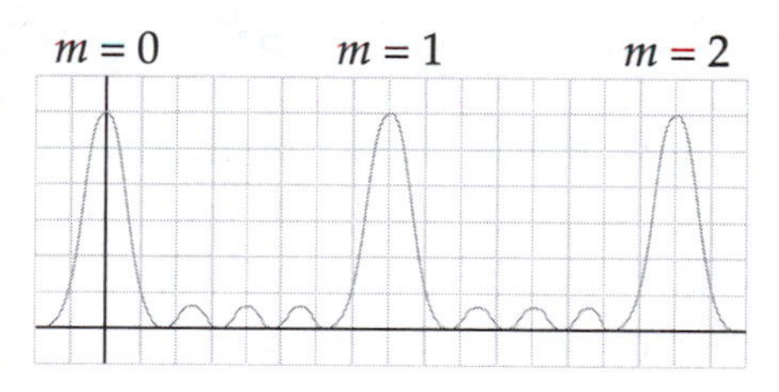

The zeroth-, first-, and second-order principal maxima of a single-wavelength diffraction pattern.

The figure illustrates the case for which the incident light contains a single wavelength component. Although not shown in the figure above, there will also be maxima corresponding to $m = -1, -2, \ldots$. These maxima occur at angles θ, measured from the line perpendicular to the grating, where $m\lambda = d \sin\theta$.

If the incident light contains a second wavelength component, a second series of principal maxima (in general with a different intensity) will be present in the diffraction pattern.

In a given spectral order, denoted by the number m, there will be principal maxima corresponding to each wavelength component incident on the grating.

24.9 Polarization of Light Waves

The electric field vector of a light wave vibrates in a plane perpendicular to the direction of propagation. In the figure on the right, the direction of propagation is out of the page. As illustrated in the top figure, for **unpolarized light,** the electric field vector vibrates along all directions in the plane perpendicular to the direction of travel with equal probability. However, at a given point and at a particular instant, there is only one resultant electric field direction (shown by the electric field vector $\vec{\mathbf{E}}$).

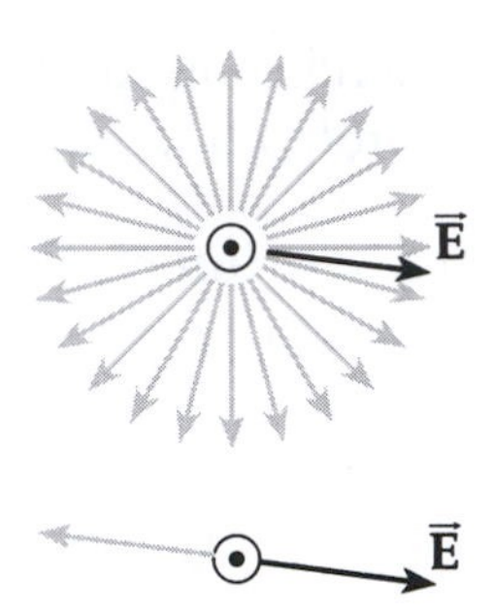

In **linearly polarized light** the electric field vibrates along the same direction at all times at a particular point, as illustrated in the bottom figure.

It is possible to obtain a linearly polarized wave from an unpolarized wave by removing from the unpolarized wave all components except those whose electric field vectors oscillate in a single plane. The plane formed by the direction of vibration of the electric field and the direction of propagation is called the **plane of polarization.**

Three important processes for producing linearly polarized light are: (1) selective absorption, (2) reflection, and (3) scattering.

Optical activity is the property of certain materials to cause rotation of the plane of polarization as a light beam travels through the material.

EQUATIONS AND CONCEPTS

In **Young's double-slit experiment,** two slits, $\mathbf{S}_1$ and $\mathbf{S}_2$, separated by a distance d, serve as monochromatic coherent sources. The light intensity at any point on the screen is the resultant of light reaching the screen from both slits. As illustrated in the figure, a point P on the screen can be identified by the angle θ or by the distance y from the center of the screen: $y = L\tan\theta \approx L\sin\theta$.

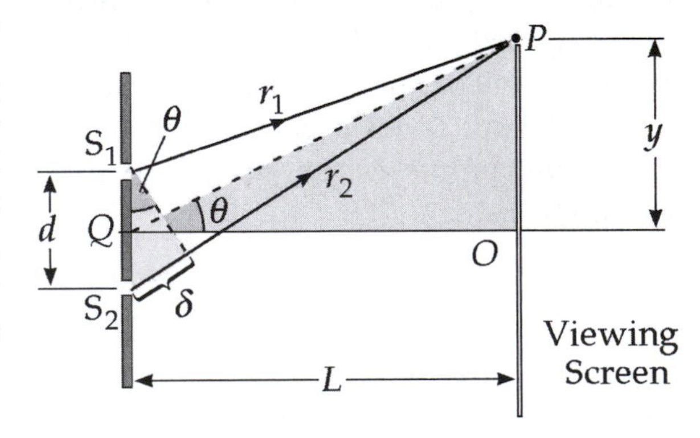

Geometry of Young's double-slit experiment

A **path difference** δ arises because waves from S_1 and S_2 travel unequal distances to reach a point on the screen (except the center point, O). *The value of δ determines whether the waves from the two slits arrive in phase or out of phase.*

$$\delta = r_2 - r_1 = d\sin\theta \qquad (24.1)$$

Constructive interference (bright fringes) will appear at points on the screen for which the path difference is equal to an integral multiple of the wavelength. The positions of bright fringes can also be located by calculating their vertical distance from the center of the screen (y). *In each case, the number m is called the order number of the fringe.* The central bright fringe ($\theta = 0°$, $m = 0$) is called the **zeroth-order maximum**, and the first maximum on either side of the central maximum is called the **first-order maximum.** Equations 24.5 and 24.6 assume that the angle θ in the figure on the previous page is small (less than 5°).

Conditions for constructive interference (two slits):

$$\delta = d\sin\theta_{\text{bright}} = m\lambda \qquad (24.2)$$

$m = 0, \pm 1, \pm 2, \ldots$

$$y_{\text{bright}} = \left(\frac{\lambda L}{d}\right)m \qquad (24.5)$$

$m = 0, \pm 1, \pm 2, \ldots$

m is called the order number

Dark fringes (destructive interference) will appear at points on the screen which correspond to path differences of an odd multiple of half wavelengths. For these points of destructive interference, waves which leave the two slits in phase arrive at the screen 180° (one-half wavelength) out of phase.

Conditions for destructive interference (two slits):

$$\delta = d\sin\theta_{\text{dark}} = \left(m + \tfrac{1}{2}\right)\lambda \qquad (24.3)$$

$m = 0, \pm 1, \pm 2, \ldots$

$$y_{\text{dark}} = \left(\frac{\lambda L}{d}\right)\left(m + \tfrac{1}{2}\right) \qquad (24.6)$$

$m = 0, \pm 1, \pm 2, \ldots$

The **wavelength of light, λ_n, in a medium having refractive index n,** is less than the wavelength in vacuum, λ.

$$\lambda_n = \frac{\lambda}{n} \qquad (24.7)$$

Interference in thin films depends on wavelength, film thickness, and the indices of refraction of the film and surrounding media. *Differences in phase may be due to path difference, phase change upon reflection, or both.* There are two general cases as described below.

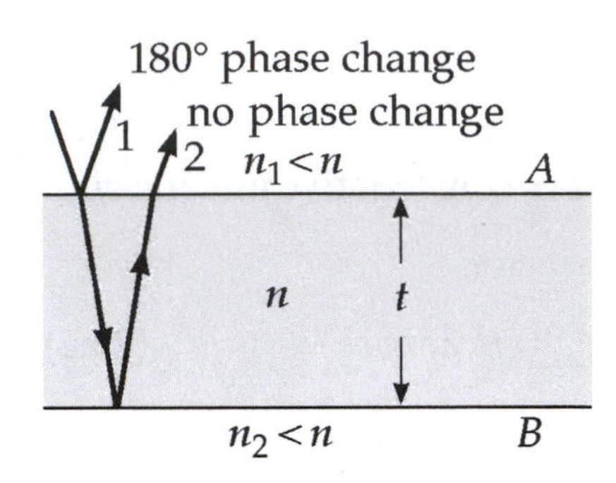

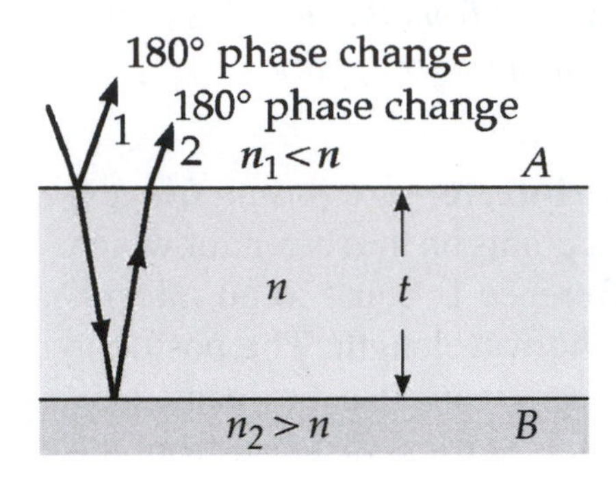

Case (I) Phase change at only one film surface

Indices of refraction of media on both sides of the film are less than that of the film ($n_1 < n$ and $n_2 < n$) as shown in the figure above left; or indices on both both sides are greater than that of the film ($n_1 > n$ and $n_2 > n$).

Constructive interference

$$2nt = \left(m + \tfrac{1}{2}\right)\lambda \quad (m = 0, 1, 2, \ldots) \tag{24.9}$$

Destructive interference

$$2nt = m\lambda \quad (m = 0, 1, 2, \ldots) \tag{24.10}$$

Case (II) Phase changes at both or neither surface

Film is between two media, either of which has an index of refraction greater than that of the film and the other a smaller index ($n_1 < n < n_2$) as shown in the figure above right, or ($n_1 > n > n_2$).

Constructive interference

$$2nt = m\lambda \quad (m = 0, 1, 2, \ldots)$$

Destructive interference

$$2nt = \left(m + \tfrac{1}{2}\right)\lambda \quad (m = 0, 1, 2, \ldots)$$

Note that the roles of the equations in Case (I) and Case (II) are reversed for constructive and destructive interference.

The diffraction pattern of a single slit consists of a broad central bright band and series of less intense and narrower side bands as illustrated in the figure on the right. The pattern is produced by interference of light waves originating from different regions of the slit.

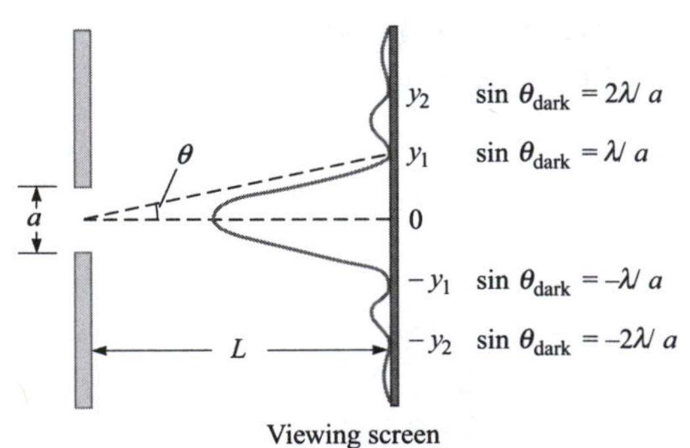

Single slit-diffraction pattern

The **general condition for destructive interference** (dark band) at a given point on the screen can be stated in terms of the angle θ_{dark}. Equation 24.11 gives the angle θ_{dark} for which the intensity will be zero (or minimum).

For destructive interference:

$$\sin\theta_{dark} = m\frac{\lambda}{a} \qquad (24.11)$$

$m = \pm 1, \pm 2, \pm 3, \ldots$

A **diffraction grating** has many narrow and equally-spaced parallel slits. The grating produces a diffraction pattern with a series of maxima (bright lines) for each different wavelength in the incident light. When the grating spacing (d) is known, the wavelength (λ) can be calculated by measuring the angle (θ_{bright}) for a given spectral line. Maxima due to wavelengths of different values comprise a spectral order denoted by an **order number** (m).

$$d\sin\theta_{bright} = m\mathtt{l} \qquad (24.12)$$

$m = 0, \pm 1, \pm 2, \ldots$

m is the order number of the diffraction pattern.

Malus's law states that the fraction of initially polarized light (from a polarizer) that will be transmitted by a second sheet of polarizing material (the analyzer) depends on the square of the cosine of the angle θ between the transmission axis of the polarizer and that of the analyzer. The transmitted intensity is a maximum when the transmission axes are parallel ($\theta = 0°$ or $180°$). When the transmission axes are perpendicular to each other, the light is completely absorbed by the analyzer, and the transmitted intensity is zero.

$$I = I_0 \cos^2\theta \qquad (24.13)$$

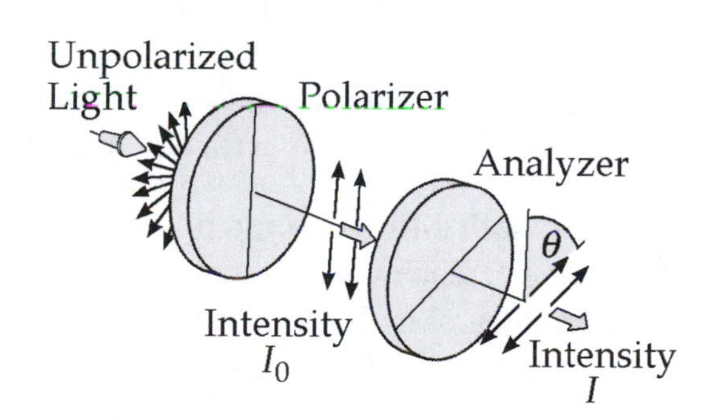

I_0 is the intensity of the beam incident on the analyzer after being transmitted by the polarizer.

The **polarizing angle** is the angle of incidence for which light incident from air and reflected from a surface of index of refraction, n, will be completely polarized with the direction of polarization parallel to the surface. Under this condition, the transmitted light will be partially polarized. Equation 24.14 is known as **Brewster's law**.

$$n = \tan\theta_p \qquad (24.14)$$

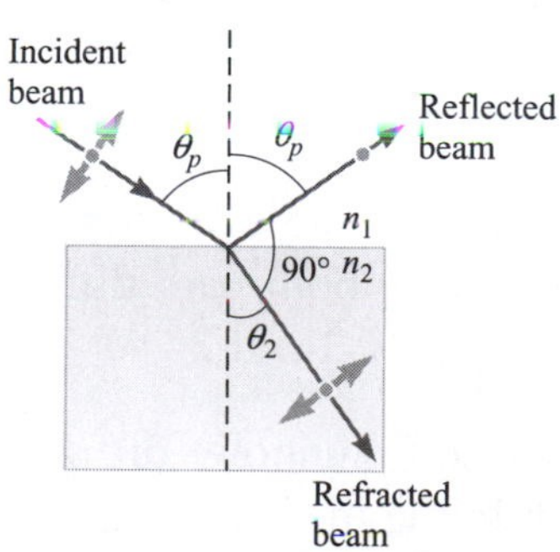

Polarization by reflection

SUGGESTIONS, SKILLS, AND STRATEGIES

THIN-FILM INTERFERENCE PROBLEMS

1. Identify the thin film from which interference effects are being observed.

2. The type of interference (constructive or destructive) that occurs in a specific situation is determined by the phase relationship between that portion of the wave reflected at the upper surface of the film and that portion reflected at the lower surface of the film.

3. The phase difference between the two portions of the reflected wave is determined by: (i) the physical path difference and (ii) any phase changes which may occur at either surface upon reflection. *You must consider both potential contributions to the total phase difference.*

   Path difference – The wave reflected from the lower surface of the film has to travel a distance equal to **twice the thickness of the film** before it returns to the upper surface of the film where it interferes with that portion of the wave reflected at the upper surface.

   Phase change upon reflection – In addition to path difference, reflections may change the predicted interference results. When a wave traveling in a particular medium reflects off a surface having a higher index of refraction than the one in which it was initially traveling, a 180° phase shift occurs. This has the same effect as if the wave lost $\frac{1}{2}\lambda$. Phase changes due to reflection must be considered in addition to the extra distance traveled by the reflected wave.

4. **When path difference and phase changes upon reflection are both taken into account,** the interference will be constructive if the two waves are out of phase ("out of step") by an integral multiple of λ_n (the wavelength as measured in the film). Destructive interference will occur when they differ in phase by an odd number of half-wavelengths.

REVIEW CHECKLIST

- Describe Young's double-slit experiment to demonstrate the wave nature of light. Account for the phase difference between light waves from the two sources as they arrive at a given point on the screen. State the conditions for constructive and destructive interference in terms of each of the following: path difference (δ), distance from center of screen (y), and the angle between the perpendicular bisector of the two sources and the location on the screen (θ). (Section 24.2)

- Account for the conditions of constructive and destructive interference in thin films considering both path difference and any expected phase changes due to reflection. (Sections 24.3 and 24.4)

- Describe Fraunhofer diffraction produced by a single slit. Determine the positions of the minima in a single-slit diffraction pattern. (Sections 24.6 and 24.7)

- Determine the positions of the principal maxima (or calculate wavelength values) in spectra formed by a diffraction grating. (Section 24.8)
- Describe qualitatively the polarization of light by selective absorption, reflection, and scattering. Also, make appropriate calculations using Brewster's law and Malus's law. (Section 24.9)

SOLUTIONS TO SELECTED END-OF-CHAPTER PROBLEMS

3. A pair of narrow, parallel slits separated by 0.250 mm is illuminated by the green component from a mercury vapor lamp ($\lambda = 546.1$ nm). The interference pattern is observed on a screen 1.20 m from the plane of the parallel slits. Calculate the distance (a) from the central maximum to the first bright region on either side of the central maximum and (b) between the first and second dark bands in the interference pattern.

Solution

When waves pass through a pair of parallel slits, the slits act as a set of coherent sources. From the sketch at the right, observe that the difference in path lengths is $\delta = d\sin\theta$, where d is the space between centers of the two slits, and θ is measured from the line perpendicular to the plane of the slits.

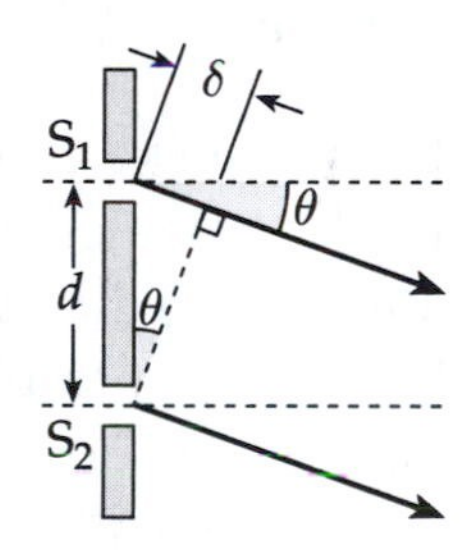

(a) To have constructive interference, the path difference δ must be an integral number of wavelengths. Thus, bright fringes occur at angles θ_m where

$$\delta = d\sin\theta_m = m\lambda \qquad m = 0, \pm1, \pm2, \pm3, \ldots$$

In most cases, with slits illuminated by visible light, the angles θ_m are very small and $\sin\theta_m \approx \tan\theta_m$. Then, the distance from the central maximum to the bright fringe of order m on a screen located distance L from the slits is given by

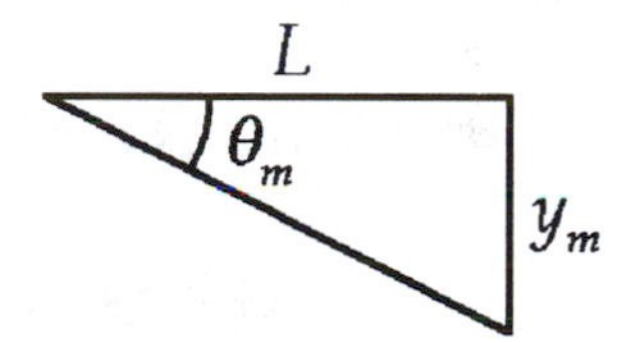

$$y_m = L\tan\theta_m \approx L\sin\theta_m = m\left(\frac{\lambda L}{d}\right)$$

Since $m = 0$ for the central maximum, and $m = \pm1$ for the first bright fringe on either side of the central maximum, the distance from the central maximum to the first bright fringe is

$$y_1 = (1)\frac{\lambda L}{d} = \frac{(546.1\times10^{-9}\ \text{m})(1.20\ \text{m})}{0.250\times10^{-3}\ \text{m}} = 2.62\times10^{-3}\ \text{m} = 2.62\ \text{mm} \qquad \Diamond$$

(b) For destructive interference and a dark fringe, the path difference δ must be an odd number of half-wavelengths. Thus, dark fringes occur where

$$\delta = d\sin\theta_m = \left(m+\tfrac{1}{2}\right)\lambda \quad \text{and} \quad y_m = \left(m+\tfrac{1}{2}\right)\frac{\lambda L}{d} \qquad m = 0, \pm1, \pm2, \pm3, \ldots$$

The spacing between the first ($m = 0$) and second ($m = 1$) dark fringes is then

$$\Delta y = y_1 - y_0 = \frac{3\lambda L}{2d} - \frac{\lambda L}{2d} = \frac{\lambda L}{d} = \frac{(546.1\times10^{-9}\text{ m})(1.20\text{ m})}{0.250\times10^{-3}\text{ m}} = 2.62\text{ mm} \quad \Diamond$$

---

15. Waves from a radio station have a wavelength of 300 m. They travel by two paths to a home receiver 20.0 km from the transmitter. One path is a direct path, and the second is by reflection from a mountain directly behind the home receiver. What is the minimum distance from the mountain to the receiver that produces destructive interference at the receiver? (Assume that no phase change occurs on reflection from the mountain.)

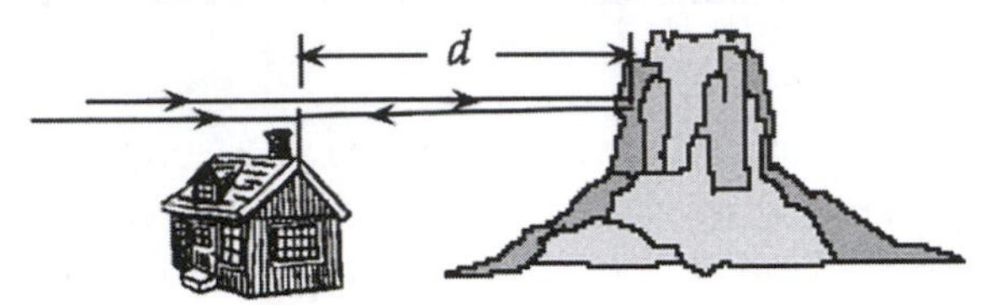

Solution

From the sketch given above, observe that one part of the radio wave is received directly by the home antenna. Another part of the wave bypasses the antenna, travels distance d to a mountain directly behind the antenna, reflects, and travels distance d back to the antenna before being received. The difference in the path lengths the two waves have traveled before being received is $\delta = 2d$. If there is no phase change upon reflection from the mountain, the waves will be received in phase and interfere constructively, if $\delta = m\lambda$, where λ is the wavelength of the waves and m is any positive integer. However, if δ is an any odd multiple of a half-wavelength [i.e., $\delta = (m + \frac{1}{2})\lambda$], the waves will be received out of phase and will interfere destructively, producing a minimum in the received signal strength.

Thus, to have destructive interference, it is necessary that

$$\delta = 2d = \left(m + \frac{1}{2}\right)\lambda, \qquad m = 0,\ 1,\ 2,\ 3,\ \ldots$$

and the distance from the home to the mountain should be

$$d = \left(m + \frac{1}{2}\right)\frac{\lambda}{2}$$

The smallest acceptable value of m is $m = 0$. Therefore, the minimum distance from the mountain to the receiver that will produce destructive interference is

$$d_{\min} = \frac{\lambda}{4} = \frac{300\text{ m}}{4} = 75.0\text{ m} \quad \Diamond$$

---

23. Astronomers observe the chromosphere of the Sun with a filter that passes the red hydrogen spectral line of wavelength 656.3 nm, called the H_α line. The filter consists of a transparent dielectric of thickness d held between two partially aluminized glass plates. The filter is kept at a constant temperature. (a) Find the minimum value of d that will produce maximum transmission of perpendicular H_α light if the dielectric has an index of refraction of 1.378. (b) If the temperature of the filter increases above the normal value increasing its thickness, what happens to the transmitted wavelength? (c) The dielectric will also pass what near-visible wavelength? One of the glass plates is colored red to absorb this light.

Solution

(a) First, we realize that maximum transmission occurs when the reflected light is a minimum. Thus, the condition for constructive interference and a maximum in the transmitted light is the same as the condition for destructive interference and a minimum in the reflected light. We assume the index of refraction of the glass on each side of the dielectric film is greater than that of the film (i.e., $n_{glass} > n_{film}$). Then light reflecting from the upper surface of the film, a glass to film transition, will not experience a phase reversal. However, light reflecting from the lower surface, a film to glass transition, will experience a phase reversal or a half-wavelength phase shift. Under these conditions, destructive interference in the reflected light occurs when $2d = m\lambda_n = m\lambda/n_{film}$, where d is the thickness of the film, n_{film} is the refractive index of the film, λ is the wavelength (in vacuum) of the light, and m is any positive integer. Therefore, the minimum nonzero value of d that will yield maximum transmission corresponds to $m = 1$, and is given by

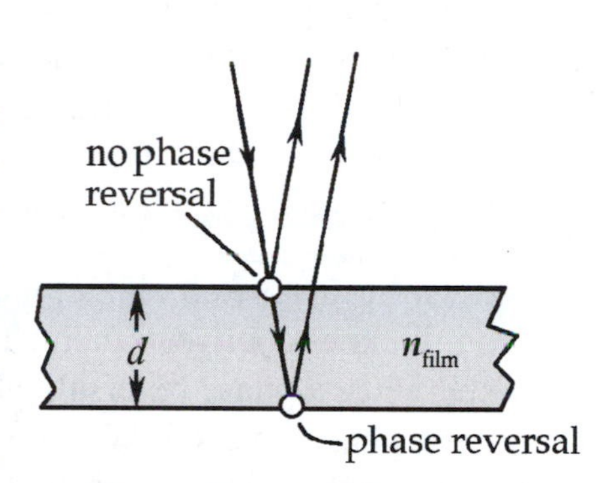

$$d_{min} = \frac{(1)\lambda}{2n_{film}} = \frac{(1)(656.3 \text{ nm})}{2(1.378)} = 238 \text{ nm}$$ ◊

(b) The filter will expand as the temperature rises, so the thickness d will increase. This means that the wavelength, $\lambda = 2n_{film}d/m$, associated with the first-order (or $m = 1$) transmission maximum, will also increase. ◊

(c) For the minimum thickness of the filter, the wavelength for which there will be a second-order (or $m = 2$) maximum in transmission is

$$\lambda = 2n_{film}d_{min}/m = \cancel{2}(1.378)(238 \text{ nm})/\cancel{2} = 328 \text{ nm}$$ ◊

27. A plano-convex lens rests with its curved side on a flat glass surface and is illuminated from above by light of wavelength 500 nm. (See Fig. 24.8.) A dark spot is observed at the center, surrounded by 19 concentric dark rings (with bright rings in between). How much thicker is the air wedge at the position of the 19th dark ring than at the center?

Solution

When light reflects while trying to cross a boundary going from one medium into a second medium with a higher index of refraction, it experiences a phase reversal or a 180° shift in phase. If the reflection occurs as the light is attempting to go from one medium into one of lower index of refraction, no such phase reversal occurs.

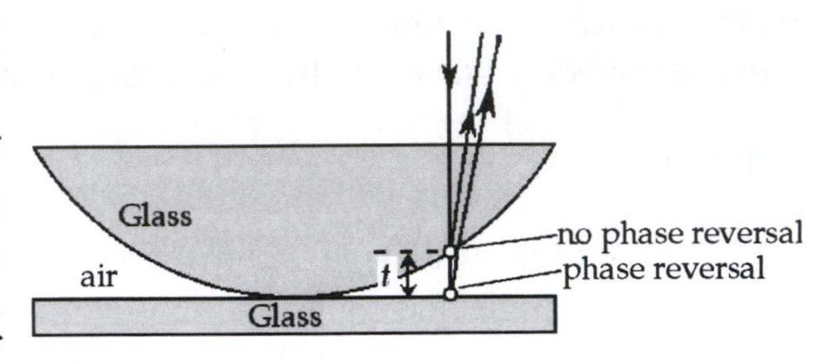

As shown in the sketch above, the plano-convex lens and glass plate have an air film of varying thickness between them. The light reflecting at the upper surface of this film does not experience a phase reversal, while light reflecting at the lower surface of the film does undergo a phase reversal. With only one reversal involved with the two reflections, Equation 24.10 in the textbook gives the condition for destructive interference in the reflected light as

$$2nt = m\lambda \qquad m = 0, 1, 2, 3, \ldots$$

where n is the index of refraction of the film (air in this case), t is the thickness of the film at the point of reflection, and λ is the wavelength of the light.

Note that $m = 0$ corresponds to a thickness of $t = 0$. Hence, there will be a dark spot in the interference pattern at the point of contact between the lens and flat plate. Higher values of m correspond to dark fringes in the interference pattern. In this case, the fringes will be circular in shape, centered on the central dark spot, and located where the air film has the thickness associated with that value of m in the above expression. The 19th dark ring corresponds to $m = 19$, and occurs where the film has thickness

$$t = \frac{m\lambda}{2n_{\text{air}}} = \frac{(19)(500\times10^{-9}\text{ m})}{2(1.00)} = 4.75\times10^{-6}\text{ m} = 4.75\ \mu\text{m} \qquad \Diamond$$

31. Light of wavelength 5.40×10^{2} nm passes through a slit of width 0.200 mm. (a) Find the width of the central maximum on a screen located 1.50 m from the slit. (b) Determine the width of the first-order bright fringe.

Solution

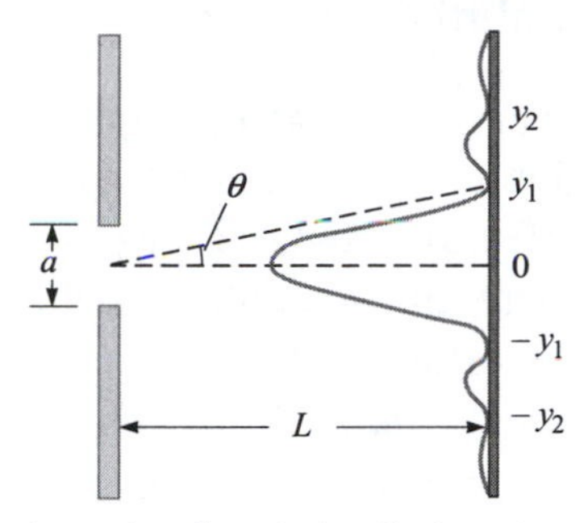

Light passing through different portions of a single slit interferes in such a way as to produce maxima and minima in intensity on a screen located at distance L beyond the slit as shown in the sketch at the right. The minima or dark fringes occur where

$$\sin\theta = m\lambda/a \quad \text{with} \quad m = \pm1, \pm2, \pm3, \ldots$$

When the slit width, a, is very large in comparison to the wavelength, λ, of the light, the angles θ are very small. In this case, $\sin\theta \approx \tan\theta = y/L$, so the locations of the minima on the screen are given by

$$y_m = mL\left(\frac{\lambda}{a}\right) \quad \text{with} \quad m = \pm1, \pm2, \pm3, \ldots$$

(a) When light having wavelength $\lambda = 5.40\times10^{2}$ nm is incident on a screen located distance $L = 1.50$ m from a slit of width $a = 0.200$ mm, the distance from the center of the central maximum to the first-order minimum $(m = 1)$ is

$$y_1 = \frac{(1)L\lambda}{a} = \frac{(1)(1.50 \text{ m})(5.40\times10^{2} \text{ nm})}{0.200\times10^{-3} \text{ m}}\left(\frac{10^{-9} \text{ m}}{1 \text{ nm}}\right) = 4.05\times10^{-3} \text{ m} = 4.05 \text{ mm}$$

The central maximum extends from the location of the first minimum on one side of the $y = 0$ position to the location of the first minimum on the other side of this position. Thus, the full width of the central maximum is

$$\Delta y_{\substack{\text{central}\\ \text{maximum}}} = 2y_1 = 2(4.05 \text{ mm}) = 8.10 \text{ mm} \qquad \Diamond$$

(b) The first-order bright fringe extends from the location of the first-order $(m = 1)$ minimum to the location of the second-order $(m = 2)$ minimum, both on the same side of the central maximum. Thus, the width of the first-order bright fringe is $\Delta y_{\substack{\text{bright}\\ \text{fringe}}} = y_2 - y_1 = (2)L(\lambda/a) - (1)L(\lambda/a) = L(\lambda/a)$, or

$$\Delta y_{\substack{\text{bright}\\ \text{fringe}}} = \frac{(1.50 \text{ m})(5.40\times10^{-7} \text{ m})}{0.200\times10^{-3} \text{ m}} = 4.05\times10^{-3} \text{ m} = 4.05 \text{ mm} \qquad \Diamond$$

35. A beam of monochromatic light is diffracted by a slit of width 0.600 mm. The diffraction pattern forms on a wall 1.30 m beyond the slit. The width of the central maximum is 2.00 mm. Calculate the wavelength of the light.

Solution

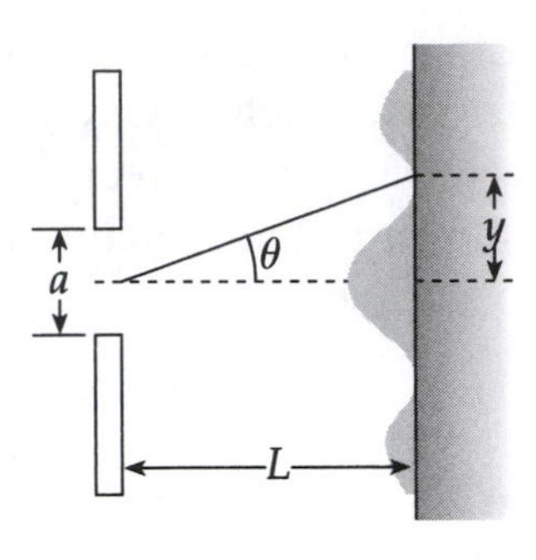

The sketch at the right illustrates the maxima and minima in intensity that make up the diffraction pattern formed when light of wavelength λ passes through a single slit of width a. The peak of the central maximum in this pattern occurs at the $y = 0$ position on the screen, and the minima in intensity occur where $\sin\theta = m\lambda/a$, with m being any nonzero integer. The width of the central maximum is $\Delta y = 2y = 2L\tan\theta$, where L is the perpendicular distance from the slit to the screen, and θ is the angular position of the first-order ($m = 1$) minimum on either side of the central maximum. When $a \gg \lambda$, we may use the small angle approximation, $\sin\theta \approx \tan\theta$, to express the width of the central maximum as

$$\Delta y = 2L\tan\theta \approx 2L\sin\theta = 2L\left[\frac{(1)\lambda}{a}\right] = \frac{2L\lambda}{a}$$

In this approximation, the wavelength of the light is given in terms of the width of the central maximum as $\lambda = a(\Delta y)/2L$. Given that $\Delta y = 2.00$ mm when $a = 0.600$ mm and $L = 1.30$ m, we find the wavelength of the light illuminating the slit to be

$$\lambda = \frac{(0.600\times10^{-3}\text{ m})(2.00\times10^{-3}\text{ m})}{2(1.30\text{ m})} = 4.62\times10^{-7}\text{ m} = 462\text{ nm} \qquad \Diamond$$

---

39. Three discrete spectral lines occur at angles of 10.1°, 13.7°, and 14.8°, respectively, in the first-order spectrum of a diffraction-grating spectrometer. (a) If the grating has 3 660 slits/cm, what are the wavelengths of the light? (b) At what angles are these lines found in the second-order spectra?

Solution

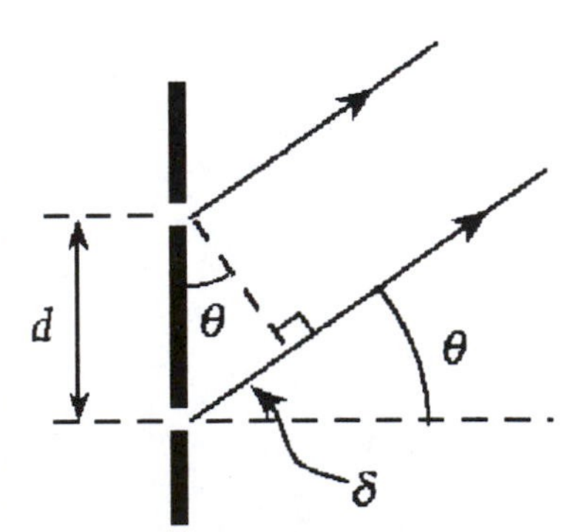

When light diffracts through a grating, constructive interference occurs, and bright lines are found when the path difference for light coming through two adjacent slits of the grating is an integer multiple of the wavelength of the light. The sketch at the right shows rays coming from two adjacent slits, separated by distance d. From this sketch, observe that the difference in path lengths for these rays is $\delta = d\sin\theta$, where θ is measured from the line perpendicular to the plane of the grating. The requirement for constructive interference, and formation of a bright line, is then

$$\delta = d\sin\theta = m\lambda \qquad \text{where} \qquad m = 0, 1, 2, 3, \ldots$$

For the described grating, the distance between adjacent slits is

$$d = \frac{1\text{ cm}}{3\,660\text{ slits}} = \frac{10^{-2}\text{ m}}{3\,660} = 2.732\times10^{-6}\text{ m} = 2\,732\text{ nm}$$

(a) The wavelength of light in a first-order ($m = 1$) bright line found at angle θ is $\lambda = d\sin\theta$. Thus,

at $\theta = 10.1°$: $\lambda_1 = d\sin\theta_1 = (2\,732\text{ nm})\sin 10.1° = 479\text{ nm}$ ◊

at $\theta = 13.7°$: $\lambda_2 = d\sin\theta_2 = (2\,732\text{ nm})\sin 13.7° = 647\text{ nm}$ ◊

at $\theta = 14.8°$: $\lambda_3 = d\sin\theta_3 = (2\,732\text{ nm})\sin 14.8° = 698\text{ nm}$ ◊

(b) In the second-order, $m = 2$, and the angle at which these wavelengths will form bright lines are:

for $\lambda_1 = 479\text{ nm}$: $\theta = \sin^{-1}(2\lambda_1/d) = \sin^{-1}[2(479\text{ nm})/2\,732\text{ nm}] = 20.5°$ ◊

for $\lambda_2 = 647\text{ nm}$: $\theta = \sin^{-1}(2\lambda_2/d) = \sin^{-1}[2(647\text{ nm})/2\,732\text{ nm}] = 28.3°$ ◊

for $\lambda_3 = 698\text{ nm}$: $\theta = \sin^{-1}(2\lambda_3/d) = \sin^{-1}[2(698\text{ nm})/2\,732\text{ nm}] = 30.7°$ ◊

---

47. Sunlight is incident on a diffraction grating that has 2 750 lines/cm. The second-order spectrum over the visible range (400–700 nm) is to be limited to 1.75 cm along a screen that is a distance L from the grating. What is the required value of L?

Solution

The spacing between adjacent slits on this grating is

$$d = \frac{1\text{ cm}}{2\,750\text{ slits}} = \frac{10^{-2}\text{ m}}{2.75\times10^{3}\text{ slits}} = 3.64\times10^{-6}\text{ m} = 3\,640\text{ nm}$$

From the grating equation, $d\sin\theta = m\lambda$, where $m = 1, 2, 3, \ldots$, the angles where the red and violet edges are located in the second-order ($m = 2$) spectrum are

$$\theta_r = \sin^{-1}\left(\frac{2\lambda_{\text{red}}}{d}\right) = \sin^{-1}\left[\frac{2(700\text{ nm})}{3\,640\text{ nm}}\right] = 22.6°$$

and

$$\theta_v = \sin^{-1}\left(\frac{2\lambda_{\text{violet}}}{d}\right) = \sin^{-1}\left[\frac{2(400\text{ nm})}{3\,640\text{ nm}}\right] = 12.7°$$

Observe from the sketch at the right, the screen positions (measured from the location of the central maximum) of the edges of this second-order spectrum are $y_r = L\tan\theta_r$ and $y_v = L\tan\theta_v$. Thus, the width of the second-order spectrum on the screen is

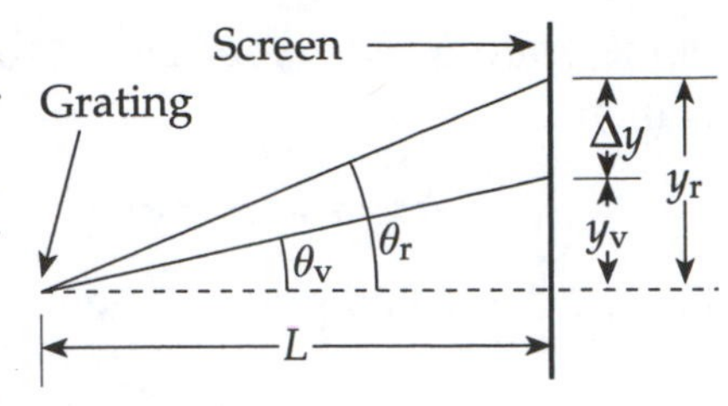

$$\Delta y = y_r - y_v = L\tan\theta_r - L\tan\theta_v = L\left(\tan\theta_r - \tan\theta_v\right)$$

If it is desired to have $\Delta y = 1.75$ cm as given, the distance between the grating and the screen must be

$$L = \frac{\Delta y}{\tan\theta_r - \tan\theta_v} = \frac{1.75 \text{ cm}}{\tan(22.6°) - \tan(12.7°)} = 9.17 \text{ cm}$$ ◊

---

51. The angle of incidence of a light beam in air onto a reflecting surface is continuously variable. The reflected ray is found to be completely polarized when the angle of incidence is 48.0°. (a) What is the index of refraction of the reflecting material? (b) If some of the incident light (at an angle of 48.0°) passes into the material below the surface, what is the angle of refraction?

Solution

If, when light is incident on a surface, the reflected beam and the refracted beam are perpendicular to each other as shown in the figure at the right, the reflected beam will be completely polarized with the electric field vector parallel to the surface. The angle of incidence, θ_p, at which this occurs is called the polarizing angle or the Brewster's angle.

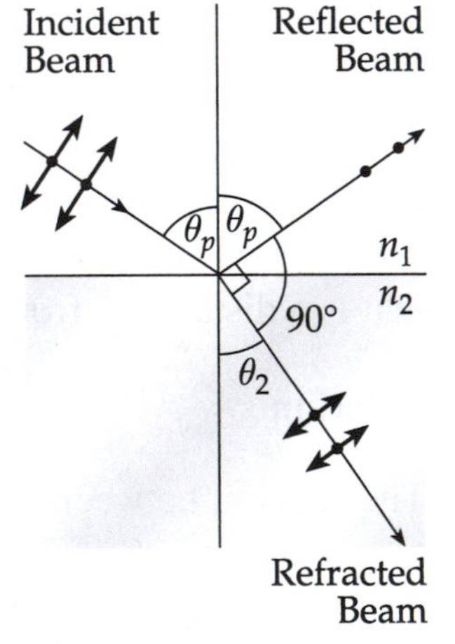

From the figure, observe that $\qquad \theta_p + 90.0° + \theta_2 = 180°$

or $\qquad \theta_2 = 90.0° - \theta_p$

That is, when the angle of incidence is the polarizing angle, the angle of refraction is the complement of that angle, and using the trigonometric identity for the sine of the difference of two angles gives

$$\sin\theta_2 = \sin\left(90.0° - \theta_p\right) = \sin 90.0° \cos\theta_p - \cos 90.0° \sin\theta_p = \cos\theta_p$$

Thus, for light incident at the polarizing angle, Snell's law, $n_1 \sin\theta_1 = n_2 \sin\theta_2$, becomes

$$n_1 \sin\theta_p = n_2 \cos\theta_p \qquad \text{or} \qquad \frac{\sin\theta_p}{\cos\theta_p} = \tan\theta_p = \frac{n_2}{n_1}$$

(a) If the incident light is in air ($n_1 = 1.00$) and the polarizing angle is found to be $\theta_p = 48.0°$, the index of refraction of the reflecting material is

$$n_2 = n_1 \tan\theta_p = (1.00)\tan 48.0° = 1.11 \qquad \lozenge$$

(b) When the angle of incidence equals the polarizing angle, $\theta_1 = \theta_p = 48.0°$, the angle of refraction for the transmitted light is

$$\theta_2 = 90.0° - \theta_p = 90.0° - 48.0° = 42.0° \qquad \lozenge$$

---

61. Light with a wavelength in vacuum of 546.1 nm falls perpendicularly on a biological specimen that is 1.000 μm thick. The light splits into two beams polarized at right angles, for which the indices of refraction are 1.320 and 1.333, respectively. (a) Calculate the wavelength of each component of the light while it is traversing the specimen. (b) Calculate the phase difference between the two beams when they emerge from the specimen.

Solution

(a) The index of refraction of a material is defined a $n = c/v$, where c is the speed of light in vacuum, and v is the speed at which light travels in this material. Thus, the wavelength of light in this material is given by

$$\lambda_n = \frac{v}{f} = \frac{c/n}{f} = \frac{c/f}{n} = \frac{\lambda_0}{n}$$

where $\lambda_0 = c/f$ is the wavelength light of this frequency would have in vacuum. Therefore, if light having wavelength $\lambda_0 = 546.1$ nm in vacuum splits into two components polarized at right angles to each other when in the material, and the indices of refraction for these components are $n_1 = 1.320$ and $n_2 = 1.333$, respectively, the wavelengths of these components in the material are

$$\lambda_1 = \frac{\lambda_0}{n_1} = \frac{546.1\text{ nm}}{1.320} = 413.7\text{ nm} \quad \text{and} \quad \lambda_2 = \frac{\lambda_0}{n_2} = \frac{546.1\text{ nm}}{1.333} = 409.7\text{ nm} \qquad \lozenge$$

(b) The number of cycles of vibration each of these components will complete as it travels through a specimen of thickness $t = 1.000\ \mu$m is

$$N_1 = \frac{t}{\lambda_1} = \frac{1.000\times10^{-6}\text{ m}}{413.7\times10^{-9}\text{ m}} = 2.417 \quad \text{and} \quad N_2 = \frac{t}{\lambda_2} = \frac{1.000\times10^{-6}\text{ m}}{409.7\times10^{-9}\text{ m}} = 2.441$$

Thus, at the end of this passage, the two components will be out of phase by

$$\Delta N = N_2 - N_1 = (2.441 - 2.417)\text{ cycles} = 0.024\text{ cycles}$$

Since each full cycle of oscillation represents 360° in phase, the angular measure of the phase difference of the two components at the end of the passage is

$$\Delta\phi = (\Delta N)(360°) = (0.024 \text{ cycles})(360°/\text{cycle}) = 8.6° \quad \lozenge$$

---

64. In a Young's interference experiment, the two slits are separated by 0.150 mm, and the incident light includes two wavelengths: $\lambda_1 = 540$ nm (green) and $\lambda_2 = 450$ nm (blue). The overlapping interference patterns are observed on a screen 1.40 m from the slits. (a) Find a relationship between the orders m_1 and m_2 that determines where a bright fringe of the green light coincides with a bright fringe of the blue light. (The order m_1 is associated with λ_1, and m_2 is associated with λ_2.) (b) Find the minimum values of m_1 and m_2 such that the overlapping of the bright fringes will occur and find the position of the overlap on the screen.

Solution

(a) In a double-slit interference pattern, the distance from the position of the central maximum to the location of the bright fringe of order m for light of wavelength λ is given by $y_m = m(\lambda L/d)$. Here, L is the distance from the plane of the slits to the plane of the viewing screen, and d is the distance separating the two slits.

Therefore, if the bright line of order m_1 for wavelength λ_1 is to coincide with the bright line of order m_2 for wavelength λ_2, it is necessary that

$$m_1 \frac{\lambda_1 \cancel{L}}{\cancel{d}} = m_2 \frac{\lambda_2 \cancel{L}}{\cancel{d}} \qquad \text{or} \qquad m_1\lambda_1 = m_2\lambda_2$$

where both m_1 and m_2 are integers. With $\lambda_1 = 540$ nm and $\lambda_2 = 450$ nm, this becomes $(540 \text{ nm})m_1 = (450 \text{ nm})m_2$. Dividing both sides of this equation by 90 nm, we obtain the relation of the overlapping order numbers as

$$6m_1 = 5m_2 \quad \lozenge$$

(b) Write the relation of part (a) as $m_2 = (6/5)m_1$ and look for integer values of m_1 that will yield an integer result for m_2. It is quickly discovered that the smallest such integer value for m_1 is $m_1 = 5$, which yields the integer result $m_2 = 6$. ◊

With $L = 1.40$ m and $d = 0.150$ mm, the distance on the screen between the central maximum and the position of the first overlap of bright fringes for the two given wavelengths is $y = m_1\lambda_1 L/d = m_2\lambda_2 L/d$, or

$$y = \frac{5(540\times10^{-9} \text{ m})(1.40 \text{ m})}{0.150\times10^{-3} \text{ m}} = \frac{6(450\times10^{-9} \text{ m})(1.40 \text{ m})}{0.150\times10^{-3} \text{ m}} = 2.52\times10^{-2} \text{ m} = 2.52 \text{ cm} \quad \lozenge$$

---

71. The transmitting antenna on a submarine is 5.00 m above the water when the ship surfaces. The captain wishes to transmit a message to a receiver on a 90.0-m-tall cliff at the ocean shore. If the signal is to be completely polarized by reflection off the ocean surface, how far must the ship be from the shore?

Solution

The sketch at the right shows the path the radio signal is to travel from the submarine's antenna to the receiver atop the cliff. If the signal is to be completely polarized by reflecting from the surface of the water, the angle of incidence when it strikes this surface must equal the polarizing angle, or Brewster's angle. For an air ($n_1 = 1.00$) and water ($n_2 = 1.333$) interface, this angle is

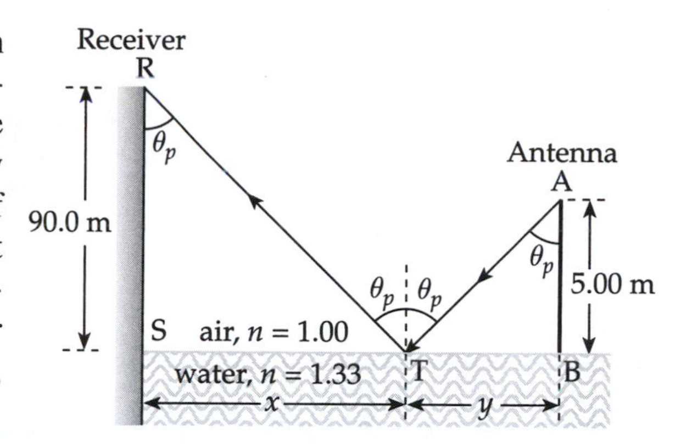

$$\theta_p = \tan^{-1}\left(\frac{n_2}{n_1}\right) = \tan^{-1}\left(\frac{1.333}{1.00}\right) = 53.1°$$

Thus, using triangle ABT in the sketch, we see that the distance from the submarine's antenna to the point where the signal must strike the water surface is

$$y = (5.00 \text{ m})\tan\theta_p = (5.00 \text{ m})\tan(53.1°) = 6.66 \text{ m}$$

Since the signal must obey the law of reflection, the angle the reflected ray will make with the vertical is $\theta_p = 53.1°$. Then, using triangle RST in the sketch, the required distance from the point of reflection to the shore is found to be

$$x = (90.0 \text{ m})\tan\theta_p = (90.0 \text{ m})\tan(53.1°) = 120 \text{ m}$$

The total distance the submarine must be from shore when the message is transmitted is now seen to be

$$d = y + x = 6.66 \text{ m} + 120 \text{ m} = 127 \text{ m}$$ ◊

25

Optical Instruments

NOTES FROM SELECTED CHAPTER SECTIONS

25.2 The Eye

The following terms are used in describing the image-forming mechanism of the eye:

- **Cornea:** transparent structure through which light enters the eye
- **Aqueous humor:** clear liquid behind the cornea
- **Pupil:** an opening in the iris, dilates and contracts to regulate light
- **Iris:** colored portion of the eye, muscular diaphragm that controls pupil size
- **Retina, rods and cones:** back surface of eye, where images are formed
- **Accommodation:** changing the focal length of the eye by action of the ciliary muscle
- **Near point:** the closest distance for which the lens can accommodate to focus a sharp image on the retina (average value of around 25 cm)
- **Far point:** the greatest distance for which the lens in a relaxed eye can focus a clear image on the retina (in a normal eye the far point is at infinity)
- **Hyperopia** (farsightedness): condition caused either when the eyeball is too short or when the ciliary muscle is unable to change the shape of the lens enough to form a properly focused image of a nearby object
- **Myopia** (nearsightedness): condition caused either when the eye is longer than normal or when the maximum focal length of the lens is insufficient to produce a clearly focused image of a distant object
- **Presbyopia:** a reduction in accommodation ability resulting in the symptoms of farsightedness
- **Astigmatism:** light from a point source produces a line image on the retina
- **Diopters:** a measure of the power of a lens which equals the inverse of the focal length in meters

25.4 The Compound Microscope

The overall magnification of a compound microscope of length L is equal to the product of the lateral magnification produced by the objective lens of focal length f_o (< 1 cm) and the angular magnification produced by the eyepiece lens of focal length f_e (a few centimeters). The two lenses, separated by a distance greater than either f_o or f_e, form an inverted, virtual image of an object.

25.5 The Telescope

There are two fundamentally different types of telescopes, both designed to aid in viewing distant objects, such as the planets in our solar system.

- The **refracting telescope** uses a combination of objective and eyepiece lenses to form an enlarged image. The two lenses are separated by a distance $L = f_e + f_o$.
- The **reflecting telescope** uses a concave mirror and an eyepiece lens to form an enlarged image.

For each type telescope, the angular magnification of the telescope equals the ratio of the focal length of the objective to the focal length of the eyepiece.

25.6 Resolution of Single-Slit and Circular Apertures

The wave nature of light limits, by diffraction, the minimum distance between objects which can be distinguished by an optical system.

Rayleigh's criterion determines the limiting condition for resolution of adjacent sources. Under this criterion two sources are just resolved when they are spaced so that the central maximum of the diffraction pattern of one source is located at the position of the first minimum of the diffraction pattern of the second source (as illustrated by the dashed vertical line). Under this condition they are separated by the minimum spacing for which they can be seen as separate sources. The curve shown in the figure as a dashed line is the sum of the diffraction patterns of the two separate sources.

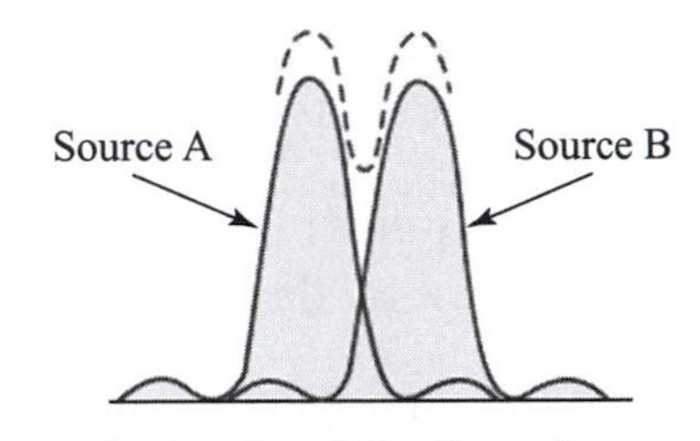

Overlapping diffraction patterns due to adjacent sources

The smallest angular separation that can be resolved is proportional to the wavelength of light used by the instrument and inversely proportional to the slit width or the diameter of a circular aperture.

The resolving power of a diffraction grating increases with the order number in the spectrum and is proportional to the number of lines illuminated in the grating. Refer to Equations 25.11 and 25.12 and consider the case of a diffraction grating in which 5 000 lines are illuminated.

In the zeroth order, $m = 0$ and all wavelengths are indistinguishable.

In the second order, $m = 2$ and the grating will produce a spectrum in which wavelengths differing in value by 1 part in 10 000 can be resolved.

25.7 The Michelson Interferometer

The Michelson interferometer can be used to measure distances to within a fraction of a wavelength of light. This is accomplished by splitting a light beam into two parts and then recombining the beams to form an interference pattern.

EQUATIONS AND CONCEPTS

The image-forming properties of several optical instruments are described in the following paragraphs and equations.

Camera

The ***f*-number** of a lens is the ratio of its focal length to its diameter. This value is a measure of the light-concentrating power of the lens and determines what is called the speed of the lens. *A fast lens has a small f-number, usually with a short focal length and a large diameter.*

$$f\text{-number} \equiv \frac{f}{D} \qquad (25.1)$$

Eye

The **power of a lens is measured in diopters** and equals the inverse of the focal length measured in meters. The correct algebraic sign must be used with *f*. *Optometrists and ophthalmologists usually prescribe lenses measured in diopters.*

$$P = \frac{1}{f}$$

f is + for a converging lens and − for a diverging lens.

Simple magnifier

The **angular magnification of a simple magnifier** is the ratio of the angle (θ) subtended by the image of an object when the object is placed near the focal point of a converging lens to the angle (θ_0) subtended by the object when it is placed at the near point of the eye (25 cm) with no lens.

$$m \equiv \frac{\theta}{\theta_0} \qquad (25.2)$$

When the **image is at the near point** of the eye (25 cm), the angular magnification of a simple magnifier is maximum.

$$m_{\text{max}} = 1 + \frac{25\text{ cm}}{f} \qquad (25.5)$$

When the **image is at infinity** (most relaxed for the eye) the angular magnification is minimum. This is the expression for the magnification of a simple lens when the object is placed at the focal point of the lens.

$$m = \frac{25\text{ cm}}{f} \qquad (25.6)$$

Compound microscope

The **overall magnification** is the product of the lateral magnification of the objective lens (M_1) and the angular magnification of the eyepiece (m_e). When an object is located just beyond the focal point of the objective, the two lenses in combination form an enlarged, virtual, and inverted image.
I_1 = image of object formed by objective lens.
I_2 = image of I_1 formed by the eyepiece.

$$M = M_1 m_e = -\frac{L}{f_o}\left(\frac{25\text{ cm}}{f_e}\right) \qquad (25.7)$$

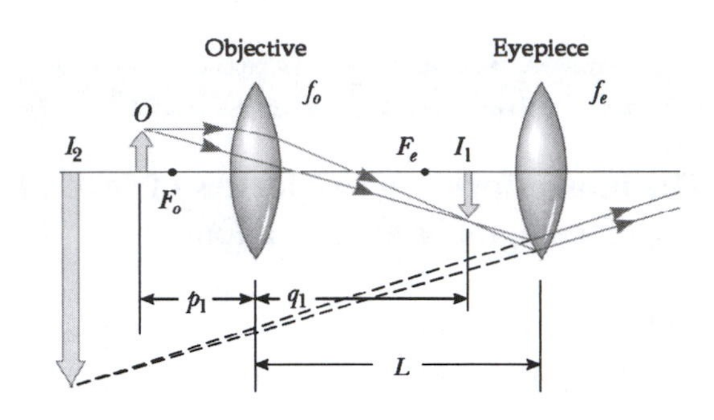

Astronomical telescope

The **angular magnification of a telescope** is equal to the ratio of the focal length of the objective to the focal length of the eyepiece. *The two converging lenses are separated by a distance equal to the sum of their focal lengths.*

$$m = \frac{f_0}{f_e} \qquad (25.8)$$

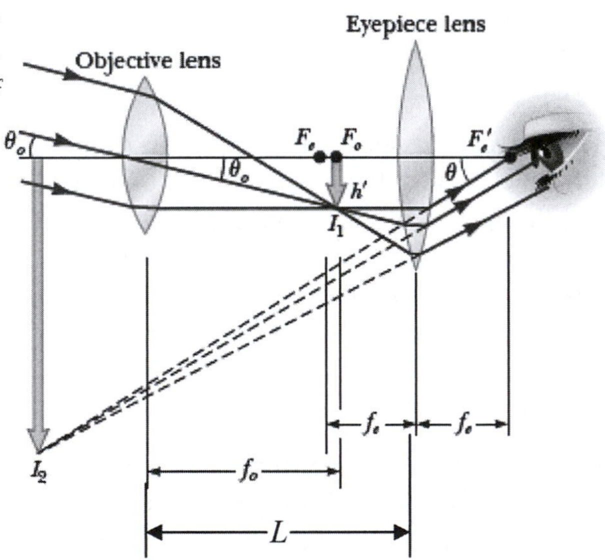

Refracting telescope; note that the two lenses are separated by a distance equal to the sum of their focal lengths.

Rayleigh's criterion states the condition for the resolution of the images of two closely spaced sources.

For a **slit**, the angular separation between sources (in radians) must be greater than the ratio of the wavelength of the light to the slit width, *a*.

$$\theta_{\text{min}} \approx \frac{\lambda}{a} \qquad (25.9)$$

For a **circular aperture,** the minimum angular separation which can be resolved depends on the wavelength of the light and the diameter of the aperture, *D*.

$$\theta_{\text{min}} = 1.22\frac{\lambda}{D} \qquad (25.10)$$

The **resolving power of a diffraction grating**, defined by Equation 25.11, increases as the number of lines illuminated, N, increases. Also, the resolving power is proportional to the order number, m, in which the spectrum is observed as stated in Equation 25.12.

$$R \equiv \frac{\lambda}{\lambda_2 - \lambda_1} = \frac{\lambda}{\Delta\lambda} \qquad (25.11)$$

where $\lambda \approx \lambda_1 \approx \lambda_2$

$$R = Nm \qquad (25.12)$$

REVIEW CHECKLIST

- Identify terms used in describing the image-forming mechanism of the eye. Describe the use of lenses in the correction of common defects of the eye. (Section 25.2)
- Define the f-number of a camera lens and relate this criterion to shutter speed. (Section 25.1)
- Describe the geometry of the optical components for each of several optical instruments: simple magnifier, compound microscope, reflecting telescope, and refracting telescope. Also, calculate the magnifying power for each instrument. (Sections 25.3, 25.4, and 25.5)
- Determine whether or not two sources under a given set of conditions are resolvable as defined by Rayleigh's criterion. (Section 25.6)
- Understand what is meant by the resolving power of a grating, and calculate the resolving power of a grating under specified conditions. (Section 25.6)
- Describe the technique employed in the Michelson interferometer for precise measurement of length based on known values for the wavelength of light. (Section 25.7)

SOLUTIONS TO SELECTED END-OF-CHAPTER PROBLEMS

3. A photographic image of a building is 0.092 0 m high. The image was made with a lens with a focal length of 52.0 mm. If the lens was 100 m from the building when the photograph was made, determine the height of the building.

Solution

From the thin-lens equation, we find

$$q = \frac{pf}{p-f} = \frac{(100\text{ m})(52.0\times 10^{-3}\text{ m})}{100\text{ m} - 52.0\times 10^{-3}\text{ m}} = 5.203\times 10^{-2}\text{ m} \approx f$$

Consider the figure below, showing two rays passing undeviated through the center of the lens to the top and bottom of the image.

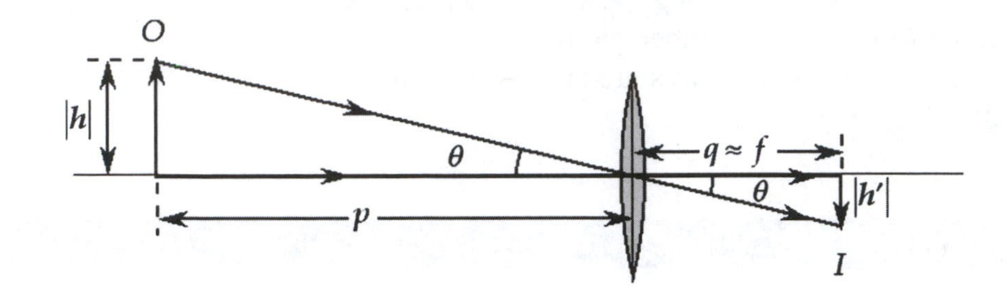

Using the properties of similar triangles, we write $|h|/|h'| = p/f$, or the size of the object must be

$$|h| = |h'|\left(\frac{p}{f}\right) = (0.092\,0\text{ m})\left(\frac{100\text{ m}}{52.0\times 10^{-3}\text{ m}}\right) = 177\text{ m}$$ ◊

Alternatively, one could have used the lateral magnification $M = h'/h = -q/p$ to obtain

$$h = \frac{h'}{-q/p} \approx -h'\left(\frac{p}{f}\right) \quad \text{or} \quad |h| = |h'|\left(\frac{p}{f}\right)$$

which is the same as what we had above and which yields the same result. The negative sign in the magnification equation simply tells us that the image is inverted in this situation.

---

7. A certain type of film requires an exposure time of 0.010 s with an *f*/11 lens setting. Another type of film requires twice the light energy to produce the same level of exposure. What *f*-number does the second type of film need with the 0.010-s exposure time?

Solution

The energy delivered to the film by the light striking it is $\Delta E = P(\Delta t) = IA(\Delta t)$, where I is the intensity of the light, A is the area of the film that is exposed, and Δt is the exposure time.

Since the exposure time and the area of film exposed (i.e., the size of the image) is the same with the two films, we see that if we are to double the energy delivered to the film, the intensity of the light reaching the film must be doubled.

The intensity of light reaching the film is proportional to the reciprocal of the square of the *f*-stop setting. (See the statement of Problem 6 in Chapter 25 of the textbook.) That is,

$$I = \frac{\textit{constant}}{(f\text{-number})^2}$$

Thus, if the intensity I_2 of the light exposing the second film is to be double the intensity I_1 used to expose the first film, it is necessary that

$$\frac{constant}{(f\text{-number})_2^2} = 2\left[\frac{constant}{(f\text{-number})_1^2}\right]$$

or $$(f\text{-number})_2^2 = \frac{(f\text{-number})_1^2}{2} = \frac{(11)^2}{2} = 61$$

giving $$(f\text{-number})_2 = 7.8$$

This means you should use the $f/8$ setting on your camera for the new film. ◊

---

14. A patient has a near point of 45.0 cm and far point of 85.0 cm. (a) Can a single lens correct the patient's vision? Explain the patient's options. (b) Calculate the power lens needed to correct the near point, so the patient can see objects 25.0 cm away. Neglect the eye-lens distance. (c) Calculate the power lens needed to correct the patient's far point, again neglecting the eye-lens distance.

Solution

(a) This patient cannot focus on objects that are farther than 85.0 cm from the eye and needs corrective action for the distant vision. Also, the patient is unable to focus on objects that are less than 45.0 cm from the eye and requires corrective action in the near vision. Both corrective actions can be accomplished using a single bifocal lens, with the upper portion having the focal length (or optical power) needed to correct the far vision, while the lower portion of the lens has the required focal length or power to correct the near vision. Another option is to use a progressive lens, in which the focal length changes continuously as the line of sight moves from the top of the lens down toward the bottom of the lens. Alternatively, the patient must purchase two pairs of glasses, one pair for reading and a second pair for distant vision. ◊

(b) To correct the near vision, the patient should be able to clearly see objects located at the near point for a normal eye ($p = 25.0$ cm). The corrective lens must form an upright, virtual image located at the near point of the patient's eye ($q = -45.0$ cm) for such objects. The thin-lens equation gives the required focal length for the lens as

$$f = \frac{pq}{p+q} = \frac{(25.0\text{ cm})(-45.0\text{ cm})}{25.0\text{ cm} - 45.0\text{ cm}} = +56.3\text{ cm} = 0.563\text{ m}$$

The power, in diopters, of this lens would be

$$P = \frac{1}{f_{\text{in meters}}} = \frac{1}{+0.563\text{ m}} = +1.78\text{ diopters}$$ ◊

(c) To correct the distant vision, the corrective lens must form upright, virtual images located at the eye's far point ($q = -85.0$ cm) for very distant objects ($p \to \infty$). The required focal length is given by the thin-lens equation as

$$\frac{1}{f} = \frac{1}{p} + \frac{1}{q} \to \frac{1}{\infty} + \frac{1}{-85.0 \text{ cm}} \quad \text{or} \quad f = -85.0 \text{ cm} = -0.850 \text{ m}$$

and the needed power is

$$P = \frac{1}{f_{\text{in meters}}} = \frac{1}{-0.850 \text{ m}} = -1.18 \text{ diopters}$$ ◊

---

19. A stamp collector uses a lens with 7.5-cm focal length as a simple magnifier. The virtual image is produced at the normal near point (25 cm). (a) How far from the lens should the stamp be placed? (b) What is the expected angular magnification?

Solution

A simple magnifier consists of a single converging lens held very close to the eye, with the object located just inside the focal point of the lens, as shown in the textbook figure duplicated here. The position of the object is adjusted until the lens forms an upright, virtual image at the near point of the eye (normally about 25 cm from the eye). This provides the largest angular width for an image that the eye is capable of focusing on clearly.

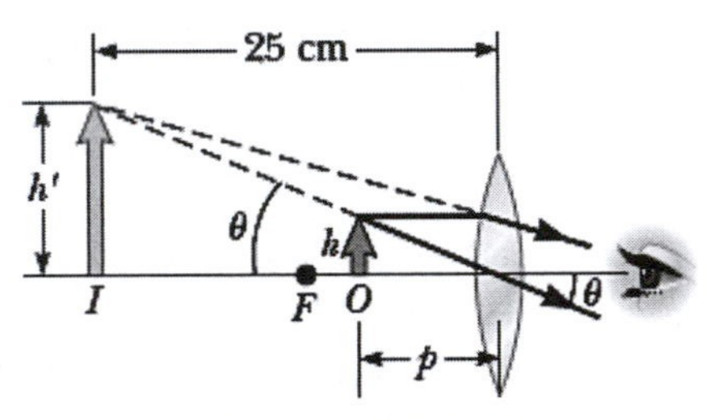

Figure 25.6(b)

(a) With a focal length of $f = +7.5$ cm, the thin-lens equation $(1/p + 1/q = 1/f)$ gives the object position required to yield a virtual image 25 cm in front of the lens $(q = -25 \text{ cm})$ as

$$p = \frac{qf}{q - f} = \frac{(-25 \text{ cm})(+7.5 \text{ cm})}{-25 \text{ cm} - 7.5 \text{ cm}} = +5.8 \text{ cm}$$ ◊

(b) When the lens is used as described above, providing maximum magnification, the angular magnification achieved is given by

$$m_{\text{max}} = 1 + \frac{25 \text{ cm}}{f} = 1 + \frac{25 \text{ cm}}{7.5 \text{ cm}} = +4.3$$ ◊

Alternatively, if the object was placed at the focal point of the lens (i.e., $p = +7.5$ cm), so the image was an infinite distance in front of the lens and parallel rays entered the eye (providing the most relaxed viewing for the eye), the angular magnification would be

$$m_{\text{relaxed}} = \frac{25 \text{ cm}}{f} = \frac{25 \text{ cm}}{+7.5 \text{ cm}} = +3.3$$

---

25. The desired overall magnification of a compound microscope is $140\times$. The objective alone produces a lateral magnification of $12\times$. Determine the required focal length of the eyepiece.

Solution

In a compound microscope, the role of the objective lens is to gather light and form an enlarged, real image of the original object. This image then serves as the object for the eyepiece, which is a converging lens acting as a simple magnifier to yield additional magnification and produce a virtual image for the eye to view. The overall magnification of the microscope is given by $m_{\text{total}} = M_1 m_e$, where M_1 is the lateral magnification produced by the objective lens and m_e is the angular magnification due to the eyepiece lens.

The given microscope is to have a magnifying power of 140×; that is, its overall magnification is to have a magnitude of $|m_{\text{total}}| = 140$. If objective element alone has a power of 12×, or $|M_1| = 12$, the additional magnification produced by the eyepiece must have a magnitude of

$$m_e = \frac{|m_{\text{total}}|}{|M_1|} = \frac{140}{12} = \frac{35}{3}$$

Since the eyepiece acts as a simple magnifier, the angular magnification it yields is given by (assuming a normal eye) $m_e = 1 + (25 \text{ cm})/f_e$ when adjusted for maximum magnification, or $m_e = (25 \text{ cm})/f_e$ when adjusted for most comfortable viewing with a relaxed eye. Because microscopes are often used for extended periods of time, the normal practice is to adjust them for most comfortable viewing, with the eyepiece forming its image at infinity and parallel rays entering the eye. Thus, we take the angular magnification produced by the eyepiece to be $m_e = (25 \text{ cm})/f_e$ and find the required focal length of the eyepiece in this microscope to be

$$f_e = \frac{25 \text{ cm}}{m_e} = \frac{25 \text{ cm}}{35/3} = \frac{75 \text{ cm}}{35} = 2.1 \text{ cm}$$ ◊

---

28. A microscope has an objective lens with a focal length of 16.22 mm and an eyepiece with a focal length of 9.50 mm. With the length of the barrel set at 29.0 cm, the diameter of a red blood cell's image subtends an angle of 1.43 mrad with the eye. If the final image distance is 29.0 cm from the eyepiece, what is the actual diameter of the red blood cell? *Hint:* To answer this question, go back to basics and use the thin-lens equation.

Solution

We are told that the final image of the blood cell is located 29.0 cm in front of the eye and that it intercepts an angle of 1.43 mrad. The diameter of this final image must be

$$h_e = r\theta = \left(29.0\times10^{-2}\text{ m}\right)\left(1.43\times10^{-3}\text{ rad}\right) = 4.15\times10^{-4}\text{ m}$$

At this point, one is tempted to use Equation 25.7 from the textbook for the overall magnification of a compound microscope and compute $h = h_e/m$ as the size of the blood cell serving as the object for the microscope. However, the derivation of that equation is based on several assumptions, one of which is that the eye is relaxed and viewing a final image located an infinite distance in front of the eyepiece. This is clearly not true in this case, and the use of Equation 25.7 would introduce considerable error. Instead, we return to basics and use the thin-lens equation to find the size of the original object.

The object for the eyepiece is the image formed by the objective lens, and we shall call the size of this image h'. The lateral magnification produced by the eyepiece is $M_e = h_e/h' = -q_e/p_e$, where $q_e = -29.0$ cm and p_e is the object distance for the eyepiece. The magnification produced by the objective lens is $M_1 = h'/h = -q_1/p_1$, with $q_1 = L - p_e$ (where $L = 29.0$ cm is the distance between the lenses of the microscope), and p_1 is the object distance for the objective lens. The overall magnification is

$$M = \frac{h_e}{h} = \left(\frac{h'}{h}\right)\left(\frac{h_e}{h'}\right) = M_1 M_e$$

and the actual size of the blood cell is $h = h_e/|M| = (4.15\times10^{-4}\text{ m})/|M|$.

From the thin-lens equation, the object distance for the eyepiece must be

$$p_e = \frac{q_e f_e}{q_e - f_e} = \frac{(-29.0\text{ cm})(0.950\text{ cm})}{-29.0\text{ cm} - 0.950\text{ cm}} = +0.920\text{ cm}$$

and $$M_e = -\frac{q_e}{p_e} = -\frac{-29.0\text{ cm}}{0.920\text{ cm}} = +31.5$$

The image distance for the objective lens is then $q_1 = 29.0\text{ cm} - 0.920\text{ cm} = 28.1$ cm, and the thin-lens equation gives the object distance for this lens as

$$p_1 = \frac{q_1 f_o}{q_1 - f_o} = \frac{(28.1\text{ cm})(1.622\text{ cm})}{28.1\text{ cm} - 1.622\text{ cm}} = +1.72\text{ cm}$$

Thus, $M_1 = -\frac{q_1}{p_1} = -\frac{28.1 \text{ cm}}{1.72 \text{ cm}} = -16.3$ and $M = M_1 M_e = (-16.3)(31.5) = -513$

The actual diameter of the blood cell must be

$$h = \frac{h_e}{|M|} = \frac{4.15 \times 10^{-4} \text{ m}}{513} = 8.09 \times 10^{-7} \text{ m} = 0.809 \ \mu\text{m}$$ ◊

---

31. Suppose an astronomical telescope is being designed to have an angular magnification of 34.0. If the focal length of the objective lens being used is 86.0 cm, find (a) the required focal length of the eyepiece and (b) the distance between the two lenses for a relaxed eye. *Hint:* For a relaxed eye, the image formed by the objective lens is at the focal point of the eyepiece.

Solution

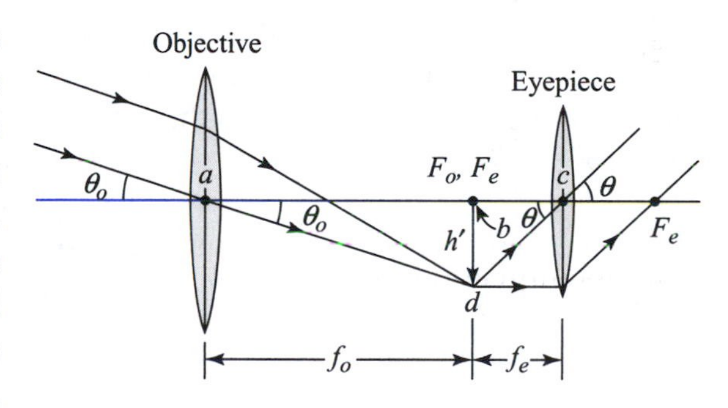

An astronomical telescope is used to observe very distant objects. This means that the light entering the objective element consists of parallel rays, and that element forms an image (of height h') at its focal point, F_o. This image serves as a real object for the eyepiece. When the telescope is adjusted for relaxed-eye viewing, this object is at the focal point F_e of the eyepiece, so parallel rays leave the eyepiece and enter the eye. The angular magnification of the telescope is $m = \theta/\theta_o$, where the angles θ and θ_o are the angles the entering and exiting rays make with the axis of the telescope as shown in the above drawing. Using the small angle approximation, $\tan\theta \approx \theta$, triangle *abd* gives $\theta_o = h'/f_o$, where f_o is the focal length of the objective. Likewise, triangle *cbd* gives $\theta = h'/f_e$, with f_e being the focal length of the eyepiece. Therefore, we see that the angular magnification produced by the telescope is given by

$$m = \frac{\theta}{\theta_o} = \frac{h'/f_e}{h'/f_o} = \frac{f_o}{f_e}$$

Also, observing the drawing given above, we see that when the telescope is adjusted for relaxed-eye viewing, its overall length (the distance between the two lenses) is given by $L = f_o + f_e$.

(a) If the telescope uses an objective lens with focal length $f_o = 86.0$ cm, and it is desired to have an angular magnification of $m = 34.0$, the required focal length of the eyepiece is

$$f_e = \frac{f_o}{m} = \frac{86.0 \text{ cm}}{34.0} = 2.53 \text{ cm}$$ ◊

(b) With the eyepiece focal length found above, the distance between the two lenses should be $L = f_o + f_e = 86.0 \text{ cm} + 2.53 \text{ cm} = 88.5 \text{ cm}$. ◊

---

41. A vehicle with headlights separated by 2.00 m approaches an observer holding an infrared detector sensitive to radiation of wavelength 885 nm. What aperture diameter is required in the detector if the two headlights are to be resolved at a distance of 10.0 km?

Solution

Any time light from a source passes through an aperture, diffraction effects occur. If the aperture is circular, the resulting diffraction pattern is a bright central spot surrounded by concentric alternating bright and dark rings. When light from two sources having a small angular separation passes through the aperture, the diffraction patterns overlap, often making it difficult to recognize that the light originated from two distinct sources.

If two headlights, sources S_1 and S_2, have a lateral separation of $\Delta s = 2.00$ m, and are viewed from a distance of $r = 10.0$ km, their angular separation is

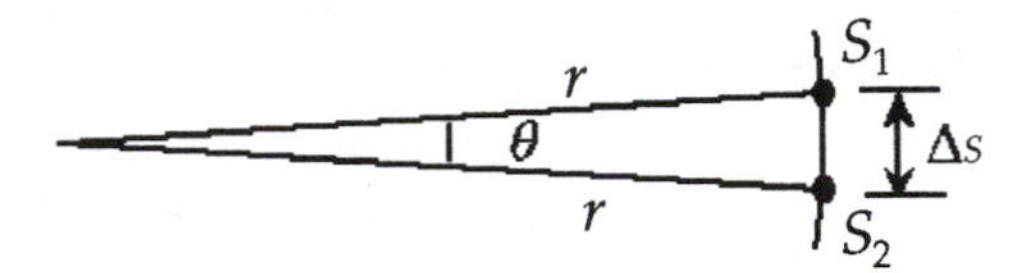

$$\theta = \frac{\Delta s}{r} = \frac{2.00 \text{ m}}{10.0 \times 10^3 \text{ m}} = 2.00 \times 10^{-4} \text{ radians}$$

After the light passes through the aperture of a detector, the images are considered just resolved, according to Rayleigh's criterion, if the central maximum in the diffraction pattern of one image falls on the first minimum of the diffraction pattern of the other image, as shown in the sketch at the right.

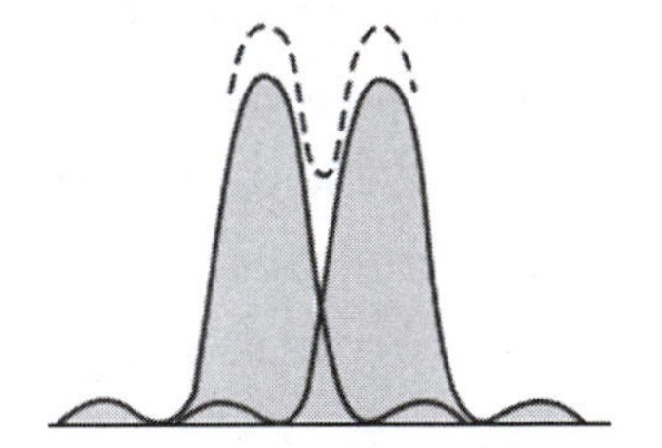

To just meet Rayleigh's criterion with a circular aperture, the angular separation of the sources must equal a minimum value given by $\theta_{\min} = 1.22\lambda/D$, where λ is the wavelength of the light and D is the diameter of the aperture. Thus, for the images of the headlights to be just resolved by the infrared detector (with $\theta = \theta_{\min}$), the required diameter of the detector aperture is

$$D = \frac{1.22\lambda}{\theta} = \frac{1.22\left(885 \times 10^{-9} \text{ m}\right)}{2.00 \times 10^{-4} \text{ rad}} = 5.40 \times 10^{-3} \text{ m} = 5.40 \text{ mm}$$ ◊

---

45. A 15.0-cm-long grating has 6 000 slits per centimeter. Can two lines of wavelengths 600.000 nm and 600.003 nm be separated with this grating? Explain.

Solution

The resolving power required to separate the two specified wavelengths is

$$R_{\text{required}} = \frac{\lambda}{\Delta\lambda} = \frac{600.000 \text{ nm}}{(600.003 - 600.000) \text{ nm}} = 2.00 \times 10^5$$

The resolving power of a diffraction grating in the *m*th order is $R = Nm$, where N is the total number of slits on the grating. The number of slits on the specified grating is

$$N = (15.0 \text{ cm})(6\,000 \text{ slits/cm}) = 9.00 \times 10^4$$

The spacing between adjacent slits on this grating is

$$d = \frac{1.00 \text{ cm}}{6\,000} = 1.67 \times 10^{-4} \text{ cm} = 1.67 \times 10^{-6} \text{ m}$$

Because the angle of diffraction cannot exceed 90°, the grating equation gives the maximum order of 600-nm light one could hope to observe with this grating as

$$m = \frac{d \sin\theta_{\text{max}}}{\lambda} = \frac{(1.67 \times 10^{-6} \text{ m})\sin 90°}{600 \times 10^{-9} \text{ m}} = 2.78$$

The order number must have integer values, and the third order cannot be reached. Therefore, when using this grating, the last observable order in the 600-nm region of the spectrum is the second order $(m_{\text{max}} = 2)$. The maximum resolving power in this region for the specified grating is then

$$R_{\text{max}} = Nm_{\text{max}} = (9.00 \times 10^4)(2) = 1.80 \times 10^5$$

Since this value is less than the required resolving power ($1.80 \times 10^5 < 2.00 \times 10^5$), it is not possible to separate the two wavelengths in question using this grating. ◊

51. A thin sheet of transparent material has an index of refraction of 1.40 and is 15.0 μm thick. When it is inserted in the light path along one arm of an interferometer, how many fringe shifts occur in the pattern? Assume the wavelength (in a vacuum) of the light used is 600 nm. *Hint:* The wavelength will change within the material.

Solution

The light used in this experiment has a wavelength in vacuum (and to a very good approximation, in air) of $\lambda_0 = 600$ nm. As the light passes through the thin sheet of transparent material, it has a shorter wavelength given by $\lambda_n = \lambda_0/n$, where n is the index of refraction of the transparent material. The number of wavelengths that will fit in a length t in vacuum or air is t/λ_0, while the number of wavelengths that will fit in the same distance t when passing through the transparent material is $t/\lambda_n = nt/\lambda_0$. Thus, when a sheet of transparent material of thickness t is inserted in the light path along one arm of an interferometer, the number of wavelengths that will fit in the length of that arm of the interferometer is increased by

$$\Delta = \frac{nt}{\lambda_0} - \frac{t}{\lambda_0} = (n-1)\frac{t}{\lambda_0}$$

In the fringe pattern produced by a Michelson interferometer, a fringe shift occurs each time the effective length of one arm of the interferometer changes by a quarter-wavelength. Therefore, when the transparent sheet is inserted and the effective length of the arm of the interferometer is increased by distance Δ, the number of fringe shifts that will occur is

$$N_{\substack{\text{fringe}\\ \text{shifts}}} = 4\Delta = 4(n-1)\frac{t}{\lambda_0}$$

In this experiment, $t = 15.0\ \mu\text{m}$, $\lambda_0 = 600$ nm, and the transparent sheet has an index of refraction of $n = 1.40$. The number of fringe shifts one should expect to occur when the sheet is inserted in the light path is therefore

$$N_{\substack{\text{fringe}\\ \text{shifts}}} = 4(1.40-1)\frac{15.0\times10^{-6}\ \text{m}}{600\times10^{-9}\ \text{m}} = 40 \qquad \lozenge$$

56. A person with a nearsighted eye has near and far points of 16 cm and 25 cm, respectively. (a) Assuming a lens is placed 2.0 cm from the eye, what power must the lens have to correct this condition? (b) Suppose contact lenses placed directly on the cornea are used to correct the person's eyesight. What is the power of the lens required in this case, and what is the new near point? *Hint:* The contact lens and the eyeglass lens require slightly different powers because they are at different distances from the eye.

Solution

To help this nearsighted person, the corrective lens must form an upright, virtual image of very distant objects at the eye's *far* point, located 25 cm in front of this eye.

(a) If, when looking at very distant objects ($p \to \infty$), the corrective lens, positioned 2.0 cm from the eye, is to form a virtual image 25 cm in front of the eye, the needed image distance for the lens is $q = -(25\text{ cm} - 2.0\text{ cm}) = -23\text{ cm}$. The thin-lens equation then gives the required focal length of the lens as

$$\frac{1}{f} = \frac{1}{p} + \frac{1}{q} = 0 + \frac{1}{-23\text{ cm}} \qquad \text{or} \qquad f = -23\text{ cm} = -0.23\text{ m}$$

The optical power of a lens having this focal length is

$$P = \frac{1}{f} = \frac{1}{-0.23\text{ m}} = -4.3\text{ diopters} \qquad \Diamond$$

(b) If the corrective lens is to be a contact lens, worn directly on the cornea, which forms the virtual image 25 cm in front of the eye for very distant objects, the needed image distance is $q = -25$ cm. In this case, with $p \to \infty$, the thin-lens equation gives

$$\frac{1}{f} = \frac{1}{p} + \frac{1}{q} = 0 + \frac{1}{-25\text{ cm}} \qquad \text{and} \qquad f = -25\text{ cm} = -0.25\text{ m}$$

The needed power for the contact lens is then

$$P = \frac{1}{f} = \frac{1}{-0.25\text{ m}} = -4.0\text{ diopters} \qquad \Diamond$$

While wearing this contact lens, the closest object the eye can focus on (i.e., the object for which the contact lens forms a virtual image at the *near* point of the eye) is

$$p = \frac{qf}{q-f} = \frac{(-16\text{ cm})(-25\text{ cm})}{-16\text{ cm} - (-25\text{ cm})} = 44\text{ cm} \qquad \Diamond$$

61. A Boy Scout starts a fire by using a lens from his eyeglasses to focus sunlight on kindling 5.0 cm from the lens. The Boy Scout has a near point of 15 cm. When the lens is used as a simple magnifier, (a) what is the maximum magnification that can be achieved, and (b) what is the magnification when the eye is relaxed? *Caution*: The equations derived in the text for a simple magnifier assume a "normal" eye.

Solution

The scout started the fire by forming an image of the Sun (a very distant object for which $p \to \infty$) on the kindling with an image distance of $q = 5.0$ cm. The thin-lens equation then gives the focal length of the lens as

$$\frac{1}{f} = \frac{1}{p} + \frac{1}{q} \approx 0 + \frac{1}{q} \qquad \text{or} \qquad f \approx q = +5.0 \text{ cm}$$

When the object is viewed most clearly with the unaided eye, it is located at the near point (at a distance of $d = 15$ cm for this eye) and has an angular size of

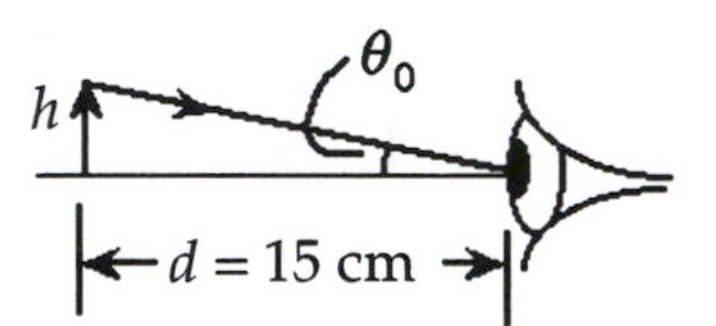

$$\theta_0 \approx \tan\theta_0 = \frac{h}{d} \qquad \text{or} \qquad \theta_0 = \frac{h}{15 \text{ cm}}$$

(a) For maximum magnification, the magnifier must form a magnified, upright, virtual image at the near point of the eye ($q = -15$ cm for this eye). To achieve this, the thin-lens equation gives the reciprocal of the required object distance as

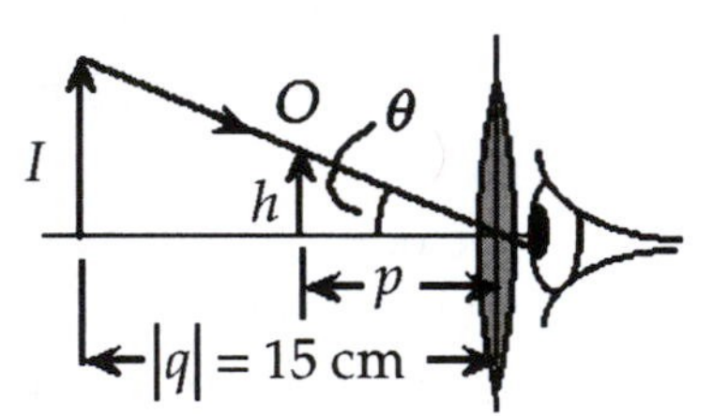

$$\frac{1}{p} = \frac{1}{f} - \frac{1}{q} = \frac{1}{f} + \frac{1}{15 \text{ cm}}$$

and the angular size of the magnified image viewed by the eye is

$$\theta \approx \tan\theta = \frac{h}{p} = h\left(\frac{1}{p}\right) = h\left(\frac{1}{f} + \frac{1}{15 \text{ cm}}\right)$$

The maximum magnification is then

$$m_{\text{max}} = \frac{\theta}{\theta_0} = \frac{h(1/f + 1/15 \text{ cm})}{h/15 \text{ cm}} = \cancel{h}\left(\frac{1}{f} + \frac{1}{15 \text{ cm}}\right)\frac{15 \text{ cm}}{\cancel{h}} = \frac{15 \text{ cm}}{f} + 1$$

or $$m_{\text{max}} = \frac{15 \text{ cm}}{5.0 \text{ cm}} + 1 = 4.0$$ ◊

(b) When the magnifier is positioned for most comfortable viewing with a relaxed eye, the object is at the focal point ($p = f$) and the virtual image recedes to infinity with parallel rays entering the eye.

Consider the second drawing given above and recognize that, when $p = f$, the angular size of the image viewed by the eye is $\theta \approx \tan\theta = h/f$ and the magnification is

$$m = \frac{\theta}{\theta_0} = \frac{h/f}{h/15\text{ cm}} = \frac{15\text{ cm}}{f} = \frac{15\text{ cm}}{5.0\text{ cm}} = 3.0$$ ◊

26

Relativity

NOTES FROM SELECTED CHAPTER SECTIONS

26.1 Galilean Relativity

According to the principle of Galilean relativity, the laws of mechanics are the same in all inertial frames of reference. *Inertial frames of reference are those coordinate systems which are at rest with respect to one another or which move at constant velocity with respect to one another. There is no preferred frame of reference for describing the laws of mechanics.*

26.2 The Speed of Light

The Michelson-Morley experiment was designed to detect the velocity of the Earth with respect to the hypothetical luminiferous ether and thereby confirm the prediction of an absolute frame of reference. The instrument used is called an interferometer (refer to Chapter 25, Section 25.7), and the measurement involved an attempt to observe a fringe shift in the interference pattern of two light beams. The light beams were directed along perpendicular paths, one parallel to and one perpendicular to the "ether wind" and then combined to form the interference pattern. *The outcome of the experiment was negative,* contradicting the ether hypothesis. *Light is now understood to be an electromagnetic wave that requires no medium for its propagation.*

26.3 Einstein's Principle of Relativity

Einstein's special theory of relativity is based upon two postulates:

1. The laws of physics are the same in all inertial frames of reference. *An inertial frame of reference is a non-accelerated frame.*

2. The speed of light in a vacuum has the same value as measured by all observers in all inertial reference frames. *The measured value for the speed of light is independent of the motion of the observer or of the motion of the light source.*

26.4 Consequences of Special Relativity

In relativistic mechanics length and time are not absolute. Following are some of the terms used to describe the consequences of special relativity:

- **Simultaneity:** Two events that are simultaneous in one reference frame are, in general, not simultaneous in another frame which is moving with respect to the first.

- **Proper time:** The proper time interval between two events is the time interval measured by an observer for whom the two events occur at the same location.

- **Time dilation:** A moving clock runs slower than an identical clock at rest with respect to the observer.

- **Proper length:** The length of an object is the length measured by an observer at rest relative to the object (moving with the object).

- **Length contraction:** The length of an object measured in a reference frame that is moving with respect to the object is always less than the proper length if this length is oriented parallel to the motion. *Lengths oriented perpendicular to the motion do not exhibit length contraction.*

26.5 Relativistic Momentum

To account for relativistic effects, it is necessary to modify the classical definition of momentum to satisfy the following conditions:

- The relativistic momentum must be conserved in all inertial frames.

- The relativistic momentum must approach the classical value, mv, as the ratio v/c approaches zero.

26.7 Relativistic Energy and the Equivalence of Mass and Energy

A particle has a rest energy proportional to the value of its inertial mass ($E_R = mc^2$). When the particle is in motion, it has a total energy equal to the sum of its kinetic energy and its rest energy.

26.8 General Relativity

Einstein's postulates of general relativity are:

- All the laws of nature have the same form for observers in any reference frame (inertial or non-inertial).

- In the vicinity of any given point, a gravitational field is equivalent to an accelerated frame of reference without a gravitational field (the principle of equivalence).

EQUATIONS AND CONCEPTS

Basic postulates of the special theory of relativity are:

- The laws of physics are the same in all inertial frames of reference.

- The speed of light has the same value in all inertial frames.

Important consequences of the theory of special relativity:

- **Time dilation** — A time interval Δt measured by an observer moving with respect to a clock is longer than the time interval Δt_p (the proper time) measured by an observer at rest with respect to the clock. *Moving clocks run slower than clocks at rest with respect to an observer.*

$$\Delta t = \frac{\Delta t_p}{\sqrt{1 - v^2/c^2}} = \gamma \, \Delta t_p \quad (26.2)$$

$$\gamma = \frac{1}{\sqrt{1 - v^2/c^2}} \quad (26.3)$$

- **Length contraction** — If an object has a length L_p (the proper length) when measured by an observer at rest with respect to the object, the length L measured by an observer moving with respect to the object will be less than L_p. *Length contraction occurs only along the direction of motion.*

$$L = \frac{L_p}{\gamma} = L_p\sqrt{1 - \frac{v^2}{c^2}} \tag{26.4}$$

- **Simultaneity** — events observed as simultaneous in one frame of reference are not necessarily observed as simultaneous in a second frame moving relative to the first.

Some important equations of relativistic mechanics:

- The **relativistic linear momentum** of a particle with mass m and moving with speed v satisfies the following conditions:

  (1) Momentum is conserved in all collisions.

  (2) The relativistic value of momentum approaches the classical value (mv) as v approaches zero.

$$p \equiv \frac{mv}{\sqrt{1 - v^2/c^2}} = \gamma mv \tag{26.5}$$

- The **relativistic kinetic energy** of a particle of mass m moving with a speed v includes the **rest energy** term mc^2.

$$KE = \gamma mc^2 - mc^2 \tag{26.9}$$

$$KE = (\gamma - 1)mc^2$$

- The **rest energy** of a particle is independent of the speed of the particle. The mass m must have the same value in all inertial frames.

$$E_R = mc^2 \tag{26.10}$$

- The **total energy** E of a particle is the sum of the kinetic energy and the rest energy. *This expression shows that mass is a form of energy.* Equation 26.13 is useful in cases where the speed of an object is not known.

$$E = KE + mc^2 = \gamma mc^2 \tag{26.11}$$

$$E = \frac{mc^2}{\sqrt{1 - v^2/c^2}} \tag{26.12}$$

$$E^2 = p^2c^2 + (mc^2)^2 \tag{26.13}$$

Photons ($m = 0$) travel with the speed of light. Equation 26.14 is an exact expression relating energy and momentum for particles which have zero mass.

$$E = pc \qquad (26.14)$$

The **electron volt** (eV) is a convenient energy unit when stating the energies of electrons and other subatomic particles.

$$1\text{ eV} = 1.60 \times 10^{-19}\text{ J}$$

$$m_e c^2 = 0.511\text{ MeV}$$

$$= 8.20 \times 10^{-14}\text{ J}$$

REVIEW CHECKLIST

- Understand the Michelson-Morley experiment, its objectives, results, and the significance of its outcome. (Section 26.2)
- State Einstein's two postulates of the special theory of relativity. (Section 26.3)
- Understand the idea of simultaneity and the fact that simultaneity is not an absolute concept. That is, two events which are simultaneous in one reference frame are not simultaneous when viewed from a second frame moving with respect to the first. (Section 26.4)
- Make calculations using the equations for time dilation and length contraction. (Section 26.4)
- State the correct relativistic expressions for the momentum, kinetic energy, and total energy of a particle. Make calculations using these equations. (Sections 26.5 and 26.7)

SOLUTIONS TO SELECTED END-OF-CHAPTER PROBLEMS

1. If astronauts could travel at $v = 0.950c$, we on Earth would say it takes $(4.20/0.950) = 4.42$ years to reach Alpha Centauri, 4.20 light-years away. The astronauts disagree. (a) How much time passes on the astronauts' clocks? (b) What is the distance to Alpha Centauri as measured by the astronauts?

Solution

(a) Observers on Earth measure the distance to Alpha Centauri as $d = 4.20$ ly, where $1\text{ ly} = c(1\text{ year})$. These observers see the ship moving toward Alpha Centauri at a constant speed of $v = 0.950c$ and compute the required travel time to be $\Delta t = d/v = 4.20\text{ ly}/0.950c = 4.42$ yr. But these observers are in motion relative to the astronauts' internal biological clocks and hence, experience a dilated version of the proper time interval Δt_p measured on those clocks. From the time dilation equation, $\Delta t = \gamma\,\Delta t_p$, where $\gamma = 1/\sqrt{1-(v/c)^2}$, we find the time that seems to have passed on the astronauts' clock during the trip to be

$$\Delta t_p = \frac{\Delta t}{\gamma} = \Delta t\sqrt{1-(v/c)^2} = (4.42\text{ yr})\sqrt{1-(0.950)^2} = 1.38\text{ yr} \qquad \Diamond$$

(b) Since any motion of Earth occurs at a speed very small in comparison to the speed of light, we consider observers on Earth at rest relative to the span of space separating Earth and Alpha Centauri. Then, the length they measure for the required travel distance is the proper length, or $L_p = d = 4.20$ ly. The astronauts are moving at $v = 0.950c$ relative to this span, and their measure of the travel distance is length contracted as given by

$$L = L_p/\gamma = L_p\sqrt{1-(v/c)^2} = (4.20\text{ ly})\sqrt{1-(0.950)^2} = 1.31\text{ ly} \qquad \Diamond$$

Now, we can note that the astronauts looking out the front window of the spacecraft see Alpha Centauri approaching them at $v = 0.950c$, and they detect an elapsed time $t = L/v = 1.31\text{ ly}/0.950c = 1.38$ yr as being required to complete the trip. This is consistent with the answer from part (a).

---

6. An astronaut is traveling in a space vehicle that has a speed of $0.500c$ relative to Earth. The astronaut measures his pulse rate at 75.0 per minute. Signals generated by the astronaut's pulse are radioed to Earth when the vehicle is moving perpendicular to a line that connects the vehicle with an Earth observer. (a) What rate does the Earth observer measure? (b) What would be the pulse rate if the speed of the space vehicle were increased to $0.990c$?

Solution

(a) As measured by the astronaut, and all others at rest relative to him, the time interval during which his heart beats 75.0 times is $\Delta t_p = 1.00$ min. According to an Earth observer, moving at a speed $v = 0.500c$ relative to the astronaut and his beating heart, the time interval during which these 75.0 beats occur is given by $\Delta t = \gamma\,\Delta t_p$, where $\gamma = 1/\sqrt{1-(v/c)^2}$. Thus, the pulse rate of the astronaut, as measured by Earth-based observers, is

$$rate_{\text{Earth}} = \frac{75.0\text{ beats}}{\Delta t} = \frac{75.0\text{ beats}}{\gamma\,\Delta t_p} = \frac{(75.0\text{ beats})\sqrt{1-(v/c)^2}}{\Delta t_p}$$

or $$rate_{\text{Earth}} = \frac{(75.0\text{ beats})\sqrt{1-(0.500)^2}}{1.00\text{ min}} = 65.0\text{ beats/min} \qquad \Diamond$$

(b) If the speed of the space vehicle relative to Earth were increased to $v = 0.990c$, the pulse rate Earth-based observers would measure for the astronaut would be

$$rate_{\text{Earth}} = \frac{(75.0\text{ beats})\sqrt{1-(0.990)^2}}{1.00\text{ min}} = 10.6\text{ beats/min} \qquad \Diamond$$

---

9. The proper length of one spaceship is three times that of another. The two spaceships are traveling in the same direction and, while both are passing overhead, an Earth observer measures the two spaceships to have the same length. If the slower spaceship has a speed of $0.350c$ with respect to Earth, determine the speed of the faster spaceship.

Solution

The faster object is the one that appears to have undergone the most length contraction to the Earth-based observer. Since this observer measures the two moving ships to have the same length, the faster one must have the greater proper length, or $L_{pf} = 3L_{ps}$.

The contracted lengths that the Earth-based observer measures for the two ships are

$$L_f = \frac{L_{pf}}{\gamma_f} = L_{pf}\sqrt{1-(v_f/c)^2}$$

and

$$L_s = \frac{L_{ps}}{\gamma_s} = L_{ps}\sqrt{1-(v_s/c)^2} = L_{ps}\sqrt{1-(0.350)^2} = 0.937L_{ps}$$

Since this observer determines that $L_f = L_s$, and we have already determined that $L_{pf} = 3L_{ps}$, we find that

$$(3L_{ps})\sqrt{1-(v_f/c)^2} = 0.937L_{ps} \qquad \text{or} \qquad \sqrt{1-(v_f/c)^2} = 0.312$$

Thus,

$$v_f = c\sqrt{1-(0.312)^3} = 0.950c \qquad \Diamond$$

---

15. An unstable particle at rest breaks up into two fragments of *unequal mass*. The mass of the lighter fragment is equal to 2.50×10^{-28} kg and that of the heavier fragment is 1.67×10^{-27} kg. If the lighter fragment has a speed of $0.893c$ after the breakup, what is the speed of the heavier fragment?

Solution

The unstable particle is initially at rest, so the total momentum is zero before breakup. To conserve momentum, the total momentum after breakup must also be zero, meaning that the momenta of the two fragments must have equal magnitudes and opposite directions. The relativistic expression for the momentum of a particle of mass m, traveling at speed v, is $p=\gamma mv$, where $\gamma = 1\big/\sqrt{1-(v/c)^2}$. Thus, equating magnitudes of momenta gives

$$p_2 = \frac{m_2 v_2}{\sqrt{1-(v_2/c)^2}} = \frac{m_1 v_1}{\sqrt{1-(v_1/c)^2}} = p_1$$

If particle 1 is the lighter fragment, moving at speed $v_1 = 0.893c$, we have

$$p_1 = \frac{(2.50\times10^{-28}\ \text{kg})(0.893c)}{\sqrt{1-(0.893)^2}} = (4.96\times10^{-28}\ \text{kg})c$$

Thus, we then have

$$p_2 = \frac{(1.67\times10^{-27}\ \text{kg})v_2}{\sqrt{1-(v_2/c)^2}} = (4.96\times10^{-28}\ \text{kg})c$$

This reduces to

$$(v_2/c) = 0.297\sqrt{1-(v_2/c)^2}$$

Squaring both sides of this equation gives

$$(v/c)^2 = 0.088\,2 - 0.088\,2\,(v/c)^2$$

and

$$v_2 = c\sqrt{\frac{0.088\,2}{1+0.088\,2}} = 0.285c \qquad \Diamond$$

17. An electron moves to the right with a speed of $0.90c$ relative to the laboratory frame. A proton moves to the left with a speed of $0.70c$ relative to the electron. Find the speed of the proton relative to the laboratory frame.

Solution

We choose toward the right as the positive direction, so the velocity of the electron relative to the laboratory frame is $v_{EL} = +0.90c$, and the velocity of the proton relative to the electron is $v_{PE} = -0.70c$.

The relativistic addition of velocities equation (Equation 26.8 in the textbook) then gives the velocity of the proton relative to the laboratory reference frame as

$$v_{PL} = \frac{v_{PE} + v_{EL}}{1 + v_{PE} v_{EL}}$$

or

$$v_{PL} = \frac{-0.70c + 0.90c}{1 + (-0.70c)(+0.90c)} = +0.54c$$

Thus, to an observer in the laboratory reference frame, the proton is seen to be moving to the right with a speed of $0.54c$. ◊

---

20. A pulsar is a stellar object that emits light in short bursts. Suppose a pulsar with a speed of $0.950c$ approaches Earth, and a rocket with a speed of $0.995c$ heads toward the pulsar. (Both speeds are measured in Earth's frame of reference.) If the pulsar emits 10.0 pulses per second in its own frame of reference, at what rate are the pulses emitted in the rocket's frame of reference?

Solution

The proper time interval between bursts emitted by the pulsar is measured in the pulsar's own frame of reference. In this reference frame, this interval is

$$\Delta t_p = \frac{1.00}{f_P} = \frac{1.00}{10.0 \text{ Hz}} = 0.100 \text{ s}$$

The time interval between pulses measured in the rocket's frame of reference is given by the time dilation relation as

$$\Delta t = \gamma_{PR} \Delta t_p = \frac{\Delta t_p}{\sqrt{1 - (v_{PR}/c)^2}}$$

where v_{PR} is the velocity of the pulsar measured in the rocket's frame of reference.

Choosing toward Earth as the positive direction, the velocity of the pulsar in the Earth's frame of reference is $v_{PE} = +0.950c$, while the velocity of the rocket in that reference frame is $v_{RE} = -0.995c$. Therefore, the velocity of Earth in the rocket's frame of reference is $v_{ER} = +0.995c$. The needed speed of the pulsar in the rocket's frame of reference is then given by the relativistic velocity addition relation (Equation 26.8 in the textbook) as

$$v_{PR} = \frac{v_{PE} + v_{ER}}{1 + v_{PE}v_{ER}} = \frac{+0.950c + (+0.995c)}{1 + (+0.950c)(+0.995c)} = +0.999\,87c$$

The time interval between pulses as measured in the rocket's frame of reference is then

$$\Delta t = \gamma_{PR}\Delta t_p = \frac{\Delta t_p}{\sqrt{1-(v_{PR}/c)^2}} = \frac{0.100 \text{ s}}{\sqrt{1-(0.999\,87)^2}}$$

and the frequency of the pulses measured by observers in the rocket will be

$$f_R = \frac{1}{\Delta t} = \frac{\sqrt{1-(0.999\,87)^2}}{0.100 \text{ s}} = 0.161 \text{ s}^{-1} = 0.161 \text{ Hz}$$

◊

---

25. What speed must a particle attain before its kinetic energy is double the value predicted by the nonrelativistic expression $KE = \frac{1}{2}mv^2$?

Solution

The nonrelativistic expression for kinetic energy is $KE_{nr} = \frac{1}{2}mv^2$, while the relativistic expression is $KE_r = E - E_R = (\gamma - 1)E_R$, where $E_R = mc^2$ and $\gamma = 1/\sqrt{1-(v/c)^2}$. Thus, if we are to have $KE_r = 2KE_{nr}$, it is necessary that

$$(\gamma - 1)mc^2 = mv^2 \qquad \text{or} \qquad \frac{1}{\sqrt{1-(v/c)^2}} - 1 = (v/c)^2$$

Rearranging and squaring this equation gives

$$\left[\frac{1}{\sqrt{1-(v/c)^2}}\right]^2 = \left[1+(v/c)^2\right]^2 \qquad \text{or} \qquad 1 = \left[1+(v/c)^2\right]^2\left[1-(v/c)^2\right]$$

Expanding and then simplifying this result yields

$$1=\left[1+2(v/c)^2+(v/c)^4\right]\left[1-(v/c)^2\right]=1+(v/c)^2-(v/c)^4-(v/c)^6$$

and $$0=-(v/c)^2\left[(v/c)^4+(v/c)^2-1\right]$$

Ignoring the trivial solution $v/c=0$, we must have $(v/c)^4+(v/c)^2-1=0$. If we let $x=(v/c)^2$, this becomes $x^2+x-1=0$, and the quadratic formula gives solutions of

$$x=\frac{-1\pm\sqrt{1-4(1)(-1)}}{2}=\frac{-1\pm\sqrt{5}}{2}$$

Since $x=(v/c)^2\geq 0$, we must reject the negative solution and obtain

$$x=(v/c)^2=\frac{-1+\sqrt{5}}{2}=0.618$$

which yields

$$v=c\sqrt{0.618}=0.786c$$ ◊

---

29. Starting with the definitions of relativistic energy and momentum, show that $E^2=p^2c^2+m^2c^4$ (Equation 26.13).

Solution

The definition of the total relativistic energy of a particle having mass m is $E=\gamma E_R$, where $\gamma=1/\sqrt{1-(v/c)^2}$, and the rest energy is given by $E_R=mc^2$. The definition of the relativistic momentum is $p=\gamma mv$.

Therefore, we may write $E^2=\gamma^2m^2c^4$ and $p^2=\gamma^2m^2v^2$, or

$$\begin{aligned}\frac{E^2}{c^2}-p^2&=\gamma^2m^2c^2-\gamma^2m^2v^2\\&=\gamma^2m^2\left(c^2-v^2\right)=\gamma^2m^2c^2\left(1-v^2/c^2\right)\\&=\left(\frac{1}{\cancel{1-v^2/c^2}}\right)m^2c^2\left(\cancel{1-v^2/c^2}\right)\end{aligned}$$

and we now have

$$\frac{E^2}{c^2} - p^2 = m^2c^2$$

Multiplying through by c^2 and rearranging slightly yields the relativistic relation between total energy and momentum (Equation 26.13).

$$E^2 - p^2c^2 = m^2c^4$$

or

$$E^2 = p^2c^2 + m^2c^4 = (pc)^2 + E_R^2 \qquad \Diamond$$

---

33. The rest energy of an electron is 0.511 MeV. The rest energy of a proton is 938 MeV. Assume both particles have kinetic energies of 2.00 MeV. Find the speed of (a) the electron and (b) the proton. (c) By how much does the speed of the electron exceed that of the proton? *Note*: perform the calculations in MeV; don't convert the energies to joules. The answer is sensitive to rounding.

Solution

The relativistic expression for kinetic energy is

$$KE = E - E_R = (\gamma - 1)E_R$$

where $\gamma = 1/\sqrt{1-(v/c)^2}$, and the rest energy is $E_R = mc^2$. Thus, we may write

$$\gamma = 1 + KE/E_R \qquad \text{and} \qquad \gamma^2 = \frac{1}{1-(v/c)^2} = (1 + KE/E_R)^2$$

Solving for the speed v of the particle then gives

$$\frac{1}{(1+KE/E_R)^2} = 1-(v/c)^2 \qquad \text{or} \qquad (v/c)^2 = 1 - \frac{1}{(1+KE/E_R)^2}$$

Thus, $$v = c\sqrt{1 - \frac{1}{(1+KE/E_R)^2}}$$

(a) For an electron ($E_R = 0.511$ MeV) with $KE = 2.00$ MeV,

$$v_e = c\sqrt{1 - \frac{1}{(1+2.00\text{ MeV}/0.511\text{ MeV})^2}} = 0.979c \qquad \Diamond$$

(b) For a proton ($E_R = 938$ MeV) with $KE = 2.00$ MeV,

$$v_p = c\sqrt{1 - \frac{1}{(1 + 2.00\text{ MeV}/938\text{ MeV})^2}} = 0.065\,2\,c$$ ◊

(c) $v_e - v_p = 0.979c - 0.065\,2\,c = 0.914c$ ◊

---

39. Owen and Dina are at rest in frame S′, which is moving with a speed of $0.600c$ with respect to frame S. They play a game of catch while Ed, at rest in frame S, watches the action (Fig. P26.39). Owen throws the ball to Dina with a speed of $0.800c$ (according to Owen) and their separation (measured in S') is equal to 1.80×10^{12} m. (a) According to Dina, how fast is the ball moving? (b) According to Dina, what time interval is required for the ball to reach her? According to Ed, (c) how far apart are Owen and Dina, and (d) how fast is the ball moving?

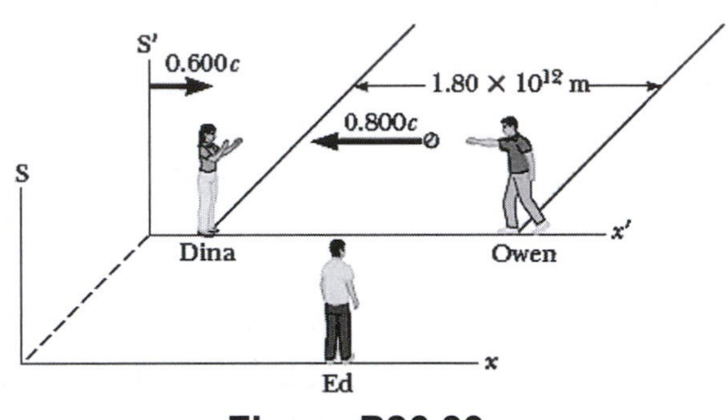

Figure P26.39

Solution

(a) Since Dina and Owen are both at rest in the same reference frame (S′), they measure the same velocity for a moving object. If we choose toward the right as the positive direction for x and x', they agree that the velocity of the ball is given by $v'_{\text{ball}} = -0.800c$. The speed Dina detects for the ball is $0.800c$. ◊

(b) Both Dina and Owen, being at rest relative to the distance separating them, measure the proper length for this distance as $L_p = 1.80 \times 10^{12}$ m. Therefore, Dina sees the ball having to travel a distance L_p at speed $|v'_{\text{ball}}|$ and computes the required travel time as

$$\Delta t_p = \frac{L_p}{|v'_{\text{ball}}|} = \frac{1.80 \times 10^{12}\text{ m}}{0.800(3.00 \times 10^8\text{ m/s})} = 7.50 \times 10^3\text{ s} = 125\text{ min}$$ ◊

(c) The length separating Dina and Owen is moving at velocity $v = +0.600c$ relative to Ed, who is stationary in frame S. Thus, he will detect the contracted length $L = L_p/\gamma = L_p\sqrt{1-(v/c)^2}$ for this distance of separation. According to Ed, Dina and Owen are separated by a distance

$$L = (1.80 \times 10^{12}\text{ m})\sqrt{1-(0.600)^2} = 1.44 \times 10^{12}\text{ m}$$ ◊

(d) The ball moves with velocity $v_{\text{BD}} = v'_{\text{ball}} = -0.800c$ relative to Dina, while Dina moves at velocity $v_{\text{DE}} = v = +0.600c$ relative to Ed. The relativistic velocity addition relation (Equation 26.8 in the textbook) then gives the velocity of the ball relative to Ed as

$$v_{\text{BE}} = \frac{v_{\text{BD}} + v_{\text{DE}}}{1 + \frac{v_{\text{BD}} v_{\text{DE}}}{c^2}} = \frac{-0.800c + 0.600c}{1 + \frac{(-0.800c)(+0.600c)}{c^2}} = -0.385c$$

or Ed sees the ball moving to the left with a speed of $0.385c$. ◊

---

45. The muon is an unstable particle that spontaneously decays into an electron and two neutrinos. In a reference frame in which the muons are stationary, if the number of muons at $t = 0$ is N_0, the number at time t is given by $N = N_0 e^{-t/\tau}$ where τ is the mean lifetime, equal to 2.2 μs. Suppose the muons move at a speed of $0.95c$ and there are 5.0×10^4 muons at $t = 0$. (a) What is the observed lifetime of the muons? (b) How many muons remain after traveling a distance of 3.0 km?

Solution

(a) The proper lifetime of the muon is measured in the reference frame where the muon is at rest. This mean lifetime is $\Delta t_p = 2.2\ \mu\text{s}$. Observers moving at speed $v = 0.95c$ relative to the muons will measure a dilated lifetime given by $\Delta t = \gamma\, \Delta t_p$, or

$$\Delta t = \frac{\Delta t_p}{\sqrt{1-(v/c)^2}} = \frac{2.2\ \mu\text{s}}{\sqrt{1-(0.95)^2}} = 7.0\ \mu\text{s}$$ ◊

(b) In the reference frame where observers see the muons move at velocity $v = 0.95c$, the time required for them to travel a distance of $L_p = 3.0$ km, measured in the observer's rest frame, is

$$t = \frac{L_p}{v} = \frac{3.00 \times 10^3\ \text{m}}{0.95\left(3.00 \times 10^8\ \text{m/s}\right)} = 1.05 \times 10^{-5}\ \text{s} = 10.5\ \mu\text{s}$$

Thus, if $N_0 = 5.0 \times 10^4$ muons are present at time $t = 0$, the number remaining after traveling 3.0 km (with the measured mean lifetime for muons in this reference frame being $\tau = \Delta t = 7.0\ \mu\text{s}$) will be

$$N = N_0 e^{-t/\tau} = \left(5.0 \times 10^4\right) e^{-10.5\ \mu\text{s}/7.0\ \mu\text{s}} = 1.1 \times 10^4$$ ◊

---

50. An alien spaceship traveling $0.600c$ toward Earth launches a landing craft with an advance guard of purchasing agents. The lander travels in the same direction with a velocity $0.800c$ relative to the spaceship. As observed on Earth, the spaceship is 0.200 light-years from Earth when the lander is launched. (a) With what velocity is the lander observed to be approaching by observers on Earth? (b) What is the distance to Earth at the time of lander launch, as observed by the aliens on the mother ship? (c) How long does it take the lander to reach Earth as observed by the aliens on the mother ship? (d) If the lander has a mass of 4.00×10^5 kg, what is its kinetic energy as observed in Earth's reference frame?

Solution

(a) We take toward Earth as the positive direction, so the velocity of the mother ship as observed on Earth is $v_{SE} = +0.600c$, and the velocity of the lander as seen from the mother ship is $v_{LS} = +0.800c$. The relativistic velocity addition relation then gives the velocity of the lander relative to Earth as

$$v_{LE} = \frac{v_{LS} + v_{SE}}{1 + v_{LS}v_{SE}/c^2} = \frac{+0.800c + 0.600c}{1 + (+0.800c)(0.600c)/c^2} = +0.946c \qquad \lozenge$$

(b) Observers at rest on Earth measure the proper distance between Earth and the spaceship at the instance of lander launch as $L_p = 0.200$ ly. Observers on the mother ship, moving relative to this span of space with velocity $v_{SE} = +0.600c$, measure the contracted length for this distance to be

$$L = L_p/\gamma = L_p\sqrt{1-(v_{SE}/c)^2} = (0.200\text{ ly})\sqrt{1-(0.600)^2} = 0.160\text{ ly} \qquad \lozenge$$

(c) Observers on the mother ship see the distance separating the lander and Earth being nibbled away from both ends as they see the lander moving toward Earth at $v_{LS} = +0.800c$, and Earth appearing to move toward the lander with a velocity $v_{ES} = -v_{SE} = -0.600c$. Thus, the time they believe will be required for their measured initial distance of separation, $L = 0.160$ ly, to shrink to zero (so lander and Earth meet) is

$$t = \frac{L}{|v_{LS}| + |v_{ES}|} = \frac{0.160\text{ ly}}{0.800c + 0.600c} = \frac{0.160}{1.40}\frac{\text{ly}}{c} = 0.114\text{ yr} \qquad \lozenge$$

(d) The relativistic kinetic energy for an object of mass $m = 4.00\times10^5$ kg and speed $v_{LE} = 0.946c$ is $KE = (\gamma_{LE} - 1)E_R = (\gamma_{LE} - 1)mc^2$, where $\gamma_{LE} = 1\big/\sqrt{1-(v_{LE}/c)^2}$. The kinetic energy of the lander, as measured by observers on Earth, is then

$$KE = \left(\frac{1}{\sqrt{1-(0.946)^2}} - 1\right)(4.00\times10^5\text{ kg})(3.00\times10^8\text{ m/s})^2 = 7.51\times10^{22}\text{ J} \qquad \lozenge$$

27

Quantum Physics

NOTES FROM SELECTED CHAPTER SECTIONS

27.1 Blackbody Radiation and Planck's Hypothesis

A **black body** is an ideal system that absorbs all radiant energy incident on it. The nature of the radiation emitted by a black body depends on its temperature. The spectral distribution of blackbody radiation (thermal radiation) at various temperatures is sketched in the figure below.

As the temperature increases, the total radiation emitted (area under the curve) increases, while the peak of the distribution shifts to shorter wavelengths. **Stefan's law** describes the total power radiated as a function of temperature. The shift to shorter wavelengths is consistent with the **Wien displacement law.** See Equation 27.1. Classical theories failed to explain the spectrum of blackbody radiation as observed experimentally. An empirical formula, proposed by Max Planck, is consistent with the observed distribution at all wavelengths. Planck made two basic assumptions:

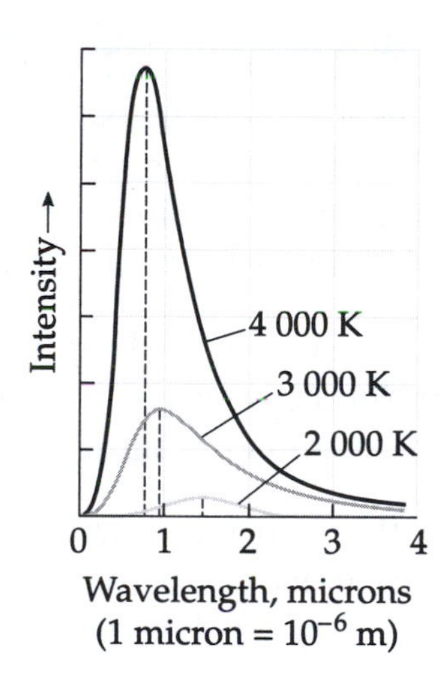

Blackbody Radiation Spectrum

(1) Blackbody radiation is caused by submicroscopic charged oscillators called **resonators.**

(2) The resonators could only have **discrete energies (quantum states)** given by $E_n = nhf$, where f is the oscillator frequency, n is a quantum number ($n = 1, 2, 3, \ldots$), and h is Planck's constant.

Subsequent developments showed that the quantum concept was necessary in order to explain several phenomena at the atomic level, including the photoelectric effect, the Compton effect, and atomic spectra.

27.2 The Photoelectric Effect and the Particle Theory of Light

When light of sufficiently high frequency is incident on certain metallic surfaces, electrons will be emitted from the surfaces. This is called the **photoelectric effect**, discovered by Hertz.

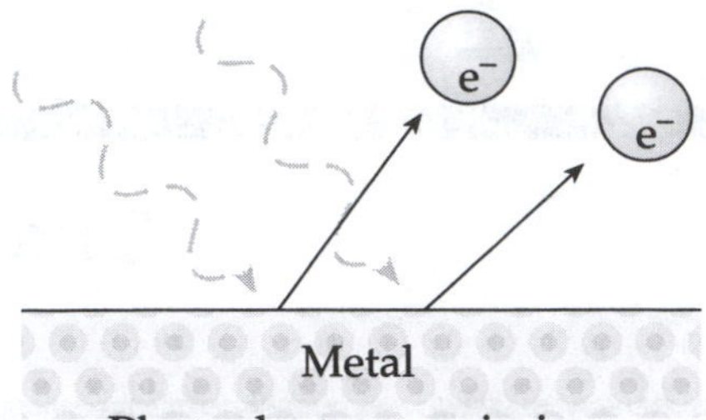

Photoelectron emission

Several features of the photoelectric effect cannot be explained with classical physics or with the wave theory of light. However, each of these features can be explained and understood on the basis of the photon theory of light. These features and their explanations include:

1. **No electrons are emitted if the frequency of the incident light is below a value called the cutoff frequency, f_c.** The value of the cutoff frequency depends on the particular material being illuminated.

   In the quantum model, the energy of electromagnetic radiation is assumed to occur in packets called photons. *Photons in the incident light interact with individual electrons.* If the energy of an incoming photon is not equal to or greater than the work function, ϕ, of the photosensitive surface, no electrons will be ejected from the surface, regardless of the intensity of the light.

2. **Above the cutoff frequency, the number of photoelectrons emitted is proportional to the light intensity, and the maximum kinetic energy of the photoelectrons is independent of light intensity.**

   If the light intensity is doubled, the number of photons is doubled, which doubles the number of photoelectrons emitted. However, their kinetic energy, which equals $hf - \phi$, depends only on the light frequency and the work function, not on the light intensity.

3. **The maximum kinetic energy of the photoelectrons increases with increasing light frequency.**

   The fact that KE_{max} increases with increasing frequency is easily understood from Equation 27.6, $KE_{max} = hf - \phi$.

4. **Electrons are emitted from the surface almost instantaneously (less than 10^{-9} s after the surface is illuminated), even at low light intensities.**

   Classically, one would expect that the electrons would require some time to absorb the incident radiation before they acquire enough kinetic energy to escape from the metal. The fact that the electrons are emitted almost instantaneously is consistent with the photon theory of light; there is a one-to-one interaction between photons and electrons.

27.3 X-Rays

X-rays are a part of the electromagnetic spectrum, characterized by frequencies higher than those of ultraviolet radiation with wavelengths in the 10 to 10^{-4} nm range. They are produced when a metal (and in some cases non-metal) target is struck by high-speed electrons. The spectrum of radiation emitted by an x-ray tube has two distinct components:

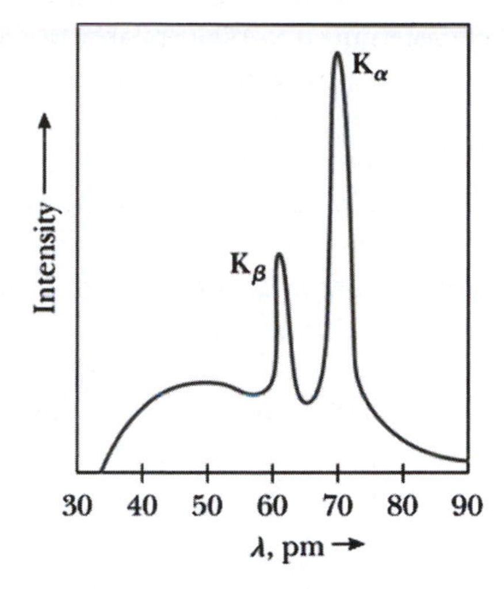

Bremsstrahlung — a continuous broad spectrum of high-energy photons produced by accelerating electrons as they pass close to large positive nuclei in the metal target.

Characteristic lines — a series of sharp lines characteristic of the target material which appear superimposed on the continuous spectrum as the accelerating voltage exceeds the threshold value. These lines represent photons emitted as electrons in atoms of the target material undergo rearrangements following collisions with the incoming electrons.

27.5 The Compton Effect

The Compton effect involves the scattering of a photon by an electron. The scattered photon undergoes a change in wavelength $\Delta\lambda$, called the **Compton shift**. An increase in wavelength for a scattered photon can be explained on the basis of a scattering event in which a portion of the energy of an incident ("bombarding") photon is transferred to an electron ("target"). *The total energy and the total momentum of the photon-electron pair are conserved during the scattering event.*

The "bombarding-photon" and "target-electron" model predicts wavelength shifts which are in excellent agreement with experimental results.

The figure at the right shows the Compton shift for x-rays scattered at 90° from graphite. In this case, the Compton shift is $\Delta\lambda = 0.002\,43$ nm, and λ_0 is the wavelength of photons in the incident x-ray beam.

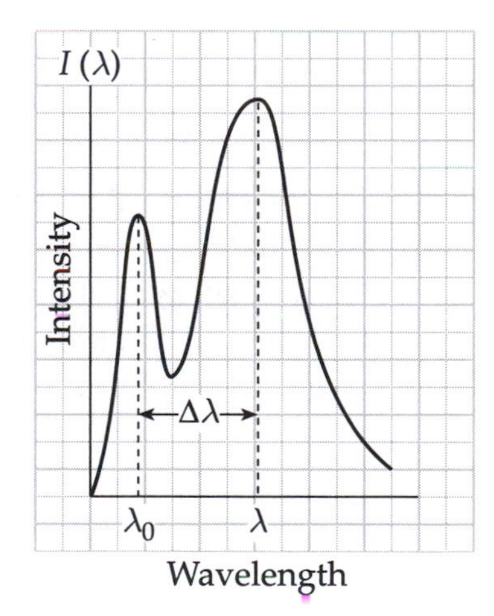

Compton shift in graphite

27.6 The Dual Nature of Light and Matter

The results of some experiments (i.e., photoelectric emission of electrons and Compton scattering) can only be explained on the basis of the photon model of light. Other phenomena (i.e., interference and diffraction of light) are consistent with the wave model. *The photon theory and the wave theory complement each other: light exhibits both wave and particle characteristics.*

De Broglie postulated that a particle in motion has wave properties and a corresponding wavelength inversely proportional to the particle's momentum. The wave nature of matter was first confirmed (Davisson and Germer) by scattering electrons from crystal targets.

27.7 The Wave Function

In the Schrödinger description of the manner in which matter waves change in space and time, each particle is represented by a wave function. *The probability of finding a particle at a specific location at some instant is given by evaluating the square of the wave function at that location and time.*

27.8 The Uncertainty Principle

It is impossible to make exact simultaneous measurements of a particle's position and linear momentum or to measure exactly the energy of a particle in an arbitrarily small time interval. These limitations are described by Equations 27.16 and 27.17.

EQUATIONS AND CONCEPTS

The **Wien displacement law** accurately describes the manner in which the peak wavelength in the spectrum of a blackbody radiator depends on temperature. *As the temperature increases, the intensity of the radiation (area under the intensity vs. wavelength curve shown in Section 27.1) increases, while the peak of the distribution shifts to shorter wavelengths.*

$$\lambda_{max} T = 0.289\,8 \times 10^{-2}\ \text{m} \cdot \text{K} \quad (27.1)$$

Discrete energy values of an atomic oscillator are determined by a quantum number, n. Each discrete energy value corresponds to a quantum state. **Planck's constant** (h) is a fundamental constant of nature.

$$E_n = nhf \quad (27.2)$$
$$(n = 1, 2, 3, \ldots)$$

$$h = 6.626 \times 10^{-34}\ \text{J} \cdot \text{s} \quad (27.3)$$

The **energy of a photon** corresponds to the energy difference between initial and final quantum states. *An oscillator emits or absorbs energy only when there is a transition between quantum states.*

$$E = hf \quad (27.5)$$

The **maximum kinetic energy of an ejected photoelectron** (typically a few eV) can be expressed in terms of the stopping potential (ΔV_s) or in terms of the work function of the metal (ϕ) and the frequency of the incident photon.

$$KE_{\max} = e\Delta V_s \qquad (27.4)$$

$$KE_{\max} = hf - \phi \qquad (27.6)$$

The **cut-off wavelength** is the maximum wavelength (corresponding to the minimum frequency) of an incident photon that will result in photoemission regardless of the intensity of the incident light.

$$\lambda_c = \frac{hc}{\phi} \qquad (27.7)$$

The **minimum wavelength x-ray** radiation, for a given accelerating voltage, is produced when an electron is completely stopped in a single collision.

$$\lambda_{\min} = \frac{hc}{e\Delta V} \qquad (27.9)$$

Bragg's law states the condition for constructive interference of x-rays diffracted by a crystal. The crystal planes are separated by a distance d; x-rays are incident at an angle θ to the crystal plane and are reflected at the same angle.

$$2d \sin\theta = m\lambda \qquad (27.10)$$
$$(m = 1, 2, 3, \ldots)$$

The **Compton shift** is the change in wavelength of a photon when scattered from an electron. *The scattered photon makes an angle θ with the direction of the incident photon as shown in the diagram on the right.* The quantity, $h/m_e c$, is called the Compton wavelength of the electron.

$$\Delta\lambda = \lambda - \lambda_o = \frac{h}{m_e c}(1 - \cos\theta) \qquad (27.11)$$

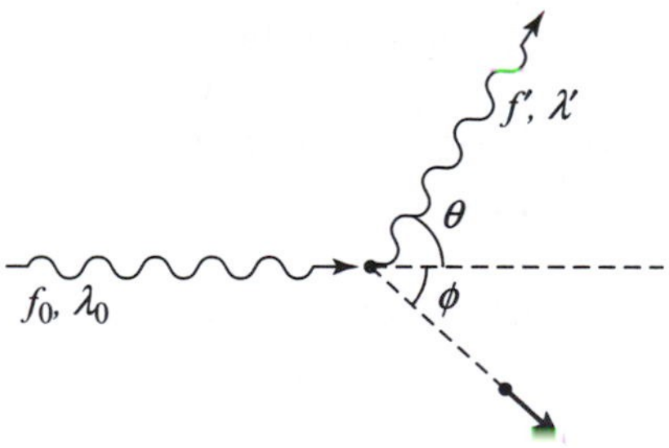

The **momentum and wavelength** of a photon are related as shown in Equation 27.13. This equation is obtained by combining two equations involving the energy of a photon: $p = E/c$ and $E = hc/\lambda$.

$$p = \frac{E}{c} = \frac{h}{\lambda} \qquad (27.13)$$

The **de Broglie wavelength** of a particle is inversely proportional to the momentum of the particle. *The waves associated with material particles are called* ***matter waves*** *with frequencies given by Equation 27.15.*

$$\lambda = \frac{h}{p} = \frac{h}{mv} \quad (27.14)$$

$$f = \frac{E}{h} \quad (27.15)$$

The **Heisenberg uncertainty principle** can be stated in two forms:

- **Simultaneous measurements of position and momentum** with respective uncertainties Δx and Δp_x.

$$\Delta x \, \Delta p_x \geq \frac{h}{4\pi} \quad (27.16)$$

- **Simultaneous measurements of energy and lifetime** with uncertainties ΔE and Δt.

$$\Delta E \, \Delta t \geq \frac{h}{4\pi} \quad (27.17)$$

REVIEW CHECKLIST

- State the assumptions made by Planck in order to explain the shape of the Intensity vs. Wavelength curve for a blackbody radiator. (Section 27.1)
- Describe the manner in which the Intensity vs. Wavelength curve for a blackbody radiator changes as the temperature of the radiator increases. (Section 27.1)
- Describe the Einstein model for the photoelectric effect and the predictions of the fundamental photoelectric effect equation for the maximum kinetic energy of photoelectrons. (Section 27.2)
- Describe the Compton effect (the scattering of x-rays by electrons) and be able to use the formula for the Compton shift equation. Recognize that the Compton effect can only be explained using the photon concept. (Section 27.5)
- Use the de Broglie equation to calculate the wavelength associated with moving particles. (Section 27.6)
- Identify experimental evidence for the wave and particle nature of light and also for the wave nature of particles. (Section 27.6)
- Discuss the manner in which the uncertainty principle makes possible a better understanding of the dual wave-particle nature of light and matter. (Section 27.8)

SOLUTIONS TO SELECTED END-OF-CHAPTER PROBLEMS

6. Suppose a star with radius 8.50×10^8 m has a peak wavelength of 685 nm in the spectrum of its emitted radiation. (a) Find the energy of a photon with this wavelength. (b) What is surface temperature of the star? (c) At what rate is energy emitted from the star in the form of radiation? Assume the star is a blackbody ($e = 1$). (d) Using the answer to part (a), estimate the rate at which photons leave the surface of the star.

Solution

(a) The energy of a photon of frequency f and wavelength $\lambda = c/f$ is $E = hf = hc/\lambda$, where $h = 6.63 \times 10^{-34}$ J·s is known as Planck's constant. The energy of a 685-nm photon is then

$$E_{peak} = \frac{hc}{\lambda_{max}} = \frac{(6.63 \times 10^{-34}\ \text{J}\cdot\text{s})(3.00 \times 10^8\ \text{m/s})}{685 \times 10^{-9}\ \text{m}} = 2.90 \times 10^{-19}\ \text{J/photon} \quad \lozenge$$

(b) Assuming the star closely approximates an ideal radiator or blackbody, the absolute temperature of its surface and the wavelength of its peak radiation are related by Wien's displacement law, $\lambda_{max} T = 0.289\,8 \times 10^{-2}$ m·K. The surface temperature must be

$$T = \frac{0.289\,8 \times 10^{-2}\ \text{m}\cdot\text{K}}{\lambda_{max}} = \frac{0.289\,8 \times 10^{-2}\ \text{m}\cdot\text{K}}{685 \times 10^{-9}\ \text{m}} = 4.23 \times 10^3\ \text{K} \quad \lozenge$$

(c) Stefan's law (see Chapter 11 in the textbook) states that the rate at which an object radiates energy is proportional to the fourth power of its absolute temperature, or $P = \sigma A e T^4$, where $\sigma = 5.669\,6 \times 10^{-8}$ W/m^2·K^4, A is the surface area of the object, and e is the emissivity. For an ideal radiator, $e = 1$. Assuming the star to be spherical, $A = 4\pi r^2$, and the radiated power must be

$$P = (5.669\,6 \times 10^{-8}\ \text{W/m}^2\cdot\text{K}^4)4\pi(8.50 \times 10^8\ \text{m})^2(1)(4.23 \times 10^3\ \text{K})^4 = 1.65 \times 10^{26}\ \text{W} \quad \lozenge$$

(d) If we assume that the average energy of the photons emitted by the star equals the energy of photons at the peak of the radiation distribution, the estimate of the number of photons emitted per second would be

$$\frac{\Delta n}{\Delta t} = \frac{P}{E_{peak}} = \frac{1.65 \times 10^{26}\ \text{J/s}}{2.90 \times 10^{-19}\ \text{J/photon}} = 5.69 \times 10^{44}\ \text{photons/s} \quad \lozenge$$

10. The work function for zinc is 4.31 eV. (a) Find the cutoff wavelength for zinc. (b) What is the lowest frequency of light incident on zinc that releases photoelectrons from its surface? (c) If photons of energy 5.50 eV are incident on zinc, what is the maximum kinetic energy of the ejected photoelectrons?

Solution

(a) From Einstein's photoelectric effect equation, the maximum kinetic energy of freed electrons in the photoelectric effect is $KE_{max} = hf - \phi = hc/\lambda - \phi$, where ϕ is the work function of the target material, f is the frequency of the incident light, and λ is the wavelength of that light. Observe that the maximum kinetic energy of the electrons decreases as the wavelength of the incident light increases. The cutoff wavelength is the longest wavelength capable of liberating electrons from this target material and hence is the wavelength for which $KE_{max} = 0$. Thus, we see that $0 = hc/\lambda_c - \phi$, and the cutoff wavelength is

$$\lambda_c = \frac{hc}{\phi} = \frac{(6.63\times10^{-34}\ \text{J}\cdot\text{s})(3.00\times10^{8}\ \text{m/s})}{(4.31\ \text{eV})(1.60\times10^{-19}\ \text{J/1 eV})} = 2.88\times10^{-7}\ \text{m} = 288\ \text{nm} \qquad \lozenge$$

(b) Light having the lowest frequency, f_c, that can free electrons from the target material has a wavelength equal to the cutoff wavelength, or

$$f_c = \frac{c}{\lambda_c} = \frac{3.00\times10^{8}\ \text{m/s}}{2.88\times10^{-7}\ \text{m}} = 1.04\times10^{15}\ \text{Hz} \qquad \lozenge$$

(c) When photons having energy $E_{photon} = hf = 5.50$ eV are incident on zinc, which has a work function $\phi = 4.31$ eV, the photoelectric effect equation gives the maximum kinetic energy of the freed electrons as

$$KE_{max} = E_{photon} - \phi = 5.50\ \text{eV} - 4.31\ \text{eV} = 1.19\ \text{eV} \qquad \lozenge$$

---

13. When light of wavelength 254 nm falls on cesium, the required stopping potential is 3.00 V. If light of wavelength 436 nm is used, the stopping potential is 0.900 V. Use this information to plot a graph like that shown in Figure 27.6, and from the graph determine the cutoff frequency for cesium and its work function.

Solution

The photoelectric effect equation,

$$KE_{max} = E_{photon} - \phi = hf - \phi$$

is a linear equation relating the maximum kinetic energy of the ejected electrons to the frequency of the incident radiation. Thus, the graph of KE_{max} versus f should be a straight line with slope h and vertical intercept $-\phi$.

We may express KE_{max} in terms of the stopping potential as $KE_{max} = V_s e$. Therefore, we have

$$KE_{max} = (3.00 \text{ V})e = 3.00 \text{ eV} \quad \text{when} \quad f = \frac{c}{\lambda} = \frac{3.00\times10^{8} \text{ m/s}}{254\times10^{-9} \text{ m}} = 11.8\times10^{14} \text{ Hz}$$

and

$$KE_{max} = (0.900 \text{ V})e = 0.900 \text{ eV} \quad \text{when} \quad f = \frac{c}{\lambda} = \frac{3.00\times10^{8} \text{ m/s}}{436\times10^{-9} \text{ m}} = 6.88\times10^{14} \text{ Hz}$$

Your graph should be similar to the sketch given at the right. At the cutoff frequency (the minimum frequency photon capable of freeing electrons from the surface), $KE_{max} = 0$.

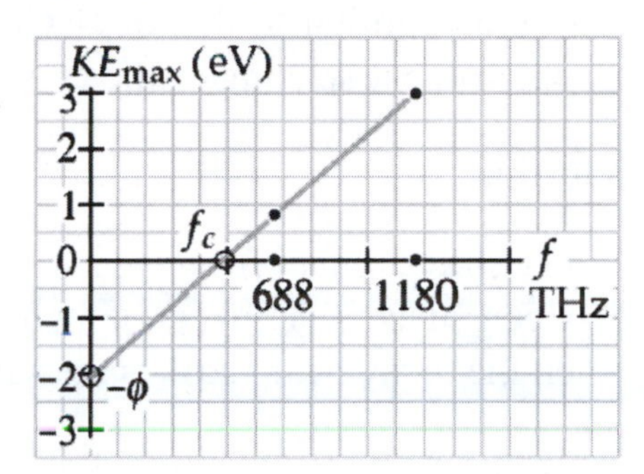

Thus, the horizontal intercept on the graph gives the cut-off frequency. The vertical intercept of the graph is equal to the negative of the work function. From your graph, you should find these values to be approximately

$$f_c = 480 \text{ THz} = 4.8\times10^{14} \text{ Hz} \qquad \text{and} \qquad \phi = 2.0 \text{ eV}$$ ◊

---

15. The extremes of the x-ray portion of the electromagnetic spectrum range from approximately 1.0×10^{-8} m to 1.0×10^{-13} m. Find the minimum accelerating voltages required to produce wavelengths at these two extremes.

Solution

When an electron, with charge $q = -e$, starts from rest and accelerates through a potential difference ΔV, its final kinetic energy is given by

$$KE = KE_i + (PE_i - PE_f) = 0 + [(-e)V_i - (-e)V_f] = e(V_f - V_i) = e\Delta V$$

When this electron strikes the target in an x-ray tube, it gives up its kinetic energy in the form of photons during collisions with atoms in the target material. The maximum-energy photons are produced when the electron gives up all its kinetic energy in a single collision. Thus, the energy and wavelength of these photons are given by $E_{photon} = hc/\lambda = KE = e\Delta V$. The minimum accelerating voltage required to produce x-rays of wavelength λ is therefore $\Delta V_{min} = hc/e\lambda$.

To produce x-rays at the long wavelength end ($\lambda = 1.0\times10^{-8}$ m) of the x-ray region in the electromagnetic spectrum, the minimum accelerating voltage required is

$$\Delta V_{min} = \frac{hc}{e\lambda} = \frac{(6.63\times10^{-34} \text{ J}\cdot\text{s})(3.00\times10^{8} \text{ m/s})}{(1.60\times10^{-19} \text{ C})(1.0\times10^{-8} \text{ m})} = 1.2\times10^{2} \text{ V}$$ ◊

The minimum accelerating voltage needed to produce x-rays at the short wavelength end ($\lambda = 1.0 \times 10^{-13}$ m) of this region in the electromagnetic spectrum is

$$\Delta V_{\min} = \frac{hc}{e\lambda} = \frac{(6.63 \times 10^{-34} \text{ J} \cdot \text{s})(3.00 \times 10^{8} \text{ m/s})}{(1.60 \times 10^{-19} \text{ C})(1.0 \times 10^{-13} \text{ m})} = 1.2 \times 10^{7} \text{ V} = 12 \text{ MV}$$ ◊

---

18. When x-rays of wavelength of 0.129 nm are incident on the surface of a crystal having a structure similar to that of NaCl, a first-order maximum is observed at 8.15°. Calculate the interplanar spacing of the crystal based on this information.

Solution

When x-rays reflect from the surface of a crystal, in which there exists a uniform spacing d between successive layers of atoms, constructive interference and strong reflections occur when the x-ray beam strikes the surface at certain angles.

Reflections of x-rays from the atomic layers within the crystal are shown in the sketch at the right. When the path difference, $\delta = 2d\sin\theta$, of radiation reflecting from adjacent layers equals some integral multiple of the wavelength of the incident x-rays, the reflected beams interfere constructively and produce maxima in the intensity of the reflected radiation. This condition is summarized by Bragg's law, which states that

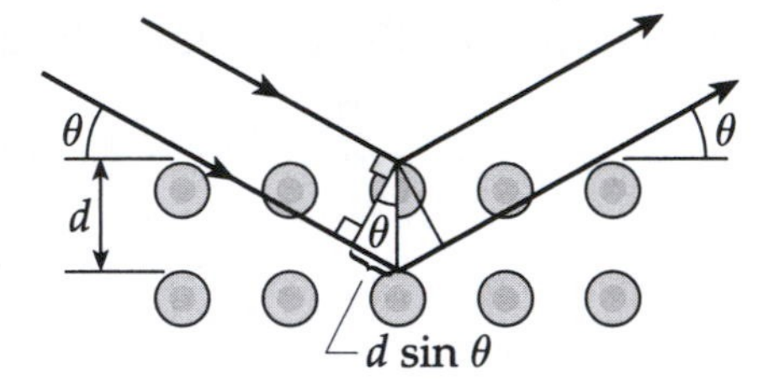

$$2d\sin\theta = m\lambda \qquad m = 1, 2, 3, \ldots$$

Thus, if a strong reflection of order m is seen at angle θ for x-rays having wavelength λ, the spacing between successive layers of atoms in this crystal is given by

$$d = \frac{m\lambda}{2\sin\theta}$$

If 0.129-nm x-rays produce the first-order ($m = 1$) maximum when the beam strikes the surface of this crystal at a grazing angle $\theta = 8.15°$, the interplanar spacing in the crystal is

$$d = \frac{(1)(0.129 \text{ nm})}{2\sin(8.15°)} = 0.455 \text{ nm}$$ ◊

---

26. In a Compton scattering experiment, an x-ray photon scatters through an angle of 17.4° from a free electron that is initially at rest. The electron recoils with a speed of 2 180 km/s. Calculate (a) the wavelength of the incident photon and (b) the angle through which the electron scatters.

Solution

This scattering event is shown in the sketch at the right, where $v = 2\,180\text{ km/s} = 2.18\times10^6\text{ m/s}$ and $\theta = 17.4°$. Note that $v \ll c$, so the electron is nonrelativistic.

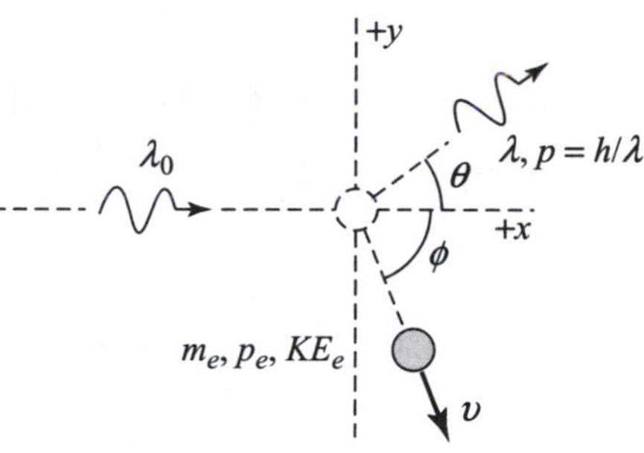

(a) The energy of a photon of wavelength λ is $E_{\text{photon}} = hc/\lambda$, so conservation of energy gives the kinetic energy of the recoiling electron as

$$KE_e = \frac{hc}{\lambda_0} - \frac{hc}{\lambda} = hc\left(\frac{\lambda-\lambda_0}{\lambda_0\lambda}\right) = \frac{(hc)\Delta\lambda}{\lambda_0\lambda}$$

Using the Compton shift formula, $\Delta\lambda = \lambda - \lambda_0 = (h/m_e c)(1-\cos\theta)$ (Equation 27.11 in the textbook), this reduces to $\frac{1}{2}m_e v^2 = (h^2/m_e\lambda_0\lambda)(1-\cos\theta)$, or $\lambda_0\lambda = 2(h/m_e v)^2$ $(1-\cos\theta)$. However,

$$\lambda_0\lambda = \lambda_0\left[\lambda_0 + \frac{h}{m_e c}(1-\cos\theta)\right] = \lambda_0^2 + \left[\frac{h}{m_e c}(1-\cos\theta)\right]\lambda_0$$

so we obtain $\lambda_0^2 + \left[\frac{h}{m_e c}(1-\cos\theta)\right]\lambda_0 - 2(h/m_e v)^2(1-\cos\theta) = 0$. This is a quadratic equation of the form $\text{A}\lambda_0^2 + \text{B}\lambda_0 + \text{C} = 0$, where

$$\text{A} = 1,\ \text{B} = (h/m_e c)(1-\cos\theta),\text{ and } \text{C} = -2(h/m_e v)^2(1-\cos\theta)$$

With $\theta = 17.4°$, the numeric values of the last two constants are

$$\text{B} = \left(\frac{6.63\times10^{-31}\text{ J}\cdot\text{s}}{(9.11\times10^{-31}\text{ kg})(3.00\times10^8\text{ m/s})}\right)(1-\cos 17.4°) = 1.11\times10^{-13}\text{ m}$$

and $$\text{C} = -2\left(\frac{6.63\times10^{-31}\text{ J}\cdot\text{s}}{(9.11\times10^{-31}\text{ kg})(2.18\times10^6\text{ m/s})}\right)^2(1-\cos 17.4°) = -1.02\times10^{-20}\text{ m}$$

Since wavelength is always positive, we use only the positive solution from the quadratic formula, $\lambda_0 = \left(-\text{B}\pm\sqrt{\text{B}^2-4\text{AC}}\right)/2\text{A}$, and find the wavelength of the incident photon to be $\lambda_0 = 1.01\times10^{-10}\text{ m} = 0.101\text{ nm}$. ◊

(b) Before the scattering event, the total momentum in the y-direction was zero. Thus, to conserve momentum, we must require that the total momentum in the y-direction be zero after the event, or $\Sigma p_y = p\sin\theta - p_e\sin\phi = 0$, giving $\sin\phi = (p/p_e)\sin\theta$. The scattering angle for the electron is found to be

$$\phi = \sin^{-1}\left[\frac{p\sin\theta}{p_e}\right] = \sin^{-1}\left[\frac{(h/\lambda)\sin\theta}{m_e v}\right]$$

where $\lambda = \lambda_0 + (h/m_e c)(1-\cos\theta) = \lambda_0 + \text{B} = 1.01\times10^{-10}\text{ m} + 1.11\times10^{-13}\text{ m}$, or $\lambda = 1.011\times10^{-10}\text{ m} = 0.1011\text{ nm}$. This gives the electron scattering angle as

$$\phi = \sin^{-1}\left[\frac{(6.63\times10^{-34}\text{ J}\cdot\text{s})\sin 17.4°}{(1.011\times10^{-10}\text{ m})(9.11\times10^{-31}\text{ kg})(2.18\times10^{6}\text{ m/s})}\right] = 80.9° \quad \lozenge$$

---

29. De Broglie postulated that the relationship $\lambda = h/p$ is valid for relativistic particles. What is the de Broglie wavelength for a (relativistic) electron whose kinetic energy is 3.00 MeV?

Solution

The kinetic energy of a 3.00-MeV electron is not small in comparison to the rest energy ($E_R = 0.511$ MeV) of an electron. Thus, a relativistic calculation for the momentum of this particle is required.

The relativistic relation between the total energy of a particle and its momentum is

$$E^2 = p^2c^2 + E_R^2$$

where the total energy is $E = KE + E_R$. Thus, the momentum of a relativistic particle may be expressed as

$$p = \frac{\sqrt{E^2 - E_R^2}}{c} = \frac{\sqrt{(KE + E_R)^2 - E_R^2}}{c}$$

and its de Broglie wavelength is

$$\lambda = \frac{h}{p} = \frac{hc}{\sqrt{(KE + E_R)^2 - E_R^2}}$$

Therefore, an electron having a kinetic energy of $KE = 3.00$ MeV will have a de Broglie wavelength of

$$\lambda = \frac{(6.63\times10^{-34}\text{ J}\cdot\text{s})(3.00\times10^{8}\text{ m/s})}{\sqrt{(3.00\text{ MeV} + 0.511\text{ MeV})^2 - (0.511\text{ MeV})^2}}\left(\frac{1\text{ MeV}}{1.60\times10^{-13}\text{ J}}\right) = 3.58\times10^{-13}\text{ m} \quad \lozenge$$

---

33. In the ground state of hydrogen, the uncertainty in the position of the electron is roughly 0.10 nm. If the speed of the electron is approximately the same as the uncertainty in its speed, about how fast is it moving?

Solution

One form of the uncertainty principle states that if the position of a particle is measured with precision Δx, and a simultaneous measurement of the linear momentum is made with precision Δp_x, the product of the two uncertainties can never be less than $h/4\pi$, or $\Delta x\, \Delta p_x \geq h/4\pi$.

Thus, if we assume that the momentum of the electron is not less than the uncertainty in the momentum, the minimum momentum of the electron is

$$p_{\min} = (\Delta p_x)_{\min} = \frac{h}{4\pi\, \Delta x} = \frac{6.63\times 10^{-34}\ \text{J}\cdot\text{s}}{4\pi(0.10\ \text{nm})(10^{-9}\ \text{m}/1\ \text{nm})} = 5.3\times 10^{-25}\ \text{kg}\cdot\text{m/s}$$

We shall assume that this electron is not relativistic and try a classical calculation for the minimum speed of the particle. Hence,

$$v_{\min} = \frac{p_{\min}}{m_e} = \frac{5.3\times 10^{-25}\ \text{kg}\cdot\text{m/s}}{9.11\times 10^{-31}\ \text{kg}} = 5.8\times 10^{5}\ \text{m/s} \qquad \text{or} \qquad v_{\min} \sim 10^{6}\ \text{m/s} \quad \Diamond$$

Observe that the computed speed is $10^6\ \text{m/s} \ll c$, and our assumption that a relativistic calculation would not be necessary is seen to be valid.

---

37. The average lifetime of a muon is about 2 μs. Estimate the minimum uncertainty in the energy of a muon.

Solution

A second form of the uncertainty principle sets a limit on the accuracy with which the energy of a system E can be measured in a finite time interval Δt. This form states that the product of the uncertainty in the energy and the duration of the interval over which that energy is measured can never be less than $h/4\pi$, or $\Delta E\, \Delta t \geq h/4\pi$.

If, on average, a muon only exists for a lifetime of $\tau = 2\ \mu$s, the maximum duration of the time interval over which its energy could be monitored and measured is $\Delta t_{\max} = \tau = 2\ \mu$s. Thus, the minimum uncertainty in the measurement of its energy is

$$\Delta E_{\min} = \frac{h}{4\pi\, \Delta t_{\max}} = \frac{6.63\times 10^{-34}\ \text{J}\cdot\text{s}}{4\pi(2\times 10^{-6}\ \text{s})} = 3\times 10^{-29}\ \text{J} \quad \Diamond$$

---

40. Find the speed of an electron having a de Broglie wavelength equal to its Compton wavelength. *Hint:* This electron is relativistic.

Solution

The de Broglie wavelength of a particle is $\lambda = h/p$, where p is the linear momentum of the particle. The Compton wavelength for a particle of mass m is $\lambda_c = h/mc$. Thus, if the de Broglie wavelength for an electron is equal to its Compton wavelength, it is necessary that $h/p = h/m_e c$, or the linear momentum of the electron must be $p = m_e c$.

The relativistic relation between the total energy of a particle and its momentum is $E^2 = (pc)^2 + E_R^2$, where the rest energy is $E_R = mc^2$, and the total energy is $E = \gamma E_R = mc^2/\sqrt{1-(v/c)^2}$. Applying this to our electron having $p = m_e c$ gives

$$\left(\frac{m_e c^2}{\sqrt{1-(v/c)^2}}\right)^2 = \left(m_e c \cdot c\right)^2 + \left(m_e c^2\right)^2$$

or

$$\frac{m_e^2 c^4}{1-(v/c)^2} = m_e^2 c^4 + m_e^2 c^4 = 2m_e^2 c^4$$

which reduces to $\frac{1}{2} = 1-(v/c)^2$. Solving for the speed of the electron then gives

$$v = c\sqrt{1-\frac{1}{2}} = \frac{c}{\sqrt{2}} \qquad \text{or} \qquad v = c\sqrt{2}/2$$ ◊

---

46. An electron initially at rest recoils after a head-on collision with a 6.20-keV photon. Determine the kinetic energy acquired by the electron.

Solution

In a head-on collision between an incident photon and an electron that was initially at rest, the electron will recoil in the direction of the incident photon's motion, while the photon scatters in the back direction with a scattering angle of $\theta = 180°$. This event is shown in the sketch at the right. The Compton shift formula gives the wavelength of the scattered photon as

Incident photon | Electron at rest
Before: λ, p, E | $+x$
After: λ', p', E' | p_e, KE_e | $+x$
Scattered photon | Recoiling electron

$$\lambda' = \lambda + \frac{h}{m_e c}(1-\cos 180°) = \lambda + \frac{2h}{m_e c} = \lambda + 2\lambda_c$$

The quantity $\lambda_c = h/m_e c$ is known as the Compton wavelength and has a numerical value of $\lambda_c = 0.002\,43 \text{ nm} = 2.43 \times 10^{-12}$ m.

From conservation of energy, the kinetic energy of the recoiling electron must be $KE_e = E_{\text{incident}} - E_{\text{scattered}}$. The energy of the incident photon is $E_{\text{incident}} = 6.20\text{ keV} = hc/\lambda$, and the wavelength of this photon is $\lambda = hc/6.20\text{ keV}$. The wavelength of the scattered photon is then

$$\lambda' = \lambda + 2\lambda_c = \frac{hc}{6.20\text{ keV}} + 2\lambda_c$$

$$= \frac{(6.63\times10^{-34}\text{ J}\cdot\text{s})(3.00\times10^{8}\text{ m/s})}{(6.20\text{ keV})(1.60\times10^{-16}\text{ J/1 keV})} + 2(2.43\times10^{-12}\text{ m}) = 2.05\times10^{-10}\text{ m}$$

Thus, the energy of the scattered photon is

$$E_{\text{scattered}} = \frac{hc}{\lambda'} = \frac{(6.63\times10^{-34}\text{ J}\cdot\text{s})(3.00\times10^{8}\text{ m/s})}{2.05\times10^{-10}\text{ m}}\left(\frac{1\text{ keV}}{1.60\times10^{-16}\text{ J}}\right) = 6.06\text{ keV}$$

and the kinetic energy of the recoiling electron is

$$KE_e = E_{\text{incident}} - E_{\text{scattered}} = 6.20\text{ keV} - 6.06\text{ keV} = 0.14\text{ keV}$$ ◊

---

47. A light source of wavelength λ illuminates a metal and ejects photoelectrons with a maximum kinetic energy of 1.00 eV. A second light source of wavelength $\lambda/2$ ejects photoelectrons with a maximum kinetic energy of 4.00 eV. What is the work function of the metal?

Solution

From Einstein's model of the photoelectric effect, the maximum kinetic energy of the electrons liberated by incident light of wavelength $f = c/\lambda$ is $KE_{\text{max}} = hf - \phi$, or $hc/\lambda - \phi = KE_{\text{max}}$, where ϕ is the work function of the metal being illuminated.

For the first light source used, this equation gives

$$\frac{hc}{\lambda} - \phi = 1.00\text{ eV} \qquad [1]$$

For the light source emitting wavelength $\lambda' = \lambda/2$, Einstein's equation yields $hc/\lambda' - \phi = (KE_{\text{max}})'$, or

$$\frac{2hc}{\lambda} - \phi = 4.00\text{ eV} \qquad [2]$$

Multiplying Equation [1] by 2 and subtracting the result from Equation [2] gives

$$\frac{2hc}{\lambda} - \phi - \frac{2hc}{\lambda} + 2\phi = 4.00\text{ eV} - 2.00\text{ eV}$$

and we find the work function of the illuminated metal to be $\phi = 2.00$ eV. ◊

28

Atomic Physics

NOTES FROM SELECTED CHAPTER SECTIONS

28.2 Atomic Spectra
28.3 The Bohr Model

The basic postulates of the Bohr model of the hydrogen atom are:

- The **electron moves in circular orbits** about the nucleus under the influence of the Coulomb force of attraction between the electron and the positively charged proton in the nucleus.

- The **electron can exist only in specific stationary states** (orbits). When the electron is in one of its allowed orbits, it does not emit energy by radiation.

- The **atom radiates energy only when the electron makes a transition** ("jumps") from one allowed stationary orbit to another less energetic state. This postulate states that the energy given off by an atom is carried away by a photon of energy hf.

- The **orbital angular momentum of the electron about the nucleus is quantized** in units of $n(h/2\pi)$, where h is Planck's constant. *This condition determines the radii of the allowed orbits.*

There are several important reasons to understand the behavior of the hydrogen atom as an atomic system:

- Much of what is learned about the hydrogen atom with its single electron can be extended to such single-electron ions as He^{+} and Li^{2+}, which are hydrogen-like in their atomic structure.

- The hydrogen atom is an ideal system for performing precise tests of theory against experiment and for improving our overall understanding of atomic structure.

- The quantum numbers used to characterize the allowed states of hydrogen can be used to describe the allowed states of more complex atoms. This enables us to understand the periodic table of the elements.

- The basic ideas about atomic structure must be well understood before we attempt to deal with the complexities of molecular structures and the electronic structure of solids.

28.4 Quantum Mechanics and the Hydrogen Atom

The possible stationary energy states of an electron in an atom are determined by the values of four quantum numbers.

- **Principal quantum number (n) with integer values from 1 to ∞**

  The principal quantum number follows from the concept of quantization of angular momentum. Electron energy states with the same principal quantum number form a **shell**. Shells are identified by the letters K, L, M... corresponding to $n = 1, 2, 3 \ldots$.

- **Orbital quantum number (ℓ) with integer values from 0 to $n-1$**

  An electron in a given allowed energy state may move in different elliptical orbits determined by the value of ℓ. For each value of n, there are n possible orbits corresponding to different values of ℓ. Energy states with given values of n and ℓ form a **subshell.** Subshells are identified by the letters s, p, d, f,... corresponding to $\ell = 0, 1, 2, 3, \ldots$. The maximum number of electrons allowed in any subshell is $2(2\ell + 1)$.

- **Orbital magnetic quantum number (m_ℓ) with integer values from $-\ell$ to $+\ell$**

  The orbital magnetic quantum number accounts for the observed **Zeeman effect**: when a gas is placed in an external magnetic field, single spectral lines are split into several lines.

- **Spin magnetic quantum number (m_s) with values of $-1/2$ and $+1/2$**

  The **spin magnetic quantum number** m_s accounts for the two closely spaced energy states in spectral lines ("**doublets**") corresponding to the two possible orientations of electron spin ($+1/2$ is "up" spin and $-1/2$ is "down" spin).

28.5 The Exclusion Principle and the Periodic Table

At this point you might find it useful to review the description of the quantum numbers in Section 28.4 of the textbook.

The Pauli exclusion principle states that no two electrons in an atom can exist in identical quantum states. This means that no two electrons in a given atom can have exactly the same values for the set of quantum numbers n, ℓ, m_ℓ, and m_s.

The order in which electrons fill subshells and the application of the exclusion principle are illustrated in the following tables in your textbook:

Table 28.3 Number of Electrons in Filled Subshells and Shells

Table 28.4 Electronic Configuration of Some Elements

28.6 Characteristic X-Rays

The x-ray spectrum of a metal target consists of a broad continuous spectrum on which are superimposed a series of sharp lines. **Characteristic x-rays** are emitted by atoms when an electron undergoes a transition from an outer shell into an electron vacancy in one of the inner shells. Transitions into a vacant state in the K shell give rise to the K series of spectral lines, transitions into a vacant state in the L shell create the L series of lines, and so on. *Lines within a series are given a notation to designate the shell from which the transition originates.*

Examples from the K and L series (in order of increasing energy and decreasing intensity) are:

- K_α line; an electron transition from the L shell to the K shell
- K_β line; an electron transition from the M shell to the K shell
- L_α line; an electron transition from the M shell to the L shell
- L_β line; an electron transition from the N shell to the L shell

28.7 Atomic Transitions and Lasers

Stimulated absorption (upward atomic transition) occurs when photons incident on a gas have energies which exactly match the energy separation between two allowed states of an atom, usually the ground state and a higher energy state. The absorption process raises an atom to various higher energy states called **excited states**.

An atom in an excited state has a certain probability of returning to its original energy state by emission of a photon. This process is called **spontaneous emission**.

When a photon with an energy equal to the excitation energy of an excited atom is incident on the atom, it can increase the probability of de-excitation. This is called **stimulated emission (downward atomic transition)** and results in a second photon. *The incident and the emitted photons are identical; they have equal energies, and they are exactly in phase.*

Population inversion in a gas sample is the condition in which there are more atoms in the excited state than there are in the ground state.

Lasers are monochromatic, coherent light sources that work on the principle of stimulated emission of radiation by a system of atoms.

The following conditions must be satisfied in order to achieve laser action:

- **Population inversion:** There must be a greater number of atoms in an excited state than in the ground state.

- **Metastable states:** The lifetime of the excited states must be long compared with the usually short lifetimes of excited states. Under these conditions stimulated emission will occur before spontaneous emission.

- **Photon confinement:** The emitted photons must remain in the system long enough to allow them to stimulate further emission from other excited atoms. This is achieved by the use of reflecting mirrors at the ends of the system. One end is made totally reflecting and the other is slightly transparent to allow the laser beam to escape.

EQUATIONS AND CONCEPTS

Wavelengths in the Balmer series (the four visible lines) in the emission spectrum of hydrogen can be calculated using empirical Equation 28.1. R_H is a constant called the **Rydberg constant**. The short wavelength limit of the Balmer series is 364.6 nm.

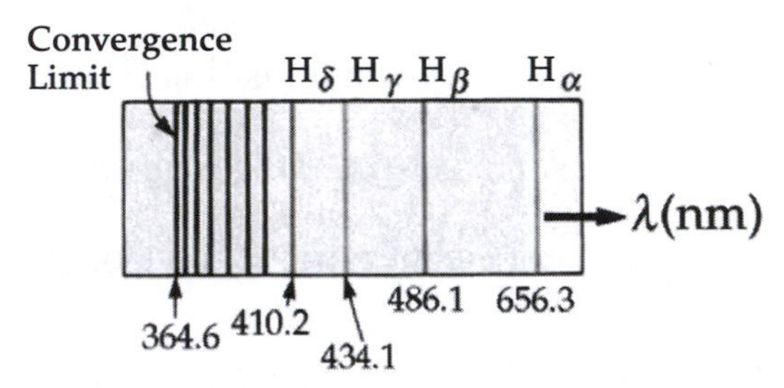

The Balmer series of spectral lines for hydrogen

$$\frac{1}{\lambda} = R_H\left(\frac{1}{2^2} - \frac{1}{n^2}\right) \quad (n = 3, 4, 5, \ldots) \quad (28.1)$$

$$R_H = 1.097\,373\,2 \times 10^7 \text{ m}^{-1} \quad (28.2)$$

An **electron transition** from an initial stationary state, E_i, to a lower energy state, E_f, results in the emission of a photon with a frequency that is proportional to the difference in energies of the initial and final states. *An electron does not radiate energy while in one of the allowed orbits (stationary states) determined by quantization of the orbital angular momentum.*

$$E_i - E_f = hf \quad (28.4)$$

The angular momentum of the electron about the nucleus must be quantized, always equal to an integral multiple of $n\hbar$. *This condition determines the radii of the allowed electron orbits.*

$$m_e vr = n\hbar \qquad n = 1, 2, 3, \ldots \quad (28.5)$$

$$\hbar = \frac{h}{2\pi} = 1.05 \times 10^{-34} \text{ J} \cdot \text{s}$$

The total energy of the hydrogen atom (*KE* plus *PE* of the proton-electron bound system) depends on the radius of the allowed orbit of the electron. *Note that the total energy of the atom is negative for all values of r except $r = \infty$ when $E = 0$.*

$$E = -\frac{k_e e^2}{2r} \qquad (28.9)$$

The **first three orbits** predicted by Bohr are shown (not to scale) in the figure at right.

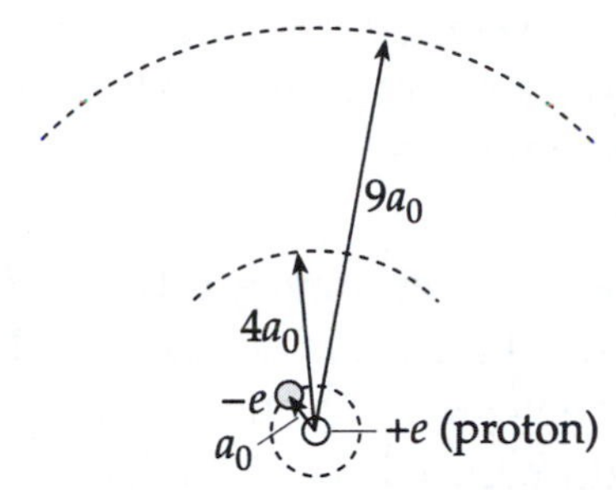

First three circular orbits predicted by the Bohr model

Radii of the allowed orbits have discrete (quantized) values that can be expressed in terms of the Bohr radius (a_0).

$$r_n = \frac{n^2\hbar^2}{m_e k_e e^2} \qquad n = 1, 2, 3, \ldots \qquad (28.10)$$

The **Bohr radius** (a_0) corresponds to $r = 1$ in Equation 28.10.

$$a_0 = \frac{\hbar^2}{m_e k_e e^2} = 0.052\,9 \text{ nm} \qquad (28.11)$$

$$r_n = n^2 a_0 = n^2(0.052\,9 \text{ nm}) \qquad (28.12)$$

Quantized energy level values can be expressed in units of electron volts (eV). The lowest allowed energy state or ground state corresponds to the principal quantum number $n = 1$. *The absolute value of the ground state energy is equal to the ionization energy of the atom. The energy level approaches $E = 0$ as r approaches infinity.*

$$E_n = -\frac{13.6}{n^2} \text{ eV} \qquad (28.14)$$

A **photon of frequency** ***f*** (and wavelength λ) is emitted when an electron undergoes a transition from an initial energy level E_i (outer orbit) to a lower level E_f (inner orbit).

$$f = \frac{E_i - E_f}{h} = \frac{m_e k_e^2 e^4}{4\pi \hbar^3}\left(\frac{1}{n_f^2} - \frac{1}{n_i^2}\right) \qquad (28.15)$$

where $n_f < n_i$

A theoretical value for the Rydberg constant can be calculated by substituting known values into Equation 28.17.

$$R_H = \frac{m_e k_e^2 e^4}{4\pi c \hbar^3} \qquad (28.17)$$

The **spectral lines** in the hydrogen spectrum can be arranged into several series. Each series corresponds to an assigned value of the principal quantum number of the final energy state (n_f).

Lyman series: $n_f=1$; $n_i=2, 3, 4, ...$
Balmer series: $n_f=2$; $n_i=3, 4, 5, ...$
Paschen series: $n_f=3$; $n_i=4, 5, 6, ...$
Brackett series: $n_f=4$; $n_i=5, 6, 7, ...$

Photon wavelengths for the various series in the hydrogen spectrum can be calculated using Equation 28.16.

$$\frac{1}{\lambda} = R_H\left(\frac{1}{n_f^2} - \frac{1}{n_i^2}\right) \tag{28.16}$$

The energy level diagram for hydrogen shows transitions for the Lyman, Balmer, and Paschen series. Within each series, the energy difference between adjacent energy levels becomes smaller as n_i increases. The **correspondence principle** states that quantum physics is in agreement with classical physics for large values of the principal quantum number (i.e., when the energy difference between quantized levels are small).

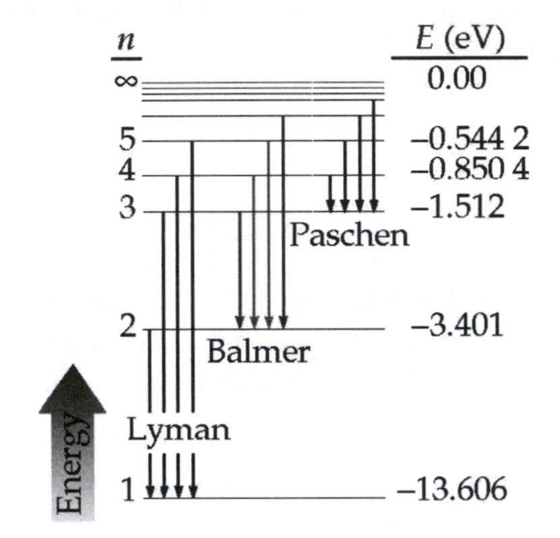

Energy level diagram for hydrogen

Quantum numbers for the hydrogen atom are listed below. *The Pauli exclusion principle states that no two electrons in an atom can have the same set of quantum number values.*

Quantum Number	Name	Allowed Values	Allowed States
n	Principal quantum number	1, 2, 3, …	any integer
ℓ	Orbital quantum number	0, 1, 2, …, $(n-1)$	n
m_ℓ	Orbital magnetic quantum number	$-\ell, -\ell+1, -\ell+2, ..., 0, 1, 2 ... \ell$	$2\ell+1$
m_s	Spin magnetic quantum number	½, −½	2

REVIEW CHECKLIST

- State the basic postulates of the Bohr model of the hydrogen atom. (Section 28.3)
- Sketch an energy level diagram for hydrogen (assign values of the principal quantum number, n), show transitions corresponding to spectral lines in the several known series, and make calculations of wavelength values. (Section 28.3)
- For each of the quantum numbers, n (the principal quantum number), ℓ (the orbital quantum number), m_ℓ (the orbital magnetic quantum number), and m_s (the spin magnetic quantum number): (i) qualitatively describe what each implies concerning atomic structure, (ii) state the allowed values which may be assigned to each, and (iii) give the number of allowed states which may exist in a particular atom corresponding to each allowed value of the principal quantum number. (Sections 28.4 and 28.5)
- State the Pauli exclusion principle and describe its relevance to the periodic table of the elements. Show how the exclusion principle leads to the known electronic ground state configuration of the light elements. (Section 28.5)

SOLUTIONS TO SELECTED END-OF-CHAPTER PROBLEMS

3. The "size" of the *atom* in Rutherford's model is about 1.0×10^{-10} m. (a) Determine the attractive electrostatic force between an electron and a proton separated by this distance. (b) Determine (in eV) the electrostatic potential energy of the atom.

Solution

(a) From Coulomb's law, the magnitude of the electrical force between two particles having charges q_1 and q_2, when separated by distance r, is

$$F=\frac{k_e|q_1||q_2|}{r^2}$$

Thus, if an electron and proton are separated by a distance of 1.0×10^{-10} m, the magnitude of the attractive force between these unlike charges is

$$F=\frac{\left(8.99\times10^9\ \text{N}\cdot\text{m}^2/\text{C}^2\right)\left(1.60\times10^{-19}\ \text{C}\right)^2}{\left(1.0\times10^{-10}\ \text{m}\right)^2}=2.3\times10^{-8}\ \text{N} \qquad \Diamond$$

(b) The electrostatic potential energy of an atom consisting of an electron and a proton separated by distance $r=1.0\times10^{-10}$ m is given by

$$PE=\frac{k_e q_1 q_2}{r}=\frac{\left(8.99\times10^9\ \text{N}\cdot\text{m}^2/\text{C}^2\right)\left(-1.60\times10^{-19}\ \text{C}\right)\left(+1.60\times10^{-19}\ \text{C}\right)}{1.0\times10^{-10}\ \text{m}}$$

or

$$PE = -2.3 \times 10^{-18} \text{ J}\left(\frac{1 \text{ eV}}{1.60 \times 10^{-19} \text{ J}}\right) = -14 \text{ eV}$$ ◊

---

8. For a hydrogen atom in its ground state, use the Bohr model to compute (a) the orbital speed of the electron, (b) the kinetic energy of the electron, and (c) the electrical potential energy of the atom.

Solution

(a) In the Bohr model of the hydrogen atom, the radii of the allowed orbits are $r_n = n^2 a_0$, where $a_0 = 0.052\,9$ nm and $n = 1, 2, 3, \ldots$. Thus, in the ground state ($n = 1$), the radius of the orbit is $r_1 = (1)^2(0.052\,9 \text{ nm}) = 5.29 \times 10^{-11}$ m.

As the electron orbits the nucleus, the required centripetal acceleration is supplied by the attractive electrical force between an electron and a proton. Thus, $m_e v_n^2 / r_n = k_e e^2 / r_n^2$, or the speed of the electron in the nth allowed orbit is $v_n = \sqrt{k_e e^2 / m_e r_n}$. In the ground state, this orbital speed will be

$$v_1 = \sqrt{\frac{k_e e^2}{m_e r_1}} = \sqrt{\frac{(8.99 \times 10^9 \text{ N}\cdot\text{m}^2/\text{C}^2)(1.60 \times 10^{-19} \text{ C})^2}{(9.11 \times 10^{-31} \text{ kg})(5.29 \times 10^{-11} \text{ m})}} = 2.19 \times 10^6 \text{ m/s}$$ ◊

(b) Since $v_1 \ll c$, the electron in the ground state of hydrogen is nonrelativistic. This means that its kinetic energy is given $KE_1 = \frac{1}{2} m_e v_1^2$, with a numeric value of

$$KE_1 = \frac{1}{2}(9.11 \times 10^{-31} \text{ kg})(2.19 \times 10^6 \text{ m/s})^2 = 2.18 \times 10^{-18} \text{ J}$$

or $$KE_1 = 2.18 \times 10^{-18} \text{ J}\left(\frac{1 \text{ eV}}{1.60 \times 10^{-19} \text{ J}}\right) = 13.6 \text{ eV}$$ ◊

(c) The electrical potential energy of the atom (an electron and a proton separated by the orbital radius) in the ground state is given by

$$PE_1 = \frac{k_e q_1 q_2}{r} = \frac{k_e(-e)(+e)}{r_1}$$

$$= -\frac{(8.99 \times 10^9 \text{ N}\cdot\text{m}^2/\text{C}^2)(1.60 \times 10^{-19} \text{ C})^2}{5.29 \times 10^{-11} \text{ m}} = -4.35 \times 10^{-18} \text{ J}$$

or

$$PE_1 = -4.35 \times 10^{-18}\ \text{J}\left(\frac{1\ \text{eV}}{1.60 \times 10^{-19}\ \text{J}}\right) = -27.2\ \text{eV}$$ ◊

Note that this gives the total energy of the hydrogen atom in its ground state as

$$E_1 = KE_1 + PE_1 = 13.6\ \text{eV} - 27.2\ \text{eV} = -13.6\ \text{eV}$$

---

11. A hydrogen atom emits a photon of wavelength 656 nm. From what energy orbit to what lower energy orbit did the electron jump?

Solution

The energy levels in the hydrogen atom are given by

$$E_n = -\frac{13.6\ \text{eV}}{n^2} \qquad \text{where} \qquad n = 1, 2, 3, 4, \ldots$$

Thus, the energies of some lower levels are:

$$E_1 = -13.6\ \text{eV},\ E_2 = -3.40\ \text{eV},\ E_3 = -1.51\ \text{eV},\ E_4 = -0.850\ \text{eV}, \ldots$$

When the electron in the atom makes a transition from a level having energy E_i to a level having lower energy E_f, the excess energy is emitted in the form of a photon with

$$E_{\text{photon}} = E_i - E_f$$

If the photon is observed to have wavelength $\lambda = 656$ nm, its energy is

$$E_{\text{photon}} = \frac{hc}{\lambda} = \frac{(6.63 \times 10^{-34}\ \text{J}\cdot\text{s})(3.00 \times 10^8\ \text{m/s})}{656 \times 10^{-9}\ \text{m}} = 3.03 \times 10^{-19}\ \text{J}$$

or

$$E_{\text{photon}} = 3.03 \times 10^{-19}\ \text{J}\left(\frac{1\ \text{eV}}{1.60 \times 10^{-19}\ \text{J}}\right) = 1.89\ \text{eV}$$

Observe that this energy equals the difference in the energies of the $n = 3$ and $n = 2$ levels in the atom, or $E_{\text{photon}} = E_3 - E_2$. Hence, the transition made by the electron was from the $E_3 = -1.51$ eV level to the $E_2 = -3.40$ eV level. ◊

---

15. The Balmer series for the hydrogen atom corresponds to electronic transitions that terminate in the state with quantum number $n = 2$ as shown in Figure P28.15. Consider the photon of longest wavelength corresponding to a transition shown in the figure. Determine (a) its energy and (b) its wavelength. Consider the spectral line of shortest wavelength corresponding to a transition shown in the figure. Find (c) its photon energy and (d) its wavelength. (e) What is the shortest possible wavelength in the Balmer series?

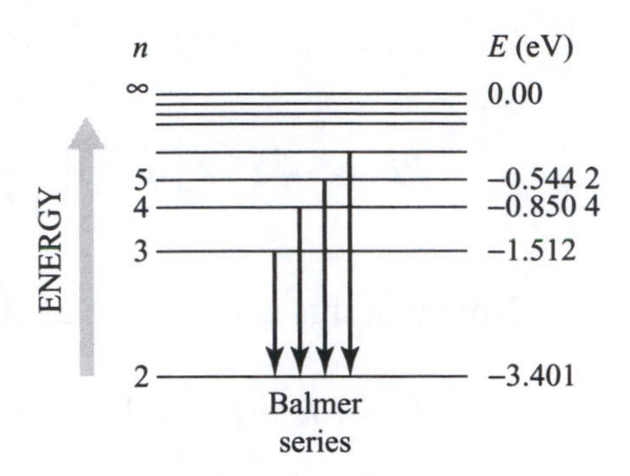

Figure P28.15

Solution

The hydrogen atom emits a photon of light when the electron makes a transition from some initial state with quantum number n_i and energy E_i to a final state with quantum number n_f and energy E_f. The energy of the emitted photon is $E_{\text{photon}} = hc/\lambda = E_i - E_f$, and its wavelength is $\lambda = hc/\left(E_i - E_f\right)$. Thus, the longest wavelength in the Balmer series is produced in the transition terminating on the $n = 2$ level that involves the smallest change in energy, namely the $n = 3$ to $n = 2$ transition.

(a) The energy of the photon emitted in this transition is

$$E_{\text{photon}} = E_3 - E_2 = -1.512\text{ eV} - (-3.401\text{ eV}) = 1.889\text{ eV} \qquad \Diamond$$

(b) The wavelength of this photon will be

$$\lambda = \frac{hc}{E_3 - E_2} = \frac{\left(6.63\times10^{-34}\text{ J}\cdot\text{s}\right)\left(3.00\times10^{8}\text{ m/s}\right)}{(1.889\text{ eV})\left(1.60\times10^{-19}\text{ J/1 eV}\right)} = 6.58\times10^{-7}\text{ m} = 658\text{ nm} \qquad \Diamond$$

The shortest wavelength corresponding to one of the shown transitions is produced in the transition involving the largest change in energy, namely the $n = 6$ to $n = 2$ transition.

(c) The energy of the photon emitted in the $n = 6$ to $n = 2$ transition is

$$E_{\text{photon}} = E_6 - E_2 = -0.378\text{ eV} - (-3.401\text{ eV}) = 3.02\text{ eV} \qquad \Diamond$$

(d) The wavelength of the photon produced in this transition is

$$\lambda = \frac{hc}{E_3 - E_2} = \frac{\left(6.63\times10^{-34}\text{ J}\cdot\text{s}\right)\left(3.00\times10^{8}\text{ m/s}\right)}{(3.02\text{ eV})\left(1.60\times10^{-19}\text{ J/1 eV}\right)} = 4.12\times10^{-7}\text{ m} = 412\text{ nm} \qquad \Diamond$$

(e) The shortest wavelength in the Balmer series is produced when the electron gives up the maximum energy and terminates on the $n = 2$ level. This occurs when the initial state is a free electron with $n_i = \infty$ and energy $E_i = 0$. When this free electron is captured into the $n = 2$ state of the hydrogen atom, the wavelength produced is

$$\lambda = \frac{hc}{0 - E_2} = \frac{\left(6.63\times10^{-34}\ \text{J}\cdot\text{s}\right)\left(3.00\times10^{8}\ \text{m/s}\right)}{(3.401\ \text{eV})\left(1.60\times10^{-19}\ \text{J/1 eV}\right)} = 3.66\times10^{-7}\ \text{m} = 366\ \text{nm}$$ ◊

---

21. An electron is in the second excited orbit of hydrogen, corresponding to $n = 3$. Find (a) the radius of the orbit and (b) the wavelength of the electron in this orbit.

Solution

(a) In the Bohr model of the hydrogen atom, the radii of the allowed orbits are given by $r_n = n^2 a_0$, where $a_0 = 0.052\,9$ nm and $n = 1, 2, 3, \ldots$. The radius of the second excited orbit is then

$$r_3 = 3^2 (0.052\,9\ \text{nm}) = 9(0.052\,9\ \text{nm}) = 0.476\ \text{nm}$$ ◊

(b) According to de Broglie's interpretation of one of Bohr's original postulates, the allowed orbits are those in which the electron can form a standing wave pattern. That is, they are the orbits for which the circumference is an integer multiple of the de Broglie wavelength of the electron in those orbits, or the orbits for which

$$\text{Circumference} = 2\pi r_n = n\lambda_n$$

Thus, the wavelength of the electron in the second excited orbit, corresponding to $n = 3$, should be

$$\lambda_3 = \frac{2\pi r_3}{3} = \frac{2\pi(0.476\ \text{nm})}{3} = 0.997\ \text{nm}$$ ◊

---

26. Using the concept of standing waves, de Broglie was able to derive Bohr's stationary orbit postulate. He assumed that a confined electron could exist only in states where its de Broglie waves form standing-wave patterns, as in Figure 28.6. Consider a particle confined in a box of length L to be equivalent to a string of length L and fixed at both ends. Apply de Broglie's concept to show that (a) the linear momentum of this particle is quantized with $p = mv = nh/2L$ and (b) the allowed states correspond to particle energies of $E_n = n^2 E_0$, where $E_0 = h^2/(8mL^2)$.

Solution

(a) In order to form a standing wave in a string fixed at both ends, it is necessary that the length of the string be an integer multiple of a half wavelength. That is,

$$L = n\left(\frac{\lambda}{2}\right) \qquad \text{or} \qquad \lambda = \frac{2L}{n}$$

According to de Broglie's hypothesis, the wavelength of a particle of mass m is given by

$$\lambda = \frac{h}{p} = \frac{h}{mv} \qquad \text{or} \qquad p = mv = \frac{h}{\lambda}$$

Thus, if we consider the particle confined in a box of length L to be equivalent to a string of length L fixed at both ends, the requirement to form standing waves would be

$$p_n = mv_n = \frac{h}{\lambda} = \frac{h}{2L/n} = \frac{nh}{2L} \qquad \text{where} \qquad n = 1, 2, 3, \ldots$$ ◊

(b) Considering the particle to be nonrelativistic, the total energy of a particle in a box would be

$$E_n = \frac{1}{2}mv_n^2 = \frac{p_n^2}{2m} = \frac{(nh/2L)^2}{2m} = \frac{n^2h^2}{8mL^2}$$

or $E_n = n^2 E_0$ where $E_0 = \dfrac{h^2}{8mL^2}$ ◊

29. The ρ-meson has a charge of $-e$, a spin quantum number of 1, and a mass 1 507 times that of the electron. If the electrons in atoms were replaced by ρ-mesons, list the possible sets of quantum numbers for ρ-mesons in the 3d subshell.

Solution

When a particle of spin quantum number s exists in a bound state within an atom, that state is characterized by a set of four quantum numbers. These are: the principal quantum number n, which may have any integer value in the range $1 \le n < \infty$; the orbital quantum number ℓ, which (for a given value of n) may take on any integer value in the range $0 \le \ell \le n-1$; the orbital magnetic quantum number m_ℓ, which (for a given value of ℓ) may have any integer value in the range $-\ell \le m_\ell \le +\ell$; and the spin magnetic quantum number m_s, which (for a particle with spin quantum number s) may vary from $-s$ to $+s$ in integer steps.

The 3d subshell within an atom consists of all allowed states having $n = 3$ and $\ell = 2$. With $\ell = 2$ in this subshell, states within this subshell may have any of 5 values for the quantum number m_ℓ. These 5 values are $m_\ell = -2,\ -1,\ 0,\ +1$, and $+2$. If the particle is a ρ-meson, with $s = 1$, there are 3 possible values of m_s (namely $m_s = -1, 0$, and $+1$) associated with each of the 5 possible values for m_ℓ. Thus, within the 3d subshell, there are $5 \times 3 = 15$ allowed combinations of the four quantum numbers, or there are 15 quantum states within the 3d subshell of an atom when the electrons are replaced with ρ-mesons. ◊

The 15 possible sets of quantum numbers for ρ-mesons in the 3d subshell are as follows:

$$(n=1, \ell=2, m_\ell=-2, m_s=-1);\ (n=1, \ell=2, m_\ell=-2, m_s=0);\ (n=1, \ell=2, m_\ell=-2, m_s=+1)$$
$$(n=1, \ell=2, m_\ell=-1, m_s=-1);\ (n=1, \ell=2, m_\ell=-1, m_s=0);\ (n=1, \ell=2, m_\ell=-1, m_s=+1)$$
$$(n=1, \ell=2, m_\ell=0, m_s=-1);\ (n=1, \ell=2, m_\ell=0, m_s=0);\ (n=1, \ell=2, m_\ell=0, m_s=+1)$$
$$(n=1, \ell=2, m_\ell=+1, m_s=-1);\ (n=1, \ell=2, m_\ell=+1, m_s=0);\ (n=1, \ell=2, m_\ell=+1, m_s=+1)$$
$$(n=1, \ell=2, m_\ell=+2, m_s=-1);\ (n=1, \ell=2, m_\ell=+2, m_s=0);\ (n=1, \ell=2, m_\ell=+2, m_s=+1)\ ◊$$

---

33. Zirconium ($Z = 40$) has two electrons in an incomplete d subshell. (a) What are the values of n and ℓ for each electron? (b) What are all possible values of m_ℓ and m_s? (c) What is the electron configuration in the ground state of zirconium?

Solution

(a) Zirconium, with 40 electrons, has four electrons outside of a closed krypton ($Z = 36$) core. In krypton (a noble gas), all the allowed quantum states are completely filled up through the 4p subshell. Normally, as atoms of higher atomic number are formed, one would expect the next electrons added would go into the 4d subshell. However, an exception to the rule occurs at this point in the periodic table. The 5s subshell fills with 2 electrons before the 4d subshell begins to fill. Thus, zirconium has the filled krypton core, a filled 5s subshell, and 2 electrons in an incomplete 4d subshell. The principal quantum number and orbital quantum number for each of these 2 electrons in the incomplete subshell are

$$n = 4 \qquad \text{and} \qquad \ell = 2$$ ◊

(b) Since the allowed values for the orbital magnetic quantum number m_ℓ consists of the integers in the range from $-\ell$ to $+\ell$, and the electrons in the incomplete $4d$ subshell in zirconium have the orbital quantum number $\ell = 2$, the possible values of m_ℓ for these two electrons are

$$m_\ell = -2,\ -1,\ 0,\ +1,\ \text{and } +2$$ ◊

Any electron has a spin quantum number $s = \frac{1}{2}$. Thus, all electrons, regardless of which shell or subshell they occupy within the atom, must have one of two possible values for the spin magnetic quantum number. These values are

$$m_s = +\tfrac{1}{2} \text{ and } m_s = -\tfrac{1}{2}$$ ◊

(c) The complete electron configuration for the ground state of zirconium is

$$1s^2 2s^2 2p^6 3s^2 3p^6 3d^{10} 4s^2 4p^6 4d^2 5s^2 = [\text{Kr}]4d^2 5s^2$$ ◊

---

35. A bismuth target is struck by electrons, and x-rays are emitted. Estimate (a) the M- to L-shell transitional energy for bismuth and (b) the wavelength of the x-ray emitted when an electron falls from the M shell to the L shell.

Solution

(a) A neutral bismuth atom ($Z = 83$) in its ground state has the K, L, M, N, and O shells completely filled and five electrons in an incomplete P shell. A vacancy in the L shell can be created by bombarding bismuth atoms with a beam of high energy electrons. When this occurs, the electrons in the M shell are partially screened from the nuclear charge by a total of 9 electrons, 2 in the filled K shell and the remaining 7 electrons in the L shell. Thus, the effective nuclear charge seen by the M shell electrons is $Z_{\text{eff}} = Z - 9 = 83 - 9 = 74$, and we approximate the energy of an electron in the M shell (with $n = 3$) as being

$$E_{\text{M}} \approx -\frac{m_e k_e^2 Z_{\text{eff}}^2 e^4}{2\hbar^2}\left(\frac{1}{n^2}\right) = -Z_{\text{eff}}^2 \frac{\left(m_e k_e^2 e^4 / 2\hbar^2\right)}{n^2} = -(74)^2 \frac{(13.6 \text{ eV})}{9}$$

$$= -8.27 \times 10^3 \text{ eV}$$

or $\quad E_{\text{M}} \approx -8.27 \text{ keV}$.

When an electron drops down to fill the vacancy in the L shell, this electron will continue to be partially screened from the nuclear charge by the two electrons in the filled K shell. Thus, the effective nuclear charge it sees is $Z_{\text{eff}} = Z - 2 = 81$, and our estimate of the energy of an electron in the L shell ($n = 2$) is

$$E_{\text{L}} \approx -Z_{\text{eff}}^2 \frac{\left(m_e k_e^2 e^4 / 2\hbar^2\right)}{n^2} = -(81)^2 \frac{(13.6 \text{ eV})}{4} = -2.23 \times 10^4 \text{ eV} = -22.3 \text{ keV}$$

The estimate of the energy given up by an electron as it makes this M- to L-shell transition (and hence, the energy of the x-ray photon emitted) is then

$$E_{\text{photon}} = E_{\text{M}} - E_{\text{L}} \approx -8.27 \text{ keV} - (-22.3 \text{ keV}) = +14 \text{ keV}$$ ◊

(b) The estimated wavelength of the x-ray photon produced by a M- to L-shell transition in bismuth is

$$\lambda = \frac{hc}{E_{\text{photon}}} \approx \frac{\left(6.63 \times 10^{-34} \text{ J} \cdot \text{s}\right)\left(3.00 \times 10^8 \text{ m/s}\right)}{14 \text{ keV}} \left(\frac{1 \text{ keV}}{1.60 \times 10^{-16} \text{ J}}\right)$$

$$= 8.9 \times 10^{-11} \text{ m}$$ ◊

---

38. In a hydrogen atom, what is the principal quantum number of the electron orbit with a radius closest to 1.0 μm?

Solution

In the hydrogen atom, the centripetal acceleration necessary to keep the electron in its orbit is supplied by the attractive electrical force between an electron and a proton. Thus,

$$\frac{m_e v^2}{r} = \frac{k_e e^2}{r^2} \qquad \text{which reduces to} \qquad r = \frac{k_e e^2}{mv^2}$$

From Bohr's quantization postulate, the allowed orbits are those for which $m_e vr = n\hbar$, where the principal quantum number n may be any integer greater than zero. The speed of the electron in the nth allowed orbit is therefore $v_n = n\hbar / m_e r_n$. Substituting this into our result from above gives $r_n = k_e e^2 m_e r_n^2 / n^2 \hbar^2$, which simplifies to

$$r_n = n^2 \left(\frac{\hbar^2}{m_e k_e e^2}\right) = n^2 a_0 \qquad n = 1, 2, 3, \ldots$$

The constant $a_0 = \hbar^2 / m_e k_e e^2$ is known as the Bohr radius and has a numeric value of $a_0 = 0.052\,9$ nm. It is the radius of the innermost ($n = 1$) orbit in the hydrogen atom.

If some orbit in the hydrogen atom is to have a radius closest to $r = 1.0\ \mu$m, the principal quantum number of this orbit ($n = \sqrt{r_n/a_0}$) must be the integer closest to

$$\sqrt{\frac{r}{a_0}} = \sqrt{\frac{1.0\times10^{-6}\text{ m}}{5.29\times10^{-11}\text{ m}}} = 137.49 \qquad \text{or} \qquad n = 137 \qquad \lozenge$$

---

41. An electron in chromium moves from the $n = 2$ state to the $n = 1$ state without emitting a photon. Instead, the excess energy is transferred to an outer electron (one in the $n = 4$ state), which is then ejected by the atom. In this Auger (pronounced "ohjay") process, the ejected electron is referred to as an Auger electron. (a) Find the change in energy associated with the transition from $n = 2$ into the vacant $n = 1$ state using Bohr theory. Assume that only one electron in the K shell is shielding part of the nuclear charge. (b) Find the energy needed to ionize an $n = 4$ electron, assuming 22 electrons shield the nucleus. (c) Find the kinetic energy of the ejected (Auger) electron. (All answers should be in electron volts.)

Solution

(a) A filled K shell would contain 2 electrons. With one vacancy in the K shell of chromium ($Z = 24$), an electron in the L shell has one electron shielding it from the nuclear charge, so $Z_{\text{eff}} = Z - 1 = 24 - 1 = 23$. The estimated energy the atom gives up during a transition from the L shell to the K shell vacancy is

$$\Delta E = E_i - E_f \approx -\frac{Z_{\text{eff}}^2(13.6\text{ eV})}{n_i^2} - \left[-\frac{Z_{\text{eff}}^2(13.6\text{ eV})}{n_f^2}\right] = Z_{\text{eff}}^2(13.6\text{ eV})\left[\frac{1}{n_f^2} - \frac{1}{n_i^2}\right]$$

or

$$\Delta E \approx (23)^2(13.6\text{ eV})\left[\frac{1}{1^2} - \frac{1}{2^2}\right] = 5.40\times10^3\text{ eV} = 5.40\text{ keV} \qquad \lozenge$$

(b) With a vacancy in the K shell, we assume that $Z - 2 = 24 - 2 = 22$ electrons shield the outermost electron (in a $4s$ state) from the nuclear charge. Thus, for this outer electron, $Z_{\text{eff}} = 24 - 22 = 2$, and the estimated energy required to remove this electron from the atom is

$$E_{\text{ionization}} = E_f - E_i = 0 - E_i \approx -\left[-\frac{Z_{\text{eff}}^2(13.6\text{ ev})}{n_i^2}\right] = \frac{2^2(13.6\text{ ev})}{4^2} = 3.40\text{ eV} \qquad \lozenge$$

(c) The atom transfers energy ΔE to this outer electron, which spends an amount $E_{\text{ionization}}$ of it in escaping from the atom. The remaining energy appears as kinetic energy of the now free electron. Therefore,

$$KE = \Delta E - E_{\text{ionization}} = 5.40\times10^3\text{ eV} - 3.40\text{ eV} \approx 5.40\times10^3\text{ eV} = 5.40\text{ keV} \qquad \lozenge$$

---

45. Use Bohr's model of the hydrogen atom to show that, when the atom makes a transition from the state n to the state $n-1$, the frequency of the emitted light is given by

$$f = \frac{2\pi^2 m_e k_e^2 e^4}{h^3}\left(\frac{2n-1}{(n-1)^2 n^2}\right)$$

Solution

According to the Bohr model of the hydrogen atom, the frequency of the photon emitted when the electron makes a transition from state n to state $n-1$ is

$$f = \frac{\Delta E}{h} = \frac{E_n - E_{n-1}}{h}$$

From Equation 28.13 in the textbook, the energy levels in hydrogen are given in terms of fundamental constants as

$$E_n = -\frac{m_e k_e^2 e^4}{2\hbar^2 n^2} = -\frac{2\pi^2 m_e k_e^2 e^4}{h^2 n^2} \qquad \text{where} \qquad n = 1, 2, 3, \ldots$$

Thus, the photon emitted in a transition from state n to state $n-1$ has frequency

$$f = \frac{1}{h}\left[-\frac{2\pi^2 m_e k_e^2 e^4}{h^2 n^2} - \left(-\frac{2\pi^2 m_e k_e^2 e^4}{h^2 (n-1)^2}\right)\right] = \frac{2\pi^2 m_e k_e^2 e^4}{h^3}\left[\frac{1}{(n-1)^2} - \frac{1}{n^2}\right]$$

or

$$f = \frac{2\pi^2 m_e k_e^2 e^4}{h^3}\left[\frac{n^2 - (n-1)^2}{(n-1)^2 n^2}\right] = \frac{2\pi^2 m_e k_e^2 e^4}{h^3}\left[\frac{n^2 - (n^2 - 2n + 1)}{(n-1)^2 n^2}\right]$$

which reduces to

$$f = \frac{2\pi^2 m_e k_e^2 e^4}{h^3}\left[\frac{2n-1}{(n-1)^2 n^2}\right] \qquad \Diamond$$

29

Nuclear Physics

NOTES FROM SELECTED CHAPTER SECTIONS

29.1 Some Properties of Nuclei

Important terms in the description of nuclear properties include:

- The **atomic number,** Z, equals the number of protons in the nucleus.
- The **neutron number,** N, equals the number of neutrons in the nucleus.
- The **mass number,** A, equals the number of nucleons (neutrons plus protons) in the nucleus.

Nuclei can be represented symbolically as ${}^{A}_{Z}\mathrm{X}$, where X designates the chemical symbol for a specific chemical element.

The nuclei of all atoms of a particular element contain the same number of protons but may contain different numbers of neutrons. Nuclei that are related in this way are called **isotopes. The isotopes of an element have the same Z value but different N and A values.**

The **unified mass unit, u, is defined such that the mass of the isotope ${}^{12}\mathrm{C}$ is exactly 12 u** ($1\ \mathrm{u} = 1.660\,559 \times 10^{-27}\ \mathrm{kg}$).

Based on nuclear scattering and other experiments, it is possible to conclude that:

- Most nuclei are approximately spherical, and all have nearly the same density.
- Nucleons combine to form a nucleus as though they were tightly packed spheres.
- The stability of nuclei is due to a short-range attractive nuclear force.
- The nuclear force is approximately the same for an interaction between any pair of nucleons (p-p, n-n, or p-n).

29.2 Binding Energy

The total mass of a nucleus is always less than the sum of the masses of its individual nucleons. This difference in mass is the origin of the nuclear **binding energy** and represents the energy that would be required to separate the nucleus into neutrons and protons.

29.3 Radioactivity

The decay of radioactive substances can be accompanied by the emission of three forms of radiation that vary in ability to penetrate shielding materials:

- **alpha (α) particles** — helium nuclei ($^4_2$He), which can barely penetrate a sheet of paper
- **beta (β) particles** — either electrons (e^-) or positrons (e^+) able to penetrate a few millimeters of aluminum
- **gamma (γ) rays** — high-energy photons capable of penetrating several centimeters of lead

A positron is the antiparticle of the electron; it has a mass equal to m_e and a charge of $+q_e$.

The decay rate, or activity, *R*, of a sample of a radioactive substance is defined as the number of decays occurring per second. The half-life ($T_{1/2}$) of a radioactive substance is the time during which half of an initial number of radioactive nuclei in a sample will decay. The customary unit of radioactivity is the Curie (Ci); the SI unit of activity is the **becquerel** (Bq), where 1 Bq = 1 decay/s and $1 \text{ Ci} = 3.7 \times 10^{10}$ Bq.

29.4 The Decay Processes

The overall decay process can be represented in equation form as

$$\underset{\text{Parent nucleus}}{X} \rightarrow \underset{\text{Daughter nucleus}}{Y} + \text{Emitted radiation}$$

In an equation representing a specific radioactive decay process, the sum of the mass numbers on each side of the equation must be equal, and the sum of the atomic numbers on each side of the equation must be equal.

The neutrino and antineutrino shown in the beta decay processes in the table below are required for conservation of energy and momentum.

The neutrino has the following properties:

- zero electric charge
- rest mass much smaller than that of the electron (recent experiments suggest that the neutrino mass may not be zero)
- spin of $\frac{1}{2}$ (satisfying the law of conservation of angular momentum)
- very weak interaction with matter (therefore very difficult to detect)

General characteristics of alpha, beta, and gamma decay are summarized below.

Decay	can be written as	Process
Alpha decay	${}^{A}_{Z}X \rightarrow {}^{A-4}_{Z-2}Y + {}^{4}_{2}He$	The parent nucleus emits an alpha particle and loses two protons and two neutrons.
Beta decay	${}^{A}_{Z}X \rightarrow {}^{A}_{Z+1}Y + e^{-} + \overline{\nu}$ (electron emission) ${}^{A}_{Z}X \rightarrow {}^{A}_{Z-1}Y + e^{+} + \nu$ (positron emission)	A nucleus can undergo beta decay in two ways. The parent nucleus can emit an electron (e^{-}) and an antineutrino ($\overline{\nu}$) or emit a positron (e^{+}) and a neutrino (ν).
Gamma decay	${}^{A}_{Z}X^{*} \rightarrow {}^{A}_{Z}X + \gamma$ * parent nucleus in an excited state	A nucleus in an excited state decays to a lower energy state (often the ground state) and emits a gamma ray.

29.5 Natural Radioactivity

Natural radioactivity is the decay of radioactive nuclei found in nature.

Artificial radioactivity is the decay of nuclei produced in the laboratory through nuclear reactions.

Most naturally decaying radionuclides are members of a **decay series**.

The uranium series, actinium series, and thorium series each begin with a long half-life, naturally occurring radioisotope and following a sequence of decays, ends in a stable isotope. A fourth series, the neptunium series, begins with the decay of an artificially produced radionuclide.

29.6 Nuclear Reactions

Nuclear reactions are events in which the structure or properties of colliding nuclei are altered. The ***Q*-value** is the energy required to balance the overall reaction equation.

Exothermic reactions release energy and have positive *Q*-values. The *Q*-values of endothermic reactions are negative, and the minimum energy of the incoming particle for which an endothermic reaction will occur is called the **threshold energy**.

29.7 Medical Applications of Radiation

Radiation damage to cells in biological organisms is primarily due to ionizing events and can be separated into two categories:

- **Somatic damage** is radiation damage to cells other than reproductive cells.
- **Genetic damage** affects only reproductive cells.

Several units are used to quantify radiation exposure and dose:

- **Roengten (R):** the amount of ionizing radiation that will produce 2.08×10^9 ion pairs in 1 cm^3 of air under standard conditions
- **rad:** the amount of radiation that deposits 10^{-2} J of energy into 1 kg of absorbing material
- **RBE** (relative biological effectiveness) factor: the number of rad of either x-ray or gamma radiation that will produce the same degree of biological damage as the actual radiation being used (e.g., electrons, protons, neutrons, heavy ions, etc.)
- **rem** (roentgen equivalent in man): This is the product of the dose measured in rads and the RBE factor. One rem of any two types of radiation will produce the same degree of biological damage. Dose in rem = dose in rad × RBE

Current SI units of radiation dose and exposure are the gray (Gy) and the sievert (Sv). 1 Sv = 100 rem and 1 Gy = 100 rad

EQUATIONS AND CONCEPTS

The **average nuclear radius** is proportional to the cube root of the mass number (or total number of nucleons). This means that the volume is proportional to A, nuclei are approximately spherical in shape, and all nuclei have nearly the same density.

$$r = r_0 A^{1/3} \quad (29.1)$$

$$r_0 = 1.2 \text{ fm}$$

The **fermi (fm)** is a convenient unit of length to use in stating nuclear dimensions.

$$1 \text{ fm} = 10^{-15} \text{ m}$$

The **decay constant, λ,** is characteristic of a particular isotope. The number of radioactive nuclei in a given sample which undergo decay during a time interval Δt is proportional to the number of nuclei present.

$$\Delta N = -\lambda N \Delta t \quad (29.2)$$

The decay rate or activity, R, of a sample of radioactive nuclei is defined as the number of decays per second. Activity, R, can be expressed in units of becquerels or curies.

$$R = \left| \frac{\Delta N}{\Delta t} \right| = \lambda N \quad (29.3)$$

$$1 \text{ Ci} \equiv 3.7 \times 10^{10} \text{ decays/s} \quad (29.6)$$

$$1 \text{ Bq} = 1 \text{ decay/s} \quad (29.7)$$

A **decay curve**, as illustrated in the figure on the right, is a plot of N vs. t (the number of nuclei remaining in a sample versus the elapsed time). *The number of nuclei remaining in a sample of a radioactive substance decreases exponentially with time.* N_0 is the number of nuclei in the sample at $t = 0$, and time is shown in units of half-lives.

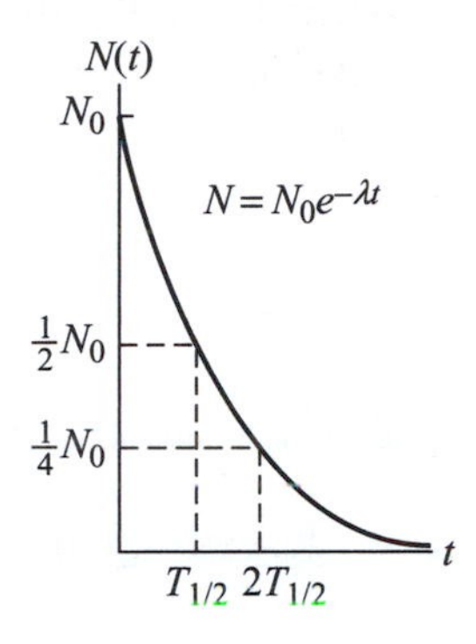

Equation 29.4b gives the number of remaining nuclei after an elapsed time of n half-lives.

$$N = N_0 e^{-\lambda t} \quad (29.4a)$$

$$N = N_0 \left(\frac{1}{2} \right)^n \quad (29.4b)$$

The **half-life of a radioactive substance** is the time required for half of the radioactive nuclei remaining in a sample of the substance to decay.

$$T_{1/2} = \frac{\ln 2}{\lambda} = \frac{0.693}{\lambda} \quad (29.5)$$

Spontaneous decay of radioactive nuclei proceeds by one the following processes: alpha decay, beta decay, electron capture, or gamma decay. The overall decay process can be represented by a decay equation.

$$\underset{\text{Parent nucleus}}{X} \rightarrow \underset{\text{Daughter nucleus}}{Y} + \text{Emitted radiation}$$

Alpha Decay

When a nucleus decays by alpha emission, the parent nucleus loses two neutrons and two protons. For alpha emission (illustrated by the scheme shown in Equation 29.8) to occur, the mass of the parent nucleus $\left({}^{A}_{Z}X\right)$ must be greater than the combined masses of the daughter nucleus $\left({}^{A-4}_{Z-2}Y\right)$ and the emitted alpha particle. Two examples of alpha decay, along with their respective half life values, are shown.

$${}^{A}_{Z}X \rightarrow {}^{A-4}_{Z-2}Y + {}^{4}_{2}He \quad (29.8)$$

$${}^{238}_{92}U \rightarrow {}^{234}_{90}Th + {}^{4}_{2}He \quad (29.9)$$

$$(T_{1/2} = 4.47 \times 10^{9} \text{ years})$$

$${}^{226}_{88}Ra \rightarrow {}^{222}_{86}Rn + {}^{4}_{2}He$$

$$(T_{1/2} = 1.60 \times 10^{3} \text{ years})$$

Beta decay

The beta decay process is accompanied by emission of a third particle that is required for conservation of energy and momentum.

Two modes of beta decay are: (1) electron emission (e^-) with antineutrino ($\bar{\nu}$) and (2) positron emission (e^+) with neutrino (ν). For each mode of decay, the equations at right show:

(i) The equation representing the process,

(ii) an example of the decay mode, and

(iii) an equation describing the origin of the electron (or proton).

Electron Emission

(i) $^{A}_{Z}X \rightarrow {}^{A}_{Z+1}Y + e^- + \bar{\nu}$

(ii) $^{14}_{6}C \rightarrow {}^{14}_{7}N + e^- + \bar{\nu}$ (29.15)

(iii) $^{1}_{1}n \rightarrow {}^{1}_{1}p + e^- + \bar{\nu}$

Positron Emission

(i) $^{A}_{Z}X \rightarrow {}^{A}_{Z-1}Y + e^+ + \nu$

(ii) $^{12}_{7}N \rightarrow {}^{12}_{6}C + e^+ + \nu$ (29.16)

(iii) $^{1}_{1}p \rightarrow {}^{1}_{1}n + e^+ + \nu$

Gamma Decay

Nuclei which undergo beta or alpha decay are often left in an excited energy state (indicated by the * symbol in Equation 29.18. The nucleus returns to the ground state by emission of one or more photons. *Gamma decay results in no change in mass number or atomic number.*

$$^{A}_{Z}X^* \rightarrow {}^{A}_{Z}X + \gamma$$

$$^{12}_{6}C^* \rightarrow {}^{12}_{6}C + \gamma \quad (29.18)$$

Nuclear reactions can occur when target nuclei (X) are bombarded with energetic particles (a) resulting in a daughter or product nucleus (Y) and an outgoing particle (b). Equation 29.21 describes the first nuclear reaction produced in the laboratory. *In this case both reactants and both products are stable isotopes. In these reactions the structure, identity, or properties of the target nuclei are changed.*

$$a + X \rightarrow Y + b$$

$$^{4}_{2}He + {}^{14}_{7}N \rightarrow {}^{17}_{8}O + {}^{1}_{1}H \quad (29.21)$$

The *Q* value of a nuclear reaction is the quantity of energy required to balance the equation representing the reaction. *Q is positive for exothermic reactions and negative for endothermic reactions.*

$$Q = KE_{\text{products}} - KE_{\text{reactants}}$$

$$Q = (m_{\text{reactants}} - m_{\text{products}})c^2$$

A **threshold energy** (minimum kinetic energy of the incident particle) is characteristic of endothermic reactions. This minimum value of kinetic energy of the incident particle is required in order for energy and momentum to be conserved.

$$KE_{\text{min}} = \left(1 + \frac{m}{M}\right)|Q| \qquad (29.24)$$

m = mass of incident particle

M = mass of target particle

SUGGESTIONS, SKILLS, AND STRATEGIES

The rest energy of a particle is given by $E = mc^2$. It is therefore often convenient to express the unified mass unit in terms of its energy equivalent, $1\text{ u} = 1.660\,559 \times 10^{-27}$ kg or $1\text{ u} = 931.50\text{ MeV}/c^2$. When masses are expressed in units of u, energy values are then $E = m(931.50\text{ MeV/u})$.

Equation 29.4a can be solved for the particular time t after which the number of remaining nuclei will be some specified fraction (N/N_0) of the original number N_0. This can be done by taking the natural log of each side of Equation 29.4a to find

$$t = \left(\frac{1}{\lambda}\right)\ln\left(\frac{N_0}{N}\right)$$

Then substitute for N the specified fraction of N_0.

REVIEW CHECKLIST

- State and apply to the solution of related problems, the formula which expresses decay rate as a function of decay constant and number of radioactive nuclei. Also apply the exponential formula which expresses the number of radioactive nuclei remaining as a function of elapsed time, decay constant, or half-life, and the initial number of nuclei. (Section 29.3)

- Identify each of the components of radiation that are emitted by nuclei through natural radioactive decay and describe the basic properties of each. Write out typical equations to illustrate the processes of alpha decay, beta decay, and gamma emission. (Section 29.4)

- Calculate the Q value of a given nuclear reaction and determine the threshold energy of an endothermic reaction. (Section 29.6)

- Given the reactants and one of the products of a nuclear reaction, be able to balance the reaction equation and identify the remaining product nuclide. (Section 29.6)

- Be familiar with the various units used to express quantities of radiation dose. (Section 29.7)

SOLUTIONS TO SELECTED END-OF-CHAPTER PROBLEMS

4. Consider the $^{65}_{29}\text{Cu}$ nucleus. Find approximate values for its (a) radius, (b) volume, and (c) density.

Solution

(a) Most nuclei tend to be approximately spherical in shape with a radius given by $r = r_0 A^{\frac{1}{3}}$, where r_0 is a constant equal to 1.2×10^{-15} m and A is the total number of nucleons in the nucleus. Thus, the approximate radius of a $^{65}_{29}\text{Cu}$ nucleus, containing 65 nucleons, will be

$$r=\left(1.2\times10^{-15}\text{ m}\right)(65)^{\frac{1}{3}}=4.8\times10^{-15}\text{ m}=4.8\text{ fm} \qquad \Diamond$$

(b) The volume of a $^{65}_{29}\text{Cu}$ is then given by

$$V_{^{65}\text{Cu}}=\frac{4}{3}\pi r^3=\frac{4}{3}\pi\left(4.8\times10^{-15}\text{ m}\right)^3=4.6\times10^{-43}\text{ m}^3 \qquad \Diamond$$

(c) The measured mass of a $^{65}_{29}\text{Cu}$ nucleus is listed in the Table of Isotopes found in Appendix B of the textbook. This mass is $m_{^{65}\text{Cu}}=64.927\,794$ u, where the unified mass unit has the value $1\text{ u}=1.66\times10^{-27}\text{ kg}=931.5\text{ MeV}/c^2$ (see the Conversion Factors on the inside cover of this manual). Thus,

$$m_{^{65}\text{Cu}}=(64.927\,794\text{ u})\left(\frac{1.66\times10^{-27}\text{ kg}}{1\text{ u}}\right)=1.08\times10^{-25}\text{ kg}$$

and the density of this nucleus is found to be

$$\rho=\frac{m_{^{65}\text{Cu}}}{V_{^{65}\text{Cu}}}=\frac{1.08\times10^{-25}\text{ kg}}{4.6\times10^{-43}\text{ m}^3}=2.3\times10^{17}\text{ kg/m}^3 \qquad \Diamond$$

This result is a typical value for atomic nuclei and illustrates the extremely high density of nuclear matter.

---

8. At the end of its life, a star with a mass of two times the Sun's mass is expected to collapse, combining its protons and electrons to form a neutron star. Such a star could be thought of as a gigantic atomic nucleus. If a star of mass $2\times1.99\times10^{30}$ kg collapsed into neutrons ($m_n=1.67\times10^{-27}$ kg), what would its radius be? Assume $r=r_0A^{\frac{1}{3}}$.

Solution

If a star containing 2 solar masses collapsed into neutrons, forming the equivalent of a gigantic atomic nucleus, the total number of nucleons (all neutrons in this case) would be

$$A=\frac{M_{\text{star}}}{m_n}=\frac{2M_{\text{sun}}}{m_n}=\frac{2\left(1.99\times10^{30}\text{ kg}\right)}{1.67\times10^{-27}\text{ kg/neutron}}=2.38\times10^{57}\text{ neutrons}$$

The approximate radius of the neutron star resulting from such a collapse is therefore given by

$$r = r_0 A^{\frac{1}{3}} = \left(1.2\times10^{-15}\ \text{m}\right)\left(2.38\times10^{57}\right)^{\frac{1}{3}} = 1.6\times10^{4}\ \text{m} = 16\ \text{km}$$ ◊

The density of this neutron star would then be

$$\rho = \frac{M_{\text{star}}}{V} = \frac{2M_{\text{Sun}}}{4\pi r^3/3} = \frac{6\left(1.99\times10^{30}\ \text{kg}\right)}{4\pi\left(1.6\times10^{4}\ \text{m}\right)^3} = 2.3\times10^{17}\ \text{kg/m}^3$$

which is the typical density of nuclear matter (see the solution to problem 4 given earlier in this chapter).

---

11. A pair of nuclei for which $Z_1 = N_2$ and $Z_2 = N_1$ are called *mirror isobars*. (The atomic and neutron numbers are interchangeable.) Binding-energy measurements on such pairs can be used to obtain evidence of the charge independence of nuclear forces. Charge independence means that the proton-proton, proton-neutron, and neutron-neutron forces are approximately equal. Calculate the difference in binding energy for the two mirror nuclei ${}^{15}_{8}\text{O}$ and ${}^{15}_{7}\text{N}$.

Solution

The binding energy of a nucleus ${}^{A}_{Z}\text{X}$ is given by $E_b = (\Delta m)c^2$, where the mass deficit is

$$\Delta m = \left[Zm_{{}^1_1\text{H}} + (A-Z)m_n\right] - m_{{}^A_Z\text{X}}$$

Here, Z is the atomic number (number of protons in the nucleus), $(A-Z)$ is the number of neutrons, m_n is the mass of a neutron, $m_{{}^1_1\text{H}}$ is the mass of a neutral hydrogen atom, and $m_{{}^A_Z\text{X}}$ is the mass of a neutral atom containing the nucleus of interest. We may use atomic masses from Appendix B of the textbook in this calculation since the included masses of Z electrons will cancel in the subtraction process.

For ${}^{15}_{8}\text{O}$, $\quad \Delta m = 8(1.007\,825\ \text{u}) + 7(1.008\,665\ \text{u}) - 15.003\,065\ \text{u} = 0.120\,190\ \text{u}$

and the binding energy is $E_b = 0.120\,190\ \text{u}\cdot c^2$. The energy equivalent of the atomic mass unit is $1\ \text{u} = 931.5\ \text{MeV}/c^2$, so the binding energy of ${}^{15}_{8}\text{O}$ is

$$E_b = (0.120\,190\ \text{u})(931.5\ \text{MeV/u}) = 112.0\ \text{MeV}$$

For ${}^{15}_{7}\text{N}$, $\quad \Delta m = 7(1.007\,825\ \text{u}) + 8(1.008\,665\ \text{u}) - 15.000\,109\ \text{u} = 0.123\,986\ \text{u}$

and $\quad E_b = (0.123\,986\ \text{u})(931.5\ \text{MeV/u}) = 115.5\ \text{MeV}$

Thus, the difference in the binding energies of these mirror isobars is

$$\Delta E_b = E_b\big|_{{}^{15}\text{N}} - E_b\big|_{{}^{15}\text{O}} = 115.5\ \text{MeV} - 112.0\ \text{MeV} = 3.5\ \text{MeV}$$ ◊

which is a small fraction of the total binding energy of either of these nuclei. This difference is primarily the result of the additional Coulomb repulsion found in the oxygen nucleus (due to the presence of its additional proton).

---

22. A building has become accidentally contaminated with radioactivity. The longest-lived material in the building is strontium-90. (The atomic mass of ${}^{90}_{38}\text{Sr}$ is 89.907 7 u.) If the building initially contained 5.0 kg of this substance and the safe level is less than 10.0 counts/min, how long will the building be unsafe?

Solution

The mass of a single ${}^{90}_{38}\text{Sr}$ atom is

$$m_{\text{atom}} = 89.907\,7 \text{ u} = (89.907\,7 \text{ u})\left(\frac{1.66\times10^{-27} \text{ kg}}{1 \text{ u}}\right) = 1.49\times10^{-25} \text{ kg}$$

so the number of ${}^{90}_{38}\text{Sr}$ nuclei initially in the building is

$$N_0 = \frac{M_{{}^{90}\text{Sr}}}{m_{\text{atom}}} = \frac{5.0 \text{ kg}}{1.49\times10^{-25} \text{ kg}} = 3.4\times10^{25}$$

Strontium-90 has a half-life of 29.1 yr (see Appendix B of the textbook), so its decay constant is $\lambda = \ln 2 / T_{1/2} = \ln 2 / 29.1$ yr, and the initial activity due to strontium-90 is

$$R_0 = \lambda N_0 = \frac{N_0 \ln 2}{T_{1/2}} = \frac{(3.4\times10^{25})\ln 2}{29.1 \text{ yr}}\left(\frac{1 \text{ yr}}{3.156\times10^{7} \text{ s}}\right)\left(\frac{60 \text{ s}}{1 \text{ min}}\right) = 1.5\times10^{18} \text{ counts/min}$$

The number of strontium-90 nuclei remaining after time t has elapsed will be $N = N_0 e^{-\lambda t}$, and the activity expected at this time is $R = \lambda N = \lambda N_0 e^{-\lambda t} = R_0 e^{-\lambda t}$. Thus, the time that must elapse before the activity in the building will reach the safe level of $R = 10.0$ counts/min is

$$t = -\frac{\ln(R/R_0)}{\lambda} = -\frac{T_{1/2}\ln(R/R_0)}{\ln 2}$$

or

$$t = -\frac{(29.1 \text{ yr})}{\ln 2}\ln\left(\frac{10.0 \text{ counts/min}}{1.5\times10^{18} \text{ counts/min}}\right) = 1.7\times10^{3} \text{ yr}$$ ◊

---

27. Determine which of the following suggested decays can occur spontaneously:

(a) ${}^{40}_{20}\text{Ca} \to e^+ + {}^{40}_{19}\text{K}$ (b) ${}^{144}_{60}\text{Nd} \to {}^{4}_{2}\text{He} + {}^{140}_{58}\text{Ce}$

Solution

For a decay to occur spontaneously, a release of energy (as indicated by a positive Q-value) must occur during the decay. The Q-value is given by $Q = (\Delta m)c^2$, where Δm is the difference between the total mass before decay and the total mass of the particles present after the decay.

(a) When the decay products include positrons (β^+ decay), the electron masses do not automatically cancel out of the Q-value calculation. Thus, one must be careful to account for electron masses on the two sides of the decay equation. To compute the Q-value for the decay ${}^{40}_{20}\text{Ca} \to e^+ + {}^{40}_{19}\text{K}$, we add 20 electrons to each side and create neutral atoms. Then we have ${}^{40}_{20}\text{Ca}_{\text{atom}} \to e^- + e^+ + {}^{40}_{19}\text{K}_{\text{atom}}$, since only 19 of the 20 added electrons on the right side were needed to create the neutral ${}^{40}_{19}\text{K}$ atom. We may now make use of the neutral atomic masses given in Appendix B of the textbook. This gives

$$Q = \left[m_{{}^{40}_{20}\text{Ca}_{\text{atom}}} - 2m_e - m_{{}^{40}_{19}\text{K}_{\text{atom}}}\right]c^2$$

$$= [39.962\,591\text{ u} - 2(0.000\,549\text{ u}) - 39.963\,999\text{ u}](931.5\text{ MeV/u})$$

$$= -2.33\text{ MeV}$$

Since $Q < 0$, this decay cannot occur spontaneously. ◊

(b) For the decay ${}^{144}_{60}\text{Nd} \to {}^{4}_{2}\text{He} + {}^{140}_{58}\text{Ce}$, we add 60 electrons to each side and obtain ${}^{144}_{60}\text{Nd}_{\text{atom}} \to {}^{4}_{2}\text{He}_{\text{atom}} + {}^{140}_{58}\text{Ce}_{\text{atom}}$. Again using neutral atomic masses from Appendix B,

$$Q = \left[m_{{}^{144}_{60}\text{Nd}_{\text{atom}}} - m_{{}^{4}_{2}\text{He}_{\text{atom}}} - m_{{}^{140}_{58}\text{Ce}_{\text{atom}}}\right]c^2$$

$$= [143.910\,083\text{ u} - 4.002\,603\text{ u} - 139.905\,434\text{ u}](931.5\text{ MeV/u})$$

$$= +1.91\text{ MeV}$$

In this case, $Q > 0$, and the decay may occur spontaneously. ◊

30. A piece of charcoal used for cooking is found at the remains of an ancient campsite. A 1.00-kg sample of carbon from the wood has an activity of 5.00×10^2 decays per minute. Find the age of the charcoal. *Hint:* Living material has an activity of 15.0 decays/minute per gram of carbon present.

Solution

Living material has an activity of 15.0 decays/minute per gram of carbon present because of the continuous intake of $^{14}_{12}C$ from the atmosphere, ground water, and other natural sources in the environment. When that material dies, its carbon-14 intake ceases, and the amount of this isotope present in the remains decreases over time due to radioactive decay. By measuring the current activity of the carbon in the remains, and knowing both the initial activity level and the half-life of carbon-14, the elapsed time since this material was part of a living organism may be determined.

When the 1.00-kg sample of carbon was part of a living organism, its activity due to carbon-14 was

$$R_0 = (1.00\text{ kg})\left(15.0\ \frac{\text{counts/minute}}{\text{g}}\right)\left(\frac{10^3\text{ g}}{1\text{ kg}}\right) = 1.50\times10^4\text{ counts/minute}$$

The activity of a radioactive material varies in time according to the equation $R = R_0e^{-\lambda t}$, where R_0 is the activity at $t=0$ and the decay constant is $\lambda = \ln 2/T_{1/2}$. The half-life of carbon-14 is $T_{1/2} = 5\,730$ yr, so if the 1.00-kg sample of carbon has a current activity of $R = 5.00\times10^2$ counts/min, the elapsed time since this charcoal was part of a living tree is given by

$$t = -\frac{\ln(R/R_0)}{\lambda} = -\frac{T_{1/2}\ln(R/R_0)}{\ln 2} = -\frac{(5\,730\text{ yr})\ln\left[\left(5.00\times10^2\text{ min}^{-1}\right)/\left(1.50\times10^4\text{ min}^{-1}\right)\right]}{\ln 2}$$

yielding $t = 2.81\times10^4$ yr, or 28,100 yrs. ◊

---

33. Identify the unknown particles X and X′ in the following nuclear reactions:

$$X + {}^4_2He \to {}^{24}_{12}Mg + {}^1_0n$$

$${}^{235}_{92}U + {}^1_0n \to {}^{90}_{38}Sr + X + 2\,{}^1_0n$$

$$2\,{}^1_1H \to {}^2_1H + X + X'$$

Solution

In any reaction, both the total number of nucleons and the total charge must be the same after the reaction as it was before. We use these criteria to identify the unknown particles.

In the reaction ${}^A_ZX + {}^4_2He \to {}^{24}_{12}Mg + {}^1_0n$, we must have $A+4 = 24+1$, giving $A = 21$. Also, it is necessary that $Z+2 = 12+0$, or $Z = 10$. Thus, the unknown particle has a charge of $Z = +10e$ and contains a total of $A = 21$ nucleons. This is the nucleus ${}^{21}_{10}Ne$. ◊

For the reaction ${}^{235}_{92}\text{U} + {}^{1}_{0}\text{n} \rightarrow {}^{90}_{38}\text{Sr} + {}^{A}_{Z}\text{X} + 2\,{}^{1}_{0}\text{n}$, we require that $235+1=90+A+2(1)$, yielding $A=144$. We also require that $92+0=38+Z+0$, or $Z=54$. Therefore, the unknown particle must be the nucleus ${}^{144}_{54}\text{Xe}$. ◊

Finally for the proton-proton reaction, $2\,{}^{1}_{1}\text{H} \rightarrow {}^{2}_{1}\text{H} + {}^{A}_{Z}\text{X} + \text{X}'$, we assume the particle X′ is a chargeless, massless particle such as a photon or neutrino. Then, it is necessary that $2(1)=2+A+0$, which gives $A=0$. Also, it is necessary that $2(1)=1+Z+0$, or $Z=+1$. Thus, the unknown particle X must be a non-nuclear particle with a charge of $+1e$. This is the positron, β^{+} or ${}^{0}_{+1}\text{e}$. The particle X′ that accompanies the production of a positron is the neutrino, ν_e. Chapter 30 will explain the reason for this when the requirement to conserve electron-lepton number is discussed. ◊

---

37. Natural gold has only one isotope, ${}^{197}_{79}\text{Au}$. If gold is bombarded with slow neutrons, e^{-} particles are emitted. (a) Write the appropriate reaction equation. (b) Calculate the maximum energy of the emitted beta particles. The mass of ${}^{198}_{80}\text{Hg}$ is 197.966 75 u.

Solution

(a) When ${}^{197}_{79}\text{Au}$ is bombarded by slow neutrons, it is possible for the gold nucleus to capture the neutron, momentarily forming an highly unstable ${}^{198}_{79}\text{Au}$ nucleus, which quickly undergoes beta decay, emitting an e^{-} particle and an antineutrino, $\bar{\nu}_e$. This process is summarized by the reaction equation:

$$ {}^{197}_{79}\text{Au} + {}^{1}_{0}\text{n} \rightarrow {}^{198}_{80}\text{Hg} + {}^{0}_{-1}\text{e} + \bar{\nu}_e $$ ◊

(b) To compute the Q-value for this reaction, we add 79 electrons to each side to form neutral atoms, yielding ${}^{197}_{79}\text{Au}_{\text{atom}} + {}^{1}_{0}\text{n} \rightarrow {}^{198}_{80}\text{Hg}_{\text{atom}} + \bar{\nu}_e$. Note that the original electron on the right side of the decay equation, along with the 79 added electrons, supply the 80 electrons needed for form a neutral mercury atom. We now use neutral atomic masses from Appendix B of the textbook to obtain

$$ Q = \left[m_{{}^{197}_{79}\text{Au}_{\text{atom}}} + m_n - m_{{}^{198}_{80}\text{Hg}_{\text{atom}}} - m_{\bar{\nu}_e} \right] c^2 $$

$$ = [196.966\,552\text{ u} + 1.008\,665\text{ u} - 197.966\,75\text{ u} - 0](931.5\text{ MeV/u}) $$

$$ = +7.89\text{ MeV} $$

Neglecting any recoil kinetic energy of the product nucleus, ${}^{198}_{80}\text{Hg}$, the 7.89 MeV of energy released in the reaction may be divided between the kinetic energies of the beta particle and the antineutrino in any proportions. Thus, the kinetic energies of the beta particles could range from a low of zero to a maximum of 7.89 MeV. ◊

---

43. A "clever" technician decides to heat some water for his coffee with an x-ray machine. If the machine produces 10 rad/s, how long will it take to raise the temperature of a cup of water by 50°C. Ignore heat losses during this time.

Solution

The radiation absorbed dose (or rad for short) is that amount of radiation which deposits 10^{-2} J of energy into 1 kg of absorbing material. Therefore, if the x-ray machine produces 10 rad/s, it would add energy to a mass m of absorbing material at a rate of

$$\frac{\Delta Q}{\Delta t} = m\left[\left(10\ \frac{\text{rad}}{\text{s}}\right)\left(\frac{10^{-2}\ \text{J/kg}}{1\ \text{rad}}\right)\right] = m(0.100\ \text{J/kg}\cdot\text{s})$$

The energy required to raise the temperature of a quantity of water having mass m by 50°C is

$$Q = mc_{\text{water}}(\Delta T) = m(4\,186\ \text{J/kg}\cdot{}^\circ\text{C})(50^\circ\text{C}) = m\left(2.1\times10^5\ \text{J/kg}\right)$$

The amount of time the x-ray machine will require to add this quantity of energy to the "clever" technician's coffee water is

$$\Delta t = \frac{Q}{\Delta Q/\Delta t} = \frac{m\left(2.1\times10^5\ \text{J/kg}\right)}{m(0.100\ \text{J/kg}\cdot\text{s})} = 2.1\times10^6\ \text{s}\left(\frac{1\ \text{d}}{86\,400\ \text{s}}\right) = 24\ \text{d}$$

◊

---

45. A patient swallows a radiopharmaceutical tagged with phosphorus-32 ($^{32}_{15}\text{P}$), a β^- emitter with a half-life of 14.3 days. The average kinetic energy of the emitted electrons is 700 keV. If the initial activity of the sample is 1.31 MBq, determine (a) the number of electrons emitted in a 10-day period, (b) the total energy deposited in the body during the 10 days, and (c) the absorbed dose if the electrons are completely absorbed in 100 g of tissue.

Solution

(a) If a radioactive sample has activity R_0 at time $t = 0$, the number of radioactive nuclei initially present in the sample is $N_0 = R_0/\lambda$, where $\lambda = \ln 2/T_{1/2}$ is the decay constant and $T_{1/2}$ is the half-life of this substance. At time t later, the number of radioactive nuclei remaining in the sample will be $N = N_0 e^{-\lambda t}$, so the number of decays that have occurred in the elapsed time is given by

$$\Delta N = N_0 - N = N_0 - N_0 e^{-\lambda t} = N_0\left(1-e^{-\lambda t}\right) = \frac{R_0\left(1-e^{-\lambda t}\right)}{\lambda} = \frac{T_{1/2}R_0\left[1-e^{-t(\ln 2/T_{1/2})}\right]}{\ln 2}$$

Since ${}^{32}_{15}\text{P}$ has a half-life of $T_{1/2} = 14.3$ days, and this sample has an initial activity of $1.31\ \text{MBq} = 1.31 \times 10^6$ decays/s, the number of decays occurring in a 10-day time period will be

$$\Delta N = \frac{(14.3\ \text{d})(1.31 \times 10^6\ \text{s}^{-1})\left[1 - e^{-(10.0\ \text{d})\ln 2/14.3\ \text{d}}\right]}{\ln 2}\left(\frac{8.64 \times 10^4\ \text{s}}{1\ \text{d}}\right) = 8.97 \times 10^{11}\ \text{decays} \quad \Diamond$$

(b) Each electron produced (one per decay) deposits its kinetic energy in the absorbing tissues as it is brought to rest. Thus, the total energy deposited in the body during this 10.0 day period is

$$E = \Delta N \cdot KE_{\text{av}} = (8.97 \times 10^{11})(700\ \text{keV})\left(\frac{1.60 \times 10^{-16}\ \text{J}}{1\ \text{keV}}\right) = 0.100\ \text{J} \quad \Diamond$$

(c) The rad (*r*adiation *a*bsorbed *d*ose) is defined as the radiation dose that deposits 10^{-2} J of energy in 1.00 kg of absorbing material. Hence, if the radiation from this sample is absorbed by $100\ \text{g} = 0.100$ kg of body tissue, the absorbed dose is

$$dose = \frac{\text{energy deposited per unit mass}}{\text{energy deposition per rad}} = \frac{0.100\ \text{J}/0.100\ \text{kg}}{\left(10^{-2}\ \frac{\text{J/kg}}{\text{rad}}\right)} = 100\ \text{rad} \quad \Diamond$$

---

53. A medical laboratory stock solution is prepared with an initial activity due to ${}^{24}\text{Na}$ of 2.5 mCi/mL, and 10.0 mL of the stock solution is diluted at $t = 0$ to a working solution whose total volume is 250 mL. After 48 h, a 5.0-mL sample of the working solution is monitored with a counter. What is the measured activity? *Note:* 1 mL = 1 milliliter.

Solution

At $t = 0$, the total activity of the working solution is

$$R_0 = (2.5\ \text{mCi/mL})(10\ \text{mL}) = 25\ \text{mCi}$$

This activity is contained in a solution having a total volume of $V_{\text{ws}} = 250$ mL. Thus, the initial concentration of the activity is

$$c_0 = \frac{R_0}{V_{\text{ws}}} = \frac{25\ \text{mCi}}{250\ \text{mL}} = 0.10\ \text{mCi/mL}$$

The initial activity of the material that will make up the 5.0-mL sample was

$$R_{0,\text{sample}} = c_0 V_{\text{sample}} = (0.10\ \text{mCi/mL})(5.0\ \text{mL}) = 0.50\ \text{mCi}$$

The half-life of ${}^{24}\text{Na}$ is $T_{1/2} = 14.96$ h, so the decay constant is

$$\lambda = \frac{\ln 2}{T_{1/2}} = \frac{\ln 2}{14.96\ \text{h}}$$

The activity of the 5.0-mL sample at $t = 48$ h is then

$$R_{\text{sample}} = R_{0,\text{sample}} e^{-\lambda t} = (0.50 \text{ mCi}) e^{-\left(\frac{\ln 2}{14.96 \text{ h}}\right)(48 \text{ h})} = 5.4 \times 10^{-2} \text{ mCi}$$

or

$$R_{\text{sample}} = \left(5.4 \times 10^{-2}\right) \times 10^{-3} \text{ Ci} = 5.4 \times 10^{-5} \text{ Ci} = 54 \times 10^{-6} \text{ Ci} = 54 \ \mu\text{Ci}$$ ◊

---

55. The theory of nuclear astrophysics is that all the heavy elements like uranium are formed in the interior of massive stars. These stars eventually explode, releasing the elements into space. If we assume that at the time of explosion there were equal amounts of ^{235}U and ^{238}U, how long ago were the elements that formed our Earth released, given that the present ^{235}U/^{238}U ratio is 0.007? (The half-lives of ^{235}U and ^{238}U are 0.70×10^9 yr and 4.47×10^9 yr, respectively.)

Solution

With the half-life of ^{235}U being $T_{1/2} = 0.70 \times 10^9$ yr, the decay constant for this isotope is

$$\lambda_{235} = \frac{\ln 2}{T_{1/2}} = \frac{\ln 2}{0.70 \times 10^9 \text{ yr}} = 9.9 \times 10^{-10} \text{ yr}^{-1}$$

Similarly, the decay constant for ^{238}U, with $T_{1/2} = 4.47 \times 10^9$ yr, is

$$\lambda_{238} = \frac{\ln 2}{T_{1/2}} = \frac{\ln 2}{4.47 \times 10^9 \text{ yr}} = 1.55 \times 10^{-10} \text{ yr}^{-1}$$

If immediately after the stellar explosion , the same number N_0 of ^{235}U and ^{238}U nuclei were present in the materials now contained in our Earth, the number of nuclei of each of these two uranium isotopes still present at time t should be $N_{235} = N_0 e^{-\lambda_{235} t}$ and $N_{238} = N_0 e^{-\lambda_{238} t}$. The ratio of the numbers of these two types of nuclei would then be

$$\frac{N_{235}}{N_{238}} = \frac{N_0 e^{-\lambda_{235} t}}{N_0 e^{-\lambda_{238} t}} = e^{-(\lambda_{235} - \lambda_{238})t}$$

Since the current value of this ratio for a sample of uranium found on Earth is $(N_{235}/N_{238})_{\text{current}} = 0.007$, the elapsed time since the stellar explosion which released the materials making up the Earth is computed to be

$$t = -\frac{\ln (N_{235}/N_{238})_{\text{current}}}{\lambda_{235} - \lambda_{238}} = -\frac{\ln (0.007)}{9.9 \times 10^{-10} \text{ yr}^{-1} - 1.55 \times 10^{-10} \text{ yr}^{-1}} = 5.9 \times 10^9 \text{ yr}$$ ◊

30

Nuclear Energy and Elementary Particles

NOTES FROM SELECTED CHAPTER SECTIONS

30.1 Nuclear Fission

Nuclear fission occurs when a heavy nucleus, such as ^{235}U, splits (fissions) into two smaller nuclei. In such a reaction, *the total rest mass of the products is less than the original rest mass.* The sequence of events in the fission process is:

- The ^{235}U nucleus captures a thermal (low energy) neutron.
- This capture results in the formation of $^{236}U^*$ (* indicates excited state), and the excess energy of this nucleus causes it to undergo violent oscillations.
- The $^{236}U^*$ nucleus becomes highly elongated, and the force of repulsion between protons in the two halves of the dumbbell-shaped nucleus tends to increase the distortion.
- The nucleus splits into two fragments, emitting several neutrons in the process.

A nuclear reactor is a system designed to maintain a **self-sustained chain reaction**. The **reproduction constant** K is defined as the average number of neutrons released from each fission event that will cause another event. In a power reactor, it is necessary to maintain a value of K close to 1. Under this condition, the reactor is said to be **critical**.

Important factors and processes relative to the design and operation of a nuclear reactor include:

- **Neutron leakage** — Some neutrons produced by a fission event will escape the volume of the reactor core before producing another fission event. The fraction of neutrons lost by leakage increases as the ratio of the surface area to the volume of the reactor vessel increases.
- **Regulating neutron energy** — The fast neutrons produced by fission are reduced in energy by allowing them to undergo collisions with a material of low atomic number (moderator). Slow neutrons have a greater probability than fast neutrons of initiating subsequent fission events.
- **Non-fissioning nuclei** — Some neutrons are captured by nuclei that do not undergo fission. The probability that a capture event will occur is reduced when the neutron energies are lower.

- **Power level** — The power level of the reactor is controlled by adjusting the number and position of control rods made of materials that are efficient neutron absorbers.

- **Reactor safety** — The primary operational concerns related to reactor safety are: containment of the fuel and radioactive fission products, flow of coolant to the core, disposal of spent fuel rods, and transportation of fuel and waste products.

30.2 Nuclear Fusion

Nuclear fusion is a process in which two light nuclei combine to form a heavier nucleus. The mass of the heavy nucleus is less than the sum of the masses of the two light nuclei; this loss of mass is the source of energy released in a fusion event. The major obstacle in obtaining useful energy from fusion is the large Coulomb repulsive force between charged nuclei as they approach close separation. Sufficient energy must be supplied to the particles to overcome this Coulomb barrier and thereby enable the nuclear attractive force to take over. Important considerations in the effort to achieve controlled fusion are:

- **Ignition temperature** — temperature at which hydrogen nuclei have sufficient kinetic energy to overcome the Coulomb force of repulsion

- **Ion density** — number of electrons and positive ions per cm^3 which are produced in the gas (plasma) at high temperature

- **Confinement time** — time that the interacting ions in the plasma are maintained at a temperature high enough for the fusion reaction to proceed

- **Lawson's criterion** — condition on plasma density and confinement time under which net power output is possible

30.3 Elementary Particles and the Fundamental Forces

There are four fundamental forces in nature:

- The **strong force** is responsible for binding quarks to form neutrons and protons and is the nuclear force that binds neutrons and protons into nuclei. *It is a very short-range force and is negligible for separations greater than the approximate size of the nucleus.*

- The **electromagnetic force** is responsible for the binding of atoms and molecules. *It is a long-range force that decreases in strength as the inverse square of the separation between interacting particles.*

- The **weak force** is a short-range nuclear force that tends to produce instability in certain nuclei. The weak interaction governs the structure of basic matter particles and is responsible for beta decay. *The electromagnetic and weak forces are now believed to be manifestations of a single force called the* ***electroweak force****.*

- The **gravitational force** is the weakest of all the fundamental forces. It is a long-range force that holds the planets, stars, and galaxies together. *The effect of the gravitational force on elementary particles is negligible.*

We model the universe as composed of field particles and matter particles. The exchange of field particles mediates the interactions of the matter particles.

- **Gluons** are the field particles for the strong force.
- **Photons** are exchanged by charged particles in electromagnetic interactions.
- **W^+, W^-, and Z bosons** mediate the weak force.
- **Gravitons** are (as yet undetected) quantum particles of the gravitational field.

30.4 Positrons and Other Antiparticles

A particle and its antiparticle (for example, an electron and a positron) have the same mass but opposite charge. Particle-antiparticle pairs may have other opposite values such as lepton number and baryon number.

Pair annihilation is an event in which a particle-antiparticle pair initially at rest, combine with each other, disappear, and produce two photons which move in opposite directions with equal energy and with the same magnitude of momentum.

30.5 Classification of Particles

All particles (other than photons) can be classified into two categories: **hadrons** and **leptons**.

Hadrons are composed of quarks and interact via all fundamental forces. They include two classes of particles grouped according to their masses and spins:

- **Mesons** — decay finally into electrons, positrons, neutrinos, and photons; they have spin quantum number 0 or 1.
- **Baryons** — (with the exception of the proton) decay in a manner leading to end products which include a proton; they have spin quantum number $\frac{1}{2}$ *or* $\frac{3}{2}$.

Leptons (e^-, μ^-, τ^-, ν_e, ν_μ, ν_τ, and their antiparticles) interact by the weak interaction, the electromagnetic interaction (if charged), and (presumably) the gravitational interaction but not by the strong force. Leptons have spin number $\frac{1}{2}$. Electrons, muons, and neutrinos are included in this group.

30.6 Conservation Laws

Conservation laws in the study of elementary particles are based on empirical evidence:

- **Conservation of baryon number** — Whenever a nuclear reaction or decay occurs, the sum of the baryon numbers before the process must equal the sum of the baryon numbers after the process. This requirement is quantified by assiging a baryon number, B ($B = +1$ for baryons, $B = -1$ for antibaryons, $B = 0$ for all other particles).

- **Conservation of lepton number** — Conservation laws are quantified by assigning lepton numbers, one for each variety of lepton: electron-lepton number (L_e), muon-lepton number (L_μ), and tau-lepton number (L_τ). *In each case the sum of the lepton numbers before a reaction or decay must equal the sum of the lepton numbers after the reaction or decay.*

- **Conservation of strangeness** — Strange particles are always produced in pairs by the strong interaction and decay very slowly, as is characteristic of the weak interaction. Strange particles are assigned a **strangeness quantum number**, S; quantitatively, $S = \pm 1, \pm 2, \pm 3$ for strange particles, and $S = 0$ for nonstrange particles. Whenever a nuclear reaction or decay occurs via the strong interaction or the electromagnetic interaction, the sum of the strangeness numbers before the process must equal the sum of the strangeness numbers after the process ($\Delta S = 0$). Decays occurring via the weak interaction include the loss of one strange particle. This process proceeds slowly and violates the law of conservation of strangeness.

30.7 The Eightfold Way
30.8 Quarks and Color

Quarks and antiquarks (of six possible types or flavors) make up baryons and mesons; the identity of each particle is determined by the particular combination of quarks. *Baryons consist of three quarks, antibaryons consist of three antiquarks, and mesons consist one quark and one antiquark.* The table at right lists the charge and baryon number for each of the six quarks and antiquarks.

Quark, Antiquark		Charge Number	Baryon Number
Up:	u, $\bar{u}$	$+\frac{2}{3}e, -\frac{2}{3}e$	$\frac{1}{3}, -\frac{1}{3}$
Down:	d, $\bar{d}$	$-\frac{1}{3}e, +\frac{1}{3}e$	$\frac{1}{3}, -\frac{1}{3}$
Strange:	s, $\bar{s}$	$-\frac{1}{3}e, +\frac{1}{3}e$	$\frac{1}{3}, -\frac{1}{3}$
Charmed:	c, $\bar{c}$	$+\frac{2}{3}e, -\frac{2}{3}e$	$\frac{1}{3}, -\frac{1}{3}$
Bottom:	b, $\bar{b}$	$-\frac{1}{3}e, +\frac{1}{3}e$	$\frac{1}{3}, -\frac{1}{3}$
Top:	t, $\bar{t}$	$+\frac{2}{3}e, -\frac{2}{3}e$	$\frac{1}{3}, -\frac{1}{3}$

The strong force between quarks is referred to as the **color force** and is mediated by massless particles called **gluons. Color charge** (red, green, blue) is a property assigned to quarks, which allows combinations of quarks to satisfy the exclusion principle.

EQUATIONS AND CONCEPTS

The **fission of an uranium nucleus** by bombardment with a low-energy neutron results in the production of **fission fragments** and typically two or three neutrons. The energy released in the fission event appears in the form of kinetic energy of the fission fragments and the neutrons.

$$ {}_0^1\text{n} + {}_{92}^{235}\text{U} \rightarrow {}_{92}^{236}\text{U}^* \rightarrow \text{X} + \text{Y} + \text{neutrons} \tag{30.1} $$

${}_{92}^{238}\text{U}^*$ is a short-lived intermediate state.

Mass and energy conservation requirements are satisfied by many different combinations of fission fragments. On the average, 2.47 neutrons are produced per fission event. Equation 30.2 shows a typical fission reaction.

$$ {}_{0}^{1}\mathrm{n} + {}_{92}^{235}\mathrm{U} \to {}_{56}^{141}\mathrm{Ba} + {}_{36}^{92}\mathrm{Kr} + 3\,{}_{0}^{1}\mathrm{n} \qquad (30.2) $$

The **fusion reactions** shown here are those most likely to be used as the basis for the design and operation of a fusion power reactor. The Q values refer to the energy released in each reaction. In the equations at right, D represents deuterium and T represents tritium.

$$ {}_{1}^{2}\mathrm{D} + {}_{1}^{2}\mathrm{D} \to {}_{2}^{3}\mathrm{He} + {}_{0}^{1}\mathrm{n} \quad (Q = 3.27\ \mathrm{MeV}) $$

$$ {}_{1}^{2}\mathrm{D} + {}_{1}^{2}\mathrm{D} \to {}_{1}^{3}\mathrm{T} + {}_{1}^{1}\mathrm{H} \quad (Q = 4.03\ \mathrm{MeV}) $$

$$ {}_{1}^{2}\mathrm{D} + {}_{1}^{3}\mathrm{T} \to {}_{2}^{4}\mathrm{He} + {}_{0}^{1}\mathrm{n} \quad (Q = 17.59\ \mathrm{MeV}) \qquad (30.4) $$

Lawson's criterion states the conditions under which a net power output of a fusion reactor is possible. In these expressions, n is the **plasma density** (number of ions per cubic cm), and τ is the plasma **confinement time** (the time during which the interacting ions are maintained at a temperature equal to or greater than that required for the reaction to proceed).

$$ n\tau \geq 10^{14}\ \mathrm{s/cm^3} \quad \text{(D-T reaction)} $$

$$ n\tau \geq 10^{16}\ \mathrm{s/cm^3} \quad \text{(D-D reaction)} \qquad (30.5) $$

Three varieties of pions correspond to three charge states: π^{+}, π^{-}, and π^{0}. Pions are very unstable particles; the equations at right show a decay mode for each type of pion.

$$ \pi^{+} \to \mu^{+} + \nu_{\mu} $$

$$ \pi^{-} \to \mu^{-} + \bar{\nu} \qquad (30.6) $$

$$ \pi^{0} \to \gamma + \gamma $$

Muons are in two varieties: μ^{-} and μ^{+} (the antiparticle). Example decay processes are shown in the equations at right.

$$ \mu^{+} \to \mathrm{e}^{+} + \nu_{\mathrm{e}} + \bar{\nu}_{\mu} $$

$$ \mu^{-} \to \mathrm{e}^{-} + \nu_{\mathrm{e}} + \bar{\nu}_{\mu} $$

REVIEW CHECKLIST

- Describe the sequence of events which occur during the fission process. Write an equation which represents a typical fission event. Use data obtained from the binding energy curve to estimate the disintegration energy of a typical fission event. (Section 30.1)
- List the major parameters influencing fission reactor design and operation which are important in maintaining a steady power level. (Section 30.1)
- Describe the basis of energy release in fusion and write out several nuclear reactions which might be used in a fusion-powered reactor. (Section 30.2)
- Calculate the minimum photon energy required to produce a given particle-antiparticle pair. (Section 30.4)
- Outline the classification of elementary particles and mention several characteristics of each group (relative mass value, spin, decay mode). (Section 30.5)
- Determine whether or not a suggested decay can occur based on the conservation of baryon number and the conservation of lepton number. (Section 30.6)
- Use conservation laws to identify reactants and products in a proposed reaction. (Section 30.6)

SOLUTIONS TO SELECTED END-OF-CHAPTER PROBLEMS

3. Find the energy released in the fission reaction

$$ {}_{0}^{1}\text{n} + {}_{92}^{235}\text{U} \rightarrow {}_{38}^{88}\text{Sr} + {}_{54}^{136}\text{Te} + 12\,{}_{0}^{1}\text{n} $$

Solution

In a fission reaction, the mass of the original heavy nucleus plus that of the incident neutron exceeds the total mass of the fission fragments. This "missing mass," or mass deficit, is converted into kinetic energy of the fragment particles during the fission process. The energy released in such a fission reaction is the Q-value of the reaction defined as $Q = (\Delta m)c^2 = (\Sigma m_{\text{before}} - \Sigma m_{\text{after}})c^2$. For the given reaction, the energy released will be

$$ Q = \left[m_{\text{n}} + m_{{}_{92}^{235}\text{U}} - m_{{}_{38}^{88}\text{Sr}} - m_{{}_{54}^{136}\text{Xe}} - 12m_{{}_{0}^{1}\text{n}} \right]c^2 $$

$$ = \left[235.043\,923 \text{ u} - 87.905\,614 - 135.907\,220 \text{ u} - 11(1.008\,665 \text{ u}) \right]c^2 $$

or $$ Q = [0.135\,774 \text{ u}]c^2 = [0.135\,774 \text{ u}]\left(\frac{931.5 \text{ MeV}/c^2}{1 \text{ u}} \right)c^2 = 126 \text{ MeV} $$ ◊

11. When a star has exhausted its hydrogen fuel, it may fuse other nuclear fuels. At temperatures above 1.0×10^8 K, helium fusion can occur. Write the equations for the following processes: (a) Two alpha particles fuse to produce a nucleus *A* and a gamma ray. What is nucleus *A*? (b) Nucleus *A* absorbs an alpha particle to produce a nucleus *B* and a gamma ray. What is nucleus *B*? (c) Find the total energy released in the reactions given in (a) and (b). *Note*: The mass of ${}^8_4\text{Be} = 8.005\,305$ u.

Solution

(a) The first reaction in this sequence is: where ${}^A_Z X$ is the unknown product nucleus A.

$${}^4_2\text{He} + {}^4_2\text{He} \rightarrow {}^A_Z X + \gamma$$

Requiring that charge be conserved gives: or $Z = 4$, meaning that A is an isotope of beryllium.

$$2 + 2 = Z + 0$$

Conserving the number of nucleons gives: or $A = 8$.

$$4 + 4 = A + 0$$

Thus, the product nucleus is ${}^8_4\text{Be}$. ◊

(b) The next reaction in the sequence is: where ${}^A_Z X$ is now the unknown product nucleus B.

$${}^8_4\text{Be} + {}^4_2\text{He} \rightarrow {}^A_Z X + \gamma$$

Requiring that charge be conserved gives: or $Z = 6$, meaning that B is an isotope of carbon.

$$4 + 2 = Z + 0$$

Conserving the number of nucleons gives: or $A = 12$.

$$8 + 4 = A + 0$$

Thus, the product nucleus is ${}^{12}_6\text{C}$. ◊

(c) The net result of this pair of reactions is to fuse three alpha particles into a carbon-12 nucleus along with the production of two gamma rays. The overall mass deficit is

$$\Delta m = 3m_{{}^4_2\text{He}} - m_{{}^{12}_6\text{C}} = 3(4.002\,603\text{ u}) - 12.000\,000\text{ u} = 0.007\,809\text{ u}$$

The total energy released in this pair of reactions is then

$$Q = (\Delta m)c^2 = (0.007\,809\text{ u})\left(\frac{931.5\text{ MeV}/c^2}{1\text{ u}}\right)c^2 = 7.27\text{ MeV}$$ ◊

13. Find the energy released in the fusion reaction

$$^2_1\text{H} + {}^2_1\text{H} \rightarrow {}^3_1\text{H} + {}^1_1\text{H}$$

Solution

In this fusion of two deuterium nuclei to form a tritium nucleus plus a proton, as in other fusion reactions, the combined mass of the two fusing nuclei exceeds the total mass of the product particles. As in the fission reaction of problem 3, this "missing mass," or mass deficit, is converted into kinetic energy of the product particles. The energy released is the Q-value of the reaction given by $Q = (\Delta m)c^2$. The energy released when two deuterons fuse as given in this reaction will be

$$Q = \left[m_{^2_1\text{H}} + m_{^2_1\text{H}} - m_{^3_1\text{H}} - m_{^1_1\text{H}}\right]c^2$$

$$= \left[2(2.014\,102\text{ u}) - 3.016\,049 - 1.007\,825\text{ u}\right]c^2$$

or $$Q = [0.004\,330\text{ u}]c^2 = [0.004\,330\text{ u}]\left(\frac{931.5\text{ MeV}/c^2}{1\text{ u}}\right)c^2 = 4.03\text{ MeV}$$ ◊

15. Assume a deuteron and a triton are at rest when they fuse according to the reaction

$$^2_1\text{H} + {}^3_1\text{H} \rightarrow {}^4_2\text{He} + {}^1_0\text{n} + 17.6\text{ MeV}$$

Neglecting relativistic corrections, determine the kinetic energy acquired by the neutron.

Solution

Since both the deuteron and the triton are at rest before the reaction, the initial total momentum is zero. Thus, the total momentum must be zero after the reaction, meaning the alpha particle and the neutron must move in opposite directions with equal magnitude momenta, $p_\alpha = p_\text{n}$.

Note that the energy released in the reaction (17.6 MeV) is very small in comparison to the rest energy of either of the product particles. Thus, these product particles will be nonrelativistic, and the relation between the kinetic energy and the momentum of either of these particles is $KE = p^2/2m$, where m is the mass of the particle. Therefore, the kinetic energies of the alpha particle and neutron after reaction are related by

$$KE_\alpha = \frac{p_\alpha^2}{2m_\alpha} = \frac{p_\text{n}^2}{2m_\alpha} = \frac{2m_\text{n}KE_\text{n}}{2m_\alpha}$$

or

$$KE_{\alpha} = \left(\frac{m_{\text{n}}}{m_{\alpha}}\right) KE_{\text{n}}$$

Since there was zero kinetic energy before the reaction, the total kinetic energy after the reaction must be

$$KE_{\text{n}} + KE_{\alpha} = Q = 17.6 \text{ MeV}$$

Thus, we have

$$\left(1 + \frac{m_{\text{n}}}{m_{\alpha}}\right) KE_{\text{n}} = 17.6 \text{ MeV}$$

or the kinetic energy of the emerging neutron is

$$KE_{\text{n}} = \frac{17.6 \text{ MeV}}{(1 + m_{\text{n}} / m_{\alpha})} = \frac{17.6 \text{ MeV}}{[1 + (1.008\,665 \text{ u})/(4.002\,603 \text{ u})]} = \underline{\underline{14.1 \text{ MeV}}}$$ ◊

17. A photon produces a proton–antiproton pair according to the reaction $\gamma \to \text{p} + \bar{\text{p}}$. What is the minimum possible frequency of the photon? What is its wavelength?

Solution

It is impossible for pair production to occur in a vacuum, since energy and momentum cannot both be conserved during pair production in such an environment. Thus, the conversion of a photon to a particle-antiparticle pair must occur in the vicinity of a massive nucleus which can absorb most (if not all) of the photon's momentum while acquiring a very small (even negligible) kinetic energy, $KE = p^2/2m$.

When pair production occurs in the vicinity of such a massive nucleus, the particles of the pair may be left at rest, with the recoiling massive nucleus being given negligible energy. In this case, the required energy of the incident photon is a minimum, and it all goes into the rest energies of the pair of particles produced. The minimum energy for the photon in the production of a proton-antiproton pair is therefore

$$E_{\text{min}} = E_{0,\text{p}} + E_{0,\bar{\text{p}}} = 2E_{0,\text{p}}$$

and the minimum frequency is

$$f_{\text{min}} = \frac{E_{\text{min}}}{h} = \frac{2E_{0,\text{p}}}{h} = \frac{2(938.3 \text{ MeV})}{6.63 \times 10^{-34} \text{ J} \cdot \text{s}} \left(\frac{1.60 \times 10^{-13} \text{ J}}{1 \text{ MeV}}\right) = 4.53 \times 10^{23} \text{ Hz}$$ ◊

The wavelength of a photon having this frequency is given by

$$\lambda_{\text{max}} = \frac{c}{f_{\text{min}}} = \frac{3.00\times10^{8}\ \text{m/s}}{4.53\times10^{23}\ \text{Hz}} = 6.62\times10^{-16}\ \text{m} = 0.662\ \text{fm}$$ ◊

---

21. Each of the following reactions is forbidden. Determine a conservation law that is violated for each reaction.

(a) $\text{p}+\bar{\text{p}}\rightarrow\mu^{+}+\text{e}^{-}$

(b) $\pi^{-}+\text{p}\rightarrow\text{p}+\pi^{+}$

(c) $\text{p}+\text{p}\rightarrow\text{p}+\pi^{+}$

(d) $\text{p}+\text{p}\rightarrow\text{p}+\text{p}+\text{n}$

(e) $\gamma+\text{p}\rightarrow\text{n}+\pi^{0}$

Solution

The basic conservation laws that must be observed in any particle reaction are: conservation of charge, Q; conservation of baryon number, B; conservation of lepton number (one for each variety of lepton), electron-lepton number L_{e}, muon-lepton number L_{μ}, and tau-lepton number L_{τ}; and conservation of strangeness. The strong and electromagnetic interactions always conserve strangeness. The weak interaction will violate conservation of strangeness but by no more that one unit.

(a) The reaction $\text{p}+\bar{\text{p}}\rightarrow\mu^{+}+\text{e}^{-}$ conserves charge, baryon number, and strangeness. It does not involve any tau-leptons but does violate both conservation of electron-lepton number ($L_{\text{e, before}}=0+0$, $L_{\text{e, after}}=0+1$) and conservation of muon-lepton number ($L_{\mu,\text{ before}}=0+0$, $L_{\mu,\text{ after}}=-1+0$). ◊

(b) The reaction $\pi^{-}+\text{p}\rightarrow\text{p}+\pi^{+}$ conserves baryon number, strangeness, and all varieties of lepton number. However, it fails to conserve charge: $Q_{\text{before}}=-e+e=0$, $Q_{\text{after}}=+e+e=2e$. ◊

(c) The reaction $\text{p}+\text{p}\rightarrow\text{p}+\pi^{+}$ conserves charge, strangeness, and all varieties of lepton number, but fails to conserve baryon number: $B_{\text{before}}=+1+1=2$, $B_{\text{after}}=+1+0=1$. ◊

(d) The reaction $\text{p}+\text{p}\rightarrow\text{p}+\text{p}+\text{n}$ conserves charge, strangeness, and all varieties of lepton number. However, it does not conserve baryon number: $B_{\text{before}}=+1+1=2$, $B_{\text{after}}=+1+1+1=3$. ◊

(e) The reaction $\gamma+\text{p}\rightarrow\text{n}+\pi^{0}$ conserves baryon number, strangeness, and all varieties of lepton number, but does not conserve charge: $Q_{\text{before}}=0+e=e$, $Q_{\text{after}}=0+0=0$. ◊

---

24. (a) Show that baryon number and charge are conserved in the following reactions of a pion with a proton:

(1) $\pi^+ + \mathrm{p} \to \mathrm{K}^+ + \Sigma^+$ (2) $\pi^+ + \mathrm{p} \to \pi^+ + \Sigma^+$

(b) The first reaction is observed, but the second never occurs. Explain these observations. (c) Could the second reaction happen if it created a third particle? If so, which particles in Table 30.2 might make it possible? Would the reaction require less energy or more energy than the reaction of Equation (1)? Why?

Solution

(a) In each reaction, the total initial charge is $Q_i = +e + e = +2e$, and the final total charge is $Q_f = +e + e = +2e$. Thus, charge is conserved. Likewise, the total baryon number is conserved in each reaction with $B_i = 0 + 1 = +1$, and $B_f = 0 + 1 = +1$.

(b) There are no leptons present either before or after the event in either reaction, so all lepton numbers are conserved. This leaves conservation of strangeness to be considered. In the first reaction, the total strangeness before is $S_i = 0 + 0 = 0$, and the total strangeness after the event is $S_f = +1 - 1 = 0$. Thus, strangeness is conserved, so all conservation laws are satisfied, and this reaction can occur.

In the second reaction, the initial total strangeness is $S_i = 0 + 0 = 0$, while the total strangeness afterwards is $S_f = 0 - 1 = -1$. Hence, $\Delta S \neq 0$, and strangeness is not conserved. This means that this reaction cannot occur via either the strong or the electromagnetic interactions.

(c) If the second reaction created a K^0 particle (with $Q = 0$, $B = 0$, and $L_e = L_\mu = L_\tau = 0$) in addition to the original π^+ and Σ^+ particles, the reaction would be $\pi^+ + \mathrm{p} \to \mathrm{K}^0 + \pi^+ + \Sigma^+$. This reaction would continue to conserve charge, baryon number, and all lepton numbers. With the inclusion of a K^0 among the particles emerging from the reaction, strangeness would now be conserved ($S_i = 0 + 0 = 0$, and $S_f = +1 + 0 - 1 = 0$). The reaction, as now modified, observes all conservation laws and could occur. However, the particles emerging from this modified reaction have a greater total rest energy than does the set of particles emerging from the reaction of Equation (1). This means that the incident particles for this reaction must have greater total energy than is required for the incident particles in the reaction of Equation (1). ◊

31. A Σ^0 particle traveling through matter strikes a proton. A Σ^+, a gamma ray, as well as a third particle, emerge. Use the quark model of each to determine the identity of the third particle.

Solution

The reaction of interest is $\Sigma^0 + \text{p} \rightarrow \Sigma^+ + \gamma + X$, where X is an unknown particle. We recognize that the Σ^0 particle and the proton are both baryons and interact via the strong interaction. Thus, we could attempt to identify the unknown particle by using the fact that the strong interaction always obeys several conservations laws: conservation of charge, conservation of baryon number, conservation of strangeness, and conservation of lepton numbers (one for each variety of lepton). However, we shall consider the quark composition (in terms of up, down, and strange quarks) of the known particles before and after the reaction to find the identity of particle X.

Consider Table 30.4 in the textbook to see that the quark composition of a Σ^0 particle is uds, that of a proton is uud, and that of a Σ^+ particle is uus. Thus, in terms of the quark composition, the reaction is

$$\text{uds} + \text{uud} \rightarrow \text{uus} + (0 \text{ quarks}) + (n_\text{u} n_\text{d} n_\text{s})$$

where n_u, n_d, and n_s represent the number of up, down, and strange quarks, respectively, contained in the mystery particle X. Note that before the reaction, there are totals of 3 up quarks, 2 down quarks, and 1 strange quark present. After the reaction, we have: $n_\text{u} + 2$ up quarks, $n_\text{d} + 0$ down quarks, and $n_\text{s} + 1$ strange quarks present.

In order to conserve the net number of each type of quark in this reaction, we must have:

$$n_\text{u} + 2 = 3 \quad \text{or} \quad n_\text{u} = 1 \quad \text{so } X \text{ must have 1 up quark}$$

$$n_\text{d} + 0 = 2 \quad \text{or} \quad n_\text{d} = 2 \quad \text{so } X \text{ must have 2 down quarks}$$

and $$n_\text{s} + 1 = 1 \quad \text{or} \quad n_\text{s} = 0 \quad \text{so } X \text{ must have 0 strange quarks}$$

Thus, the quark composition of particle X must be udd, and from Table 30.4 we see that particle X must be a neutron. ◊

37. A 2.0-MeV neutron is emitted in a fission reactor. If it loses one-half its kinetic energy in each collision with a moderator atom, how many collisions must it undergo to reach an energy associated with a gas at a room temperature of 20.0°C?

Solution

From the kinetic theory of gases (see Chapter 10 in the textbook), the average kinetic energy of a particle in a gas at room temperature ($T = 20.0°\text{C} = 293\text{ K}$) is given by

$$KE_{\text{av}} = \frac{3}{2}k_B T = \frac{3}{2}\left(1.38\times10^{-23}\text{ J/K}\right)\left(293\text{ K}\right)\left(\frac{1\text{ eV}}{1.60\times10^{-19}\text{ J}}\right) = 0.0379\text{ eV}$$

If the neutron, with an initial kinetic energy of $KE_i = 2.0$ MeV, loses one-half of its incident kinetic energy in each collision with a moderator atom, its kinetic energy after n such collisions will be

$$KE_f = \left(\frac{1}{2}\right)^n KE_i = \frac{2.0\text{ MeV}}{2^n}$$

Thus, if its final kinetic energy is to be $KE_f = KE_{\text{av}} = 0.0379\text{ eV}$, we must have $0.0379\text{ eV} = (2.0\text{ MeV})/2^n$, or $2^n = (2.0\text{ MeV})/(0.0379\text{ eV})$. To determine the number of collisions with moderator atoms required to make this true, we take the natural logarithm of both sides of this result, and obtain

$$n\ln 2 = \ln\left[\frac{2.0\text{ MeV}}{0.0379\text{ eV}}\left(\frac{10^6\text{ ev}}{1\text{ MeV}}\right)\right] = \ln\left(5.3\times10^7\right)$$

and

$$n = \frac{\ln\left(5.3\times10^7\right)}{\ln 2} = 26$$

◊

41. A π-meson at rest decays according to

$$\pi^- \rightarrow \mu^- + \bar{\nu}_\mu$$

What is the energy carried off by the neutrino? Assume the neutrino has no mass and moves off with the speed of light. Take $m_\pi c^2 = 139.6$ MeV and $m_\mu c^2 = 105.7$ MeV. *Note:* Use relativity; see Equation 26.13.

Solution

Both energy and momentum must be conserved in this pion decay. Since the π-meson was at rest before the decay, the total energy before decay was the rest energy of this particle, $E_{0,\pi^-} = 139.6$ MeV. Conservation of energy then requires that

$$E_\mu + E_\nu = 139.6 \text{ MeV} \qquad \textbf{[1]}$$

Also, with the pion initially at rest, its momentum was zero. Conservation of momentum then requires that the total momentum after the decay must be zero, meaning the muon and the antineutrino must travel in opposite directions with equal magnitude momenta, or $p_\mu = p_\nu$. For relativistic particles, their total energy and momentum are related by $E^2 = p^2c^2 + E_0^2$. For the muon, $E_{0,\mu} = 105.7$ Mev, so this becomes $E_\mu^2 = p_\mu^2 c^2 + (105.7 \text{ MeV})^2$. For the neutrino, $E_{0,\nu} = 0$, giving $E_\nu^2 = p_\nu^2 c^2$, or $p_\nu^2 = E_\nu^2 / c^2$. Since $p_\mu = p_\nu$, we must then have $p_\mu^2 = E_\nu^2 / c^2$, and the equation for the total energy of the muon is

$$E_\mu^2 = \left(E_\nu^2 / c^2\right)c^2 + (105.7 \text{ MeV})^2$$

This reduces to $E_\mu^2 - E_\nu^2 = \left(E_\mu + E_\nu\right)\left(E_\mu - E_\nu\right) = (105.7 \text{ MeV})^2$. Substituting Equation [1] into this expression yields $(139.6 \text{ MeV})\left(E_\mu - E_\nu\right) = (105.7 \text{ MeV})^2$, or

$$E_\mu - E_\nu = \frac{(105.7 \text{ MeV})^2}{139.6 \text{ MeV}} = 80.0 \text{ MeV} \qquad \textbf{[2]}$$

Finally, subtracting Equation [2] from Equation [1] gives $2E_\nu = 59.6$ MeV, so the energy carried away by the neutrino must be

$$E_\nu = 59.6 \text{ MeV}/2 = 29.8 \text{ MeV}$$

◊

43. The Sun radiates energy at the rate of 3.85×10^{26} W. Suppose the net reaction

$$4\text{p}+2\text{e}^- \rightarrow \alpha + 2\nu_\text{e} + 6\gamma$$

accounts for all the energy released. Calculate the number of protons fused per second. *Note:* Recall that an alpha particle is a helium-4 nucleus.

Solution

To determine the energy released in each occurrence of the reaction $4\text{p}+2\text{e}^- \rightarrow \alpha + 2\nu_\text{e} + 6\gamma$, we add two electrons to each side of the reaction equation to form neutral atoms and obtain

$$4\left({}^1_1\text{H}_\text{atom}\right) \rightarrow {}^4_2\text{He}_\text{atom} + 2\nu_\text{e} + 6\gamma$$

Then, using neutral atomic masses from Appendix B of the textbook, and recognizing that both the neutrinos and the photons are massless particles, we find

$$Q = (\Delta m)c^2 = \left(4m_{{}^1_1\text{H}_\text{atom}} - m_{{}^4_2\text{He}_\text{atom}}\right)c^2$$

$$= \left[4(1.007\,825\text{ u}) - 4.002\,603\text{ u}\right](931.5\text{ MeV/u})$$

$$= 26.7\text{ MeV}$$

Each occurrence of this reaction consumes four protons. Thus, the energy released per proton consumed is $E_1 = 26.7\text{ MeV}/4\text{ protons} = 6.68\text{ MeV/proton}$.

Therefore, the rate at which the Sun must be fusing protons to generate its present power output is

$$= \frac{P}{E_1} = \frac{3.85\times10^{26}\text{ J/s}}{6.68\text{ MeV/proton}}\left(\frac{1\text{ MeV}}{1.60\times10^{-13}\text{ J}}\right) = 3.60\times10^{38}\text{ protons/s} \qquad \lozenge$$